SKETCH *shewing the several lines of* ROADS *of the* WESTERN DISTRICT.

Ordered by The House of Commons to be Printed, 16th May 1833.

James and Luke G. Hansard & Sons, Printers.

S. Arrowsmith, Lithog.

Map of Ballynahinch Barony drawn by Alexander Nimmo in 1828 to illustrate works underway on the central Connemara road from the Halfway House to Emlaghdauroe

Showing the old centre road (Grand Jury road), laid out by Thomas Martin, the new centre road designed by Nimmo and then under contract to Samuel Jones, and the section of Nimmo's new coast road running from Camus Bay to Ballinaboy. The unfinished section of this road is indicated with broken line. Bridleroads are indicated with dots. (NAI CSORP 1828 865)

Alexander Nimmo & The Western District

Alexander Nimmo & The Western District

Emerging infrastructure in pre-Famine Ireland

KATHLEEN VILLIERS-TUTHILL

Connemara Girl Publications

First Published 2006 by
Connemara Girl Publications, D'Arcy Cottage, Ardbear, Clifden, Co Galway

email: kvt@connemaragirlpublications.com

ISBN 0-9530455-3-6
978-0-9530455-3-6
British Library Cataloguing in Publication Data.
A catalogue record for this book is available from the British Library.

By the same author
HISTORY OF CLIFDEN 1810-1860 (1982)

BEYOND THE TWELVE BENS:
A History of Clifden and District 1860-1923 (1986)

PATIENT ENDURANCE:
The Great Famine in Connemara (1997)

HISTORY OF KYLEMORE CASTLE & ABBEY (2002)

Layout and Typesetting: Bernard Kaye
Cover: Johan Hofsteenge
Printed by ColourBooks Ltd, Dublin

This publication received financial support from
Cold Chon (Galway) Ltd

COLD CHON (GALWAY) LTD

CONTENTS

List of Illustrations

Abbreviations

NLS – National Library of Scotland
NLI – National Library of Ireland
NA – National Archives of Ireland
CSORP – Chief Secretaries Official Registered Papers
OP – Official Papers
BPP – British Parliamentary Papers

Acknowledgements

Firstly, I want to acknowledge my sincere gratitude to Professor Gearóid Ó Tuathaigh for providing expert guidance throughout my research for this book. It was he who first suggested that, instead of Nimmo being a mere chapter in a re-write of my earlier work, *History of Clifden 1810–1860*, I should expand my research and write a book on his contribution to the development of the West of Ireland in the 1820s. Over the past two and a half years, Gearóid's advice and editorial judgement were frequently sought and always generously given. He is never sparing with his time or his knowledge, and I am always grateful for the interest he takes in my work.

I greatly appreciate the advice and assistance received from Peter and Paula Vine on matters relating to the publication of this book and to Paula for reading the final manuscript and offering valuable advice. My thanks to Paddy Barry for advice on all things nautical and engineering, for reading the manuscript and for his many helpful suggestions. I would also like to thank Bernard Kaye for designing the book. Needless to add, I take full responsibility for any errors that may become evident.

I wish to express my gratitude to the staff of the National Archives of Ireland, in particular to Gregory O'Connor, who is always so willing to provide assistance. My thanks to the staff of the National Library of Ireland for their continued friendliness, courtesy and assistance over the years. I would particularly like to thank the cloakroom staff for their sense of humour, which usually helped to put things into perspective at the end of a long day's search, with little or nothing to show for it.

I appreciate the assistance provided by Maureen Moran in the Galway County Library; John Callanan, archivist in the Institute of Engineers; Arnold Horner, for helpful advice and for the use of the Bog Commission maps; David Cabot, for generously allowing me unlimited access to his private library; Timothy Collins NUI Galway, for assistance in tracking down illustrations; Robert Preece, Academy of Inverness, for providing extracts from his unpublished manuscript on the history of the Academy.

I would also like to thank the staff in the following libraries, Mayo County Library, in particular Maureen Costello in the Maps Department; Fran Hegarty, Local Studies Department, Sligo County Library; Local Studies Dept, Leitrim County Library; Kieran Hoare, Librarian, NUI Galway; The Map Library, Trinity College Dublin; The British Library; National Library of Scotland; Mary Kelleher, RDS Dublin and Derek Cullen, Photographic Section, Failte Ireland.

For assistance in identifying particular roads and bridges in Mayo, I am indebted to the following engineers, Padraig Philbin, Executive Engineer Mayo County Council, Joe Beirne, Mayo County Engineer, and Tom McDonnell, Area Engineer Belmullet. My thanks to Paul Duffy for his help in identifying roads in the Tuam district of Galway. I would also like to express my thanks for the assistance of Dr Ciaran McLoughlan, Clifden, Marie Mannion, Galway County Heritage Officer, Eila Joyce, Clifden, John Homes, Leenaun, Shane Joyce CE, Clifden, and Dennis Sweeney, Galway.

For providing welcome relief from long hours of research in libraries and archives and for helping me to view Connemara from another perspective, my sincere thanks to Emer Joyce, Paraic Folan and Martin Barrett, the owners of the Galway Hooker, *The Truelight*, for inviting me to sail with them in the summer of 2005 and to the skipper, Mattie Casey, his crew, Sean Casey and Frank Cook, for making it a most memorable summer. My thanks also to Michael Barrett who first introduced me to the joys of sailing on a Galway Hooker and to his extended family on Foirnis Island for their continued hospitality.

As ever, I am grateful to my husband, John, and our two sons, William and Graham, for providing support and love, and for allowing me the space to pursue my great love affair with the West of Ireland.

I am deeply indebted to John Killeen, CEO of the Colas Group of companies in Ireland, for so generously providing financial support towards the publication of this book, for his interest in the development of the project and for his assistance along the way. Without such support this book would not have appeared in its present format.

Preface

The inspiration for this book came from two distinct, but interlocking, interests with which my research and writings have been concerned for some time. Firstly, my earlier work on aspects of the history of Connemara raised questions which drew me into a closer examination of the key episodes that had led to the economic and social development of the west of Ireland. The second source of my interest was Alexander Nimmo, a key agent in the story of this development. Nimmo, though his story and his significance has frequently been noted by engineers, still awaits a full biography.[1] This study does not seek to fill this void. Rather, its objective is to focus on the exertions of Nimmo in Ireland in the period 1811-31. My theme, therefore, is the infrastructural development of the geographical region encompassing counties Galway, Mayo, Sligo, Leitrim and Roscommon, with a particular focus on the districts of Connemara and of Erris, in counties Galway and Mayo, respectively. My key subject is the Scottish engineer, Alexander Nimmo, who oversaw the construction of many of the early roads, bridges and piers in this region.

This study seeks to provide the reader with the background to Nimmo's work in Ireland; to demonstrate the conditions under which he worked in the west, the infrastructure (or lack of it) that existed prior to his arrival, and the impact that his work had on the coastal regions in particular. It outlines the material available to Nimmo when he embarked on his coastal survey, and the contemporary assumptions and ideas that may well have influenced his findings and recommendations.

In the official reports that he compiled, in his correspondence throughout his years in Ireland, and in the opinions and assessments of Nimmo and his work contained in the writings of his contemporaries, we have an unusually rich commentary on the challenges posed by 'development' in the west of Ireland in the twenty years between 1811 and 1831. This study is based mainly upon this primary evidence, and it examines closely the role and legacy of Nimmo's actions in this vital period of pre-famine State 'intervention' in the improvement of the lives of the people – especially the poorer people – in the underdeveloped west.

The Western District
The Western District was established in 1822 for the purpose of administering public works during a period of distress. The district covered from Sligo Bay to Galway Bay, encompassing counties Leitrim, Roscommon, Sligo, Mayo and Galway. Alexander Nimmo was the government-appointed engineer for the Western District from June 1822 to June 1831. (*New and Correct Irish Atlas* circa pre-1825)

Introduction

In the early decades of the nineteenth century there was a considerable spate of enquiries and investigations into various aspects of 'the state of Ireland', sponsored by government or by the Westminster parliament. Ireland's absorption into the United Kingdom following the Act of Union of 1801 meant that the Westminster parliament in London now had responsibility for ruling Ireland. This new responsibility is likely to have increased the urgency of the ruling elite, in parliament and in government, in seeking to investigate those aspects of Irish conditions that posed a particular challenge. Moreover, the early nineteenth century was a time of increasing government interest in social and economic conditions in Britain as well as in Ireland.

The conduct of day-to-day government and administration in Ireland was the responsibility of the Irish Administration centred in Dublin Castle. At the head of the Irish Administration was the lord lieutenant, representing the crown and with an official residence in Phoenix Park, and the chief secretary, representing the government. Both of these were political appointments. The chief secretary was assisted by the under-secretary, the principal civil servant of the administration in Dublin Castle. Irish local government, other than in the case of urban corporations, was carried out by the county grand juries, with relatively little interference from central government. The Grand Jury was, in certain respects, the forerunner of the later county councils. Members of the Grand Jury were elected by the High Sheriff, from the local property owners of the county. The High Sheriff was himself appointed by the Crown. Funds for administering the county were raised by means of a local tax, the county cess. In the early years of the nineteenth century the Grand Jury was responsible for constructing and maintaining all roads and bridges in the county, for the building and repair of jails, courthouses and county infirmaries, as well as providing relief for the poor.

At local or parish level, in the general absence of any alternative to agriculture as a way of living, the landlord was in a position of singular power. Frequently he, or his nominee, was magistrate and member of the Grand Jury. Many of the landlords were also the members of parliament. In a myriad number of ways they exercised enormous power over the mass of the tenantry and the landless labourers.

Successive governments relied on the administration in Dublin Castle to provide them with reliable information, and advice, on the state of Ireland, in order to enable the government to formulate sensible policies for the future. One way of collecting such information was through the setting up of parliamentary committees or, in the case of areas requiring more substantial investigation, commissions of enquiry. The evidence gathered by these enquiries provides valuable data on social and economic conditions in

Ireland at this time. The principal sources from which this information was gathered were the individuals who came before the enquiries to give evidence; in the main, landlords, clergymen, members of the higher professions, and such others as were deemed to have expert or special information to contribute to the enquiry. In the early decades of the nineteenth century, with population increasing rapidly, in a country in which agriculture remained (outside of the northeast) the mainstay of the economy and of employment, it is hardly surprising that economic and social issues (including education and law and order) were the focus of particular attention.

Among the areas in which there was an understandable interest was the challenge of developing natural resources and the construction of an infrastructure for travel and transport. The need to improve communications was widely acknowledged, through the construction of roads and the cutting of canals. The construction and maintenance of roads was the responsibility of the Grand Jury and paid for by a tax levied on the county. The Grand Jury presentment system, under which the roads were constructed, came into operation in 1760. Under this system, maps, plans and estimates were brought before the magistrates by a petitioner, wishing to build or to repair a road. The petitioner was very often the landlord through whose land the road would run. On the approval of the magistrates, the proposal was then taken as a presentment before the Grand Jury of the county at one of the biannual assizes. If the Grand Jury passed the presentment, the petitioner then swore an oath as to the truth of the estimates, and received the sanction of the presiding judge. The petitioner was then responsible for the construction of the road, or the carrying out of the repair, and, on completion of the work, was paid by the county treasurer. Landlords frequently took on the construction and repair of roads as employment for their poorer tenants; but, instead of paying the labourers in money, they allowed it against their rent. This practice of jobbery, commonly referred to as 'jobbing', was strongly criticised by Nimmo, on the grounds that it produced poor quality roads, facilitated high rents, and increased the number of applications by landlords for presentments. Roads in remoter districts, where the landlord was either absent or not in the habit of attending Grand Juries, were usually neglected, thus contributing further to their isolation.

Alexander Nimmo came to Ireland in 1811 and remained in the country until his death in 1832. During this time he was employed by a number of government bodies on various projects throughout the country: he conducted surveys for the Bog Commission in Kerry and Galway, he surveyed two-thirds of the coastline for the Fishery Board, he directed the construction of two hundred and forty-three miles of road in the west, and supervised the erection of over forty piers along the west coast. Nimmo was also responsible for a number of private commissions. His opinions and observations were highly valued in his day by social commentators and those concerned with public policy. Nimmo had studied conditions in Ireland before taking up his appointment in the country, and he took every available opportunity to add to his knowledge of Ireland in the years that followed.

During the course of his work, Nimmo made many journeys around the country, and referred to himself as 'an observing traveller'. This enabled him to witness at first hand, and to give testimony to, the general lack of enterprise which prevailed among the landed proprietors of the country, and the many instances that he noted of their mismanagement of their estates, and neglect of their duties as magistrates and Grand Jury members. Nimmo also noted the neglect of the natural resources of the country, by individuals and government alike. He commented in particular on the living conditions of the poor. On all of these aspects of 'the condition of the people', Nimmo presented evidence to various enquiries, on occasion supplying suggestions as to how conditions might be improved and offering solutions to some of the many deep-seated problems that bedevilled Ireland at this time. In general, his evidence shows a distinct sympathy with the predicament of the tenant and the poor, compared to his critical tone in relation to landlords and governments.

Nimmo's evidence before commissions was much more forceful in tone than that used by him in his regular reports to Dublin Castle. Indeed the contrast is striking. In one, he is the dutiful employee, carrying out the wishes of his superiors; before commissions he is a man of principle and ideas, not afraid to share them with others and not afraid to criticise the shortcomings of the government, i.e. his employer. Several of Nimmo's recommendations to these enquiries were original, and were eventually taken up many years later: a system of relief of the poor by local authority, an arbitration court to adjudicate on rents between landlord and tenant, the development of industry in urban centres to reduce the over-dependence on land.

Nimmo's evidence to these special enquiries, coupled with the information supplied in his many reports to the Bog Commission and the Fishery Board, furnishes us with a close examination of conditions on the west coast of Ireland at a time when relatively few travelers had documented the region. The value to the historian of such material need hardly be emphasised. An interesting feature of Nimmo's reports and recommendations is the fact that he rarely remarks on, or attributes economic or social conditions to, the 'character' of the people. This is to say, he is generally free of moralistic, racial or religious explanations for economic and social conditions. At a time when the Catholic question was being hotly debated in every newspaper, as well as generally in public and private meetings, Nimmo's reports of the 1820s generally carry no material of a political nature: there is no mention of local divisions, local politics, law and order issues, or crime, except with regard to the use of corn in illicit distilling.

While the full range of Nimmo's opinions and observations is worthy of attention, the particular materials supplied in his reports to the Fishery Board, on the coast from Sligo Bay to Galway Bay, are so rich in local data that I felt compelled to reproduce them in their entirety in the appendices to this book. They still retain their interest today, almost two hundred years after they were first written.

Chapter 1

Investigating the west coast of Ireland 1811–1814

Irish history in the 1820s is dominated by the Catholic Association's campaign for Emancipation, which was eventually granted in 1829, and by the resistance to tithes, taxes and rising rents. Commentators on this period usually give brief mention to a partial failure of the potato crop in 1821. This resulted in famine in the south and west in 1822, to which the government responded with public works. Historians then quickly move on to the 1830s, which seems a far more eventful decade for agitation and reform.

However, for the western counties of Mayo and Galway, the 1820s was a time of huge structural development. It was during this period that many of today's roads were laid out, and a large number of piers were built. The piers and the roads were built in response to the 1822 famine. The works to a great extent were financed by the government and brought large-scale employment to the poor. But more significantly they left permanent improvements in their wake.

The construction of the roads and piers opened up the west as never before. Previously inaccessible regions such as Erris, the Mullet Peninsula and Achill in Mayo, and Connemara in Galway were made accessible and more land became available for cultivation. This increase in available land brought an increase in population and agricultural produce. And where the roads and piers met, villages usually developed. At this time many of the landlords, whose families prior to this had almost ignored their estates in the west, took up residency and began planning towns and villages. Markets developed, import and export trade was established, and, under the Fishery Board, the fishing industry was encouraged and ancillary industries, such as curing, sail making and boat building, began to develop.

Alexander Nimmo

The name of the Scottish engineer, Alexander Nimmo, is frequently associated with the developments that took place in Connemara at this time, but in fact his influence is to be found throughout the west. In his capacity as engineer for the Western District and engineer with the Fishery Board, Nimmo directed the entire operation. It was he who chose the site for the piers and the routes for the roads. For the first half of the 1820s he successfully combined both positions for the betterment of the region. As engineer with the Fishery Board he was responsible for overseeing the construction of over forty piers between 1822 and 1824

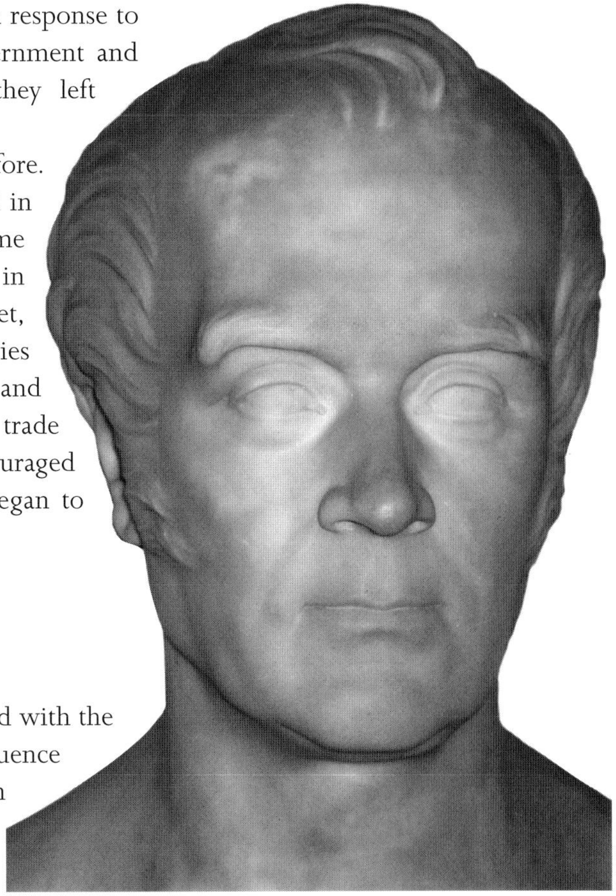

Alexander Nimmo (1783 – 1832)
Marble bust on display in the Library of the Royal Dublin Society.

and as engineer for the Western District he oversaw the construction of two hundred and forty-three miles of government roads from 1822 until 1831. Decisions made by Nimmo at this time were vital to the survival of isolated communities and districts, not to mention the future financial stability of the local landed proprietors. The choices he made determined which villages survived and which were abandoned, which districts were marginalized and which were developed. For a road to bypass an existing village meant death to that village and, when the site for a pier was chosen, there was a shift of focus to that section of the coast. How Nimmo found himself in such an influential position was no happy coincidence, but rather an indication of the high regard in which the Irish Administration at Dublin Castle held him.

Alexander Nimmo first came to Ireland in 1811 to work as an engineer for the government-appointed Bog Commission. When his term with this body was over in 1813, he remained on to take up a number of private commissions in various parts of the country. From the start Nimmo was considered a man of broad experience and it was well recognized that he was well informed on subjects far beyond the confines of his profession. This knowledge had been gathered through study and observation, both of which would remain life-long pursuits. His keen intellect and thirst for knowledge had marked him out from an early age. Born into a relatively modest family in Cupar, Fife, Scotland, in 1783, Nimmo was the son of a watchmaker who later ran a hardware store in Kirkcaldy. He attended the grammar school at Kirkcaldy and was, by all accounts, a very bright student. He was awarded the Bayne bursary, a grant of £120 per annum to students of poorer families, and this helped finance his university education.[1] He attended St. Andrews University from 1796 to 1799, where he studied Latin, Greek, mathematics, logic, ethics and natural philosophy. This was followed by a year at Edinburgh University, studying physics, ethics and mathematics. Early biographical references tell us that he read several languages, he was an accomplished scholar in Latin and Greek, and 'the higher branches of mathematics and algebra were his favourite studies'.[2]

Having completed his university education at the early age of seventeen or eighteen (1800 or 1801), Nimmo took up a teaching position as second master at Fortrose Academy in Scotland. At Fortrose he taught arithmetic, bookkeeping, and drawing; 'the elements of Euclid, navigation, land surveying and other mensurations; architecture, fortification and gunnery; also the elements of chemistry and of natural philosophy'. He was also housemaster for a number of students who boarded in a house provided for him. He left Fortrose in 1805 to take up the appointment of Rector [Headmaster] of the Academy of Inverness, where he remained until 1811. Here again Nimmo taught mathematics and natural philosophy, along with Spanish, Italian and French.[3]

While Nimmo was at Inverness there were a great many roads and bridges under construction in the region, the work being carried out under the direction of the renowned engineer, Thomas Telford. Telford had conducted a survey of Scotland for the government in 1802 and reported on ways of improving communications in the country and of promoting the fisheries on the east and west coasts. Later in the same year he was instructed to make a further survey of the interior of the Highlands. The latter report was published and became the basis for large-scale improvements carried out in the Highlands over the next eighteen years. During that period, nine hundred and twenty miles of road and one thousand two hundred bridges were constructed and financed by equal contributions from central government and local taxation.[4] Those recommending that similar work be undertaken in Ireland would later point to the example set by Telford in the Highlands.

Soon after taking up his position in Inverness, Nimmo became acquainted with Telford and began to show an interest in surveying. Telford was impressed by his ability and went on to recommend him to the commissioners for Highland Roads and Bridges. The commissioners employed Nimmo during his summer vacation of 1806 to make a survey of the boundaries of the five northern counties and to insert these on to Aaron Arrowsmith's map of Scotland. Before setting out he was issued with a skeleton map and supplied with a telescope, compass and sextant. From June to July, he travelled on horseback examining and mapping the county boundaries, usually accompanied by a guide, the agent or 'factor' from the neighbouring estate. He employed a young man 'well acquainted with the Gaelic language' as an interpreter, along with a manservant to attend to the horses.[5]

Nimmo's handwritten report to his superior took the form of a journal and is as informative on Inverness-shire in 1806 as his future surveys of the west of Ireland would prove to be. In it he mentions several topics which he would raise again in Ireland. He comments on the geography of the region, the mode of agriculture practiced by the tenants in the highlands and on the state of the roads. Some sheep farmers, he wrote, had cleared the small tenants from their lands and were using their holdings to provide winter pasture or to produce winter feed for the sheep. Nimmo recommended that, rather than reduce the population of the highlands by such disruptive measures, they should instead reclaim the bog land and convert it into 'hay pastures'. He also noted how heavy flooded mountain streams frequently damaged bridges and destroyed roads, demonstrating the importance of well-constructed roads and bridges, an observation that would, no doubt, prove useful in the future.[6]

Nimmo also conducted a survey for Telford of the proposed route for a new road 'southward from the Great Glen along Lough Treig and across Rannock Moor to Glen Lyon'. The road was never made, although Nimmo's report was published. It is suggested that Nimmo may have carried out further work for Telford in the highlands while still employed at the Academy. Certainly he was seldom, if ever, idle at Inverness, as he used his time to further advance his studies. Evidence of Nimmo's academic endeavours at this time can be found in the several articles he contributed to *The Edinburgh Encyclopaedia* displaying his knowledge of bridge building, carpentry, drainage and civil engineering. He was already winning a reputation as 'a man of scientific knowledge and skill'. He was elected an honorary Member of the Geological Society of London in 1809 and a Fellow of the Royal Society of Edinburgh in 1810.[7] In June 1811 Nimmo resigned from the Academy of Inverness and, again on the recommendation of Thomas Telford, took up a position as engineer with the Bog Commission in Ireland.[8]

The Bog Commission

The Bog Commission was one of a number of government appointed commissions set up to gather reliable information and to make practical recommendations that would enable the government of the day to take appropriate measures to 'improve' the conditions of Ireland.[9] In the case of the Bog Commission it was information collected by qualified engineers and surveyors while surveying the bogs of Ireland over a number of years.

The Bog Commission was set up in 1809 to ascertain the nature and extent of the bogs of Ireland, and the practicality of converting them into agricultural land. The commission employed ten engineers between the years 1809 and 1813, to survey and map over one million acres of bog and to report their findings.[10] At a time when accurate cartographic data on Ireland remained defective, particularly at a local level, conducting a survey on this scale was a huge undertaking. This was the first time the government had employed professional engineers and surveyors to survey and map large sections of the country and it is recognized today as a turning point in the history of cartography in Ireland.[11] The engineers – Alexander Nimmo, Richard Griffith, Richard Lovell Edgeworth, Thomas Townsend, Richard Brassington, James Alexander Jones, John Longfield, David Aher, William Bald, and Thomas Colbourne – were paid two guineas a day, plus allowances for travel and accommodation.[12] They were assisted in the field by forty surveyors and a large number of chainmen,[i] staff men and labourers. The surveyors received one guinea per day and it was left to the engineer to 'make the best bargains in their power' when hiring the chainmen and others, but this was not to exceed three shillings per day.[13]

Some of the engineers had already established practices in Ireland, such as Richard Brassington, John Longfield, Richard Lovell Edgeworth and Richard Griffith; others came specially to Ireland to work for the commission. Edgeworth, an improving landlord from County Longford, was the father of the writer Maria Edgeworth, and his son, William, later worked on the survey. Richard Griffith was born in Ireland in 1784 and had studied in London and Edinburgh before returning home in 1808. As well as being an engineer, he was a well-known geologist and a Fellow of the Royal Society of Edinburgh.[14] Griffith's name later became synonymous with the Primary Valuation carried out under his direction. William Bald had come to Ireland in 1809 to conduct the Trigonometrical Survey of County Mayo. Like Nimmo, Bald was born in Fife, Scotland (1789).[15] He was appointed to the Bog Commission on 22 May 1810 and worked with them for 598 days, receiving a salary of

[i] A chainman was responsible for the measurement of length using specially constructed accurate codified 'chains'.

£1,246.11s.[16] He went on to produce a county map and survey of Mayo, published in 1816, and later worked with Nimmo on the construction of piers in that county. Of the ten engineers employed by the commission, Griffith received the largest salary, having worked the greater number of days; he earned £2,466.4.9d for 858 days. Nimmo was next, receiving £1,719.16s for 720 days.[17]

Each engineer surveyed a district allocated to him by the commission. He drew up a map and supplied a detailed report, according to the guidelines given. The commission later published these in a series of reports to the House of Commons, accompanied by scaled down versions of the maps. The maps showed the extent and boundaries of the bogs and identified their natural drainage. They also illustrated the topographical features of the surrounding countryside and indicated any drainage and improvement works proposed by the engineer. The engineers' reports described the different types of bog. They pointed out which bogs could be drained and which, if any, should be flooded. They investigated the possibility of draining the bogs using navigation canals, which could later be used to convey fertilizer into the bog and future produce from it. They also suggested a suitable method for the future cultivation of the reclaimed bog. They outlined any new roads needed to facilitate the transport of fertilizer and crops between the bogs and existing roads in the vicinity. Estimates were given for all works proposed and projections were made for the increased value of the reclaimed land. The engineers also gave the names of the local proprietors and included details on any bogs already reclaimed in their district.[18]

At one time over three million acres of bog covered Ireland. Most of this had been brought into cultivation, but still thousands of acres were to be found among the rich arable lands in the centre of the country and in the remote mountainous regions along the coast. This resulted in almost every landowner holding some acres of bog on his estate. This bog, or wasteland, was seen by some as a reservoir of future agricultural land, an underdeveloped resource awaiting only the energy and funds to bring it into

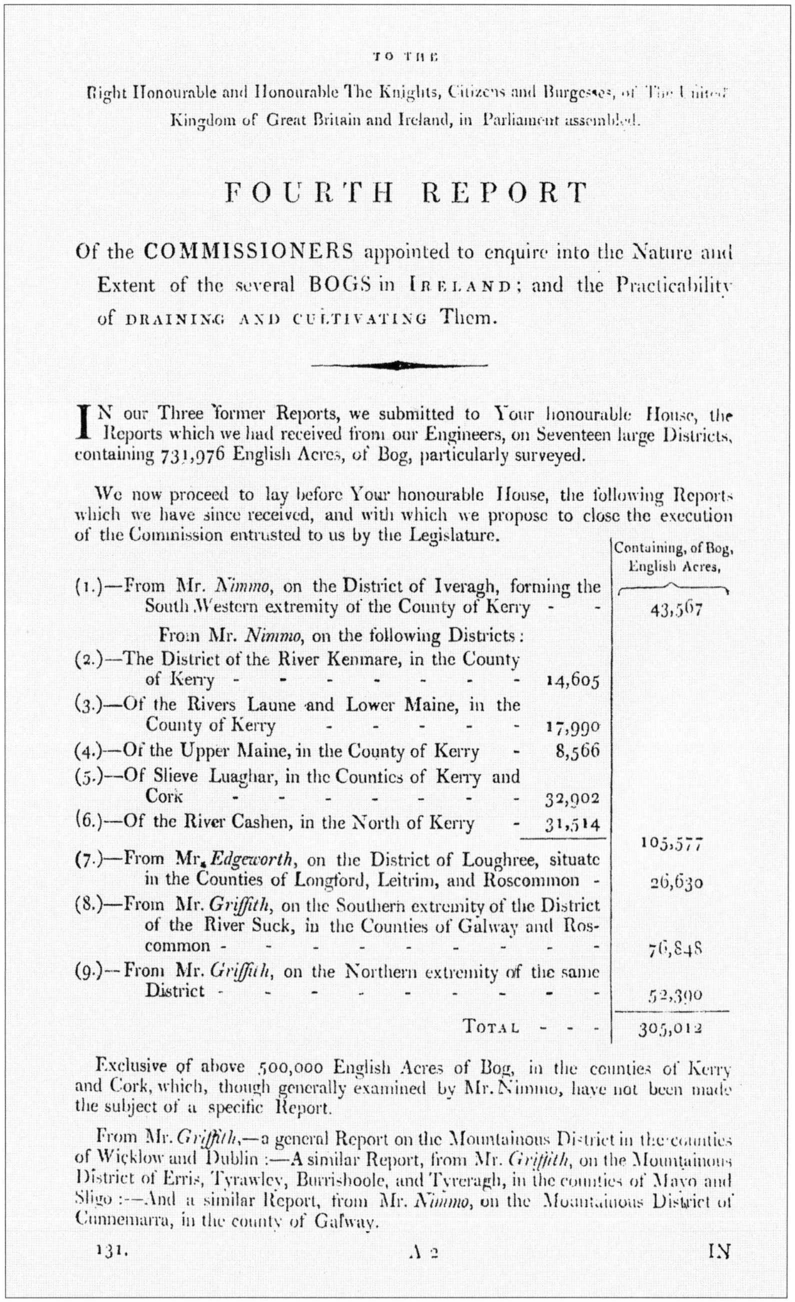
TO THE

Right Honourable and Honourable The Knights, Citizens and Burgesses, of The United Kingdom of Great Britain and Ireland, in Parliament assembled.

FOURTH REPORT

Of the COMMISSIONERS appointed to enquire into the Nature and Extent of the several BOGS in IRELAND; and the Practicability of DRAINING AND CULTIVATING Them.

IN our Three former Reports, we submitted to Your honourable House, the Reports which we had received from our Engineers, on Seventeen large Districts, containing 731,976 English Acres, of Bog, particularly surveyed.

We now proceed to lay before Your honourable House, the following Reports which we have since received, and with which we propose to close the execution of the Commission entrusted to us by the Legislature.

		Containing, of Bog, English Acres,
(1.)—From Mr. *Nimmo*, on the District of Iveragh, forming the South Western extremity of the County of Kerry - -		43,567
From Mr. *Nimmo*, on the following Districts:		
(2.)—The District of the River Kenmare, in the County of Kerry - - - - - - -	14,605	
(3.)—Of the Rivers Laune and Lower Maine, in the County of Kerry - - - - -	17,990	
(4.)—Of the Upper Maine, in the County of Kerry -	8,566	
(5.)—Of Slieve Luaghar, in the Counties of Kerry and Cork - - - - - - - -	32,902	
(6.)—Of the River Cashen, in the North of Kerry -	31,514	
		105,577
(7.)—From Mr. *Edgeworth*, on the District of Loughree, situate in the Counties of Longford, Leitrim, and Roscommon -		26,630
(8.)—From Mr. *Griffith*, on the Southern extremity of the District of the River Suck, in the Counties of Galway and Roscommon - - - - - - - - -		76,848
(9.)—From Mr. *Griffith*, on the Northern extremity of the same District - - - - - - - - -		52,390
TOTAL - - -		305,012

Exclusive of above 500,000 English Acres of Bog, in the counties of Kerry and Cork, which, though generally examined by Mr. Nimmo, have not been made the subject of a specific Report.

From Mr. *Griffith*,—a general Report on the Mountainous District in the counties of Wicklow and Dublin:—A similar Report, from Mr. *Griffith*, on the Mountainous District of Erris, Tyrawley, Burrishoole, and Tyreragh, in the counties of Mayo and Sligo:—And a similar Report, from Mr. *Nimmo*, on the Mountainous District of Connemarra, in the county of Galway.

131. A 2 IN

Bogs surveyed by A. Nimmo in counties Kerry, Cork and Galway
(Fourth Report of the Bog Commission 1813-14)

productivity. It was hoped that the government would assist the landlords by providing some form of grant by way of an incentive to help them begin reclaiming their bogs. The setting up of the commission was seen as a possible first step in that direction.

The principal bogs were already surveyed and several of the engineers' reports published by the time Nimmo joined the team. He is recorded as having been appointed to the commission on 5 January 1811, but does not appear to have taken up his position until later in the year. He was assigned the bogs of Kerry and part of Cork on 11 June 1811.[19] Immediately on his appointment, Nimmo began reading background material on Ireland to prepare himself for the task ahead: 'When I was employed as an engineer on the subject of the survey of the bogs in Ireland, which was the object for which I first went to Ireland, I made it my particular business to make myself thoroughly acquainted with the general state of agriculture, the rent of land, and the produce; at that time farming was comparatively a profitable occupation'.[20]

His first report from Kerry was dated 27 December 1811; the rest were handed in to the commission in 1812, and all were published in 1814. During his time in Kerry, Nimmo had conducted a full survey of a little over 105,000 acres of bog and produced six detailed maps, at two inches to one Irish mile (I:40,320).[21] The principal bogs surveyed were on the Iveragh peninsula, between Dingle Bay and Kenmare River. Also surveyed were bogs in the areas drained by the Kenmare, Laune, Main and Cashen rivers, and the Luachra mountains in Kerry and Cork. A general examination was made of a further 500,000 acres lying between Kerry and Cork, but no specific report was requested for this.[22]

In his reports Nimmo expanded on methods of draining and reclaiming bogs, pointing out techniques adopted in Scotland in similar terrain. He drew attention to the backward state of agriculture in the region and commented on the absence of 'the plough, cart or team', and suggested improvements to the common spade used by the people, which was totally unsuitable for bogs. Ireland was, he observed, in the same state as Scotland, before the introduction of 'English husbandry [had] led to the improvement of the older implements'. To improve communications he proposed routes for new roads, suggesting that roads were more appropriate in poor districts than canals and railways; the latter being best suited to districts where capital and produce were in abundance. Although other engineers in previously published reports had already discussed the methods and projected the cost of reclaiming bogs, the commissioners considered Nimmo's report to be a 'highly valuable' addition and his calculations for reclaiming mountain bogs, 'ingenious'.[23]

Having obtained reports on the bogs in the centre of the country the commissioners next initiated a 'general examination' of the more remote bogs in the mountain districts of Wicklow, Erris in Mayo and Connemara in Galway. William Bald had already surveyed 20,000 acres in Mayo and Jones and Griffith had covered north and east Galway. The western districts of Mayo and Galway, 'in their present state of desolation and abandonment', did not, in the opinion of the commissioners, justify the expense of a full survey and they requested instead that the engineers produce a general report and add their findings to available maps on these districts. Richard Griffith was allocated Wicklow, (which included south Dublin), and Erris, (which included north Sligo). Nimmo was allocated the 'extensive and desert district of Cunnamara (sic), occupying the west of the county of Galway'.[24]

Marginalized districts and their people

In sanctioning the surveys of Mayo and Galway the Bog Commission introduced the first 'men of science' into the west.[25] Nimmo, Griffith and Bald were the first engineers to survey the remote western parts of both counties. Earlier surveys of Ireland tended to be of a general nature and made only passing observations on the west coast of both counties. Edward Wakefield in his *An Account of Ireland Statistical and Political*, published in 1812, confessed to not having set foot in Erris: 'I was on the edge of this country but did not cross it, and few of the gentry who have property here, to use their own expression, "have ever been in it"'.[26] Dr Daniel Beaufort pleaded bad weather as his excuse for not visiting the Mullet: 'When I was in Mayo, the season was unfortunately too much advanced, for to venture so far into this difficult country'. He did, however, gives a good description of Connemara. A close study of Nimmo's bog report on Connemara, and of evidence he later gave before government inquiries, clearly indicates that he was familiar with Beaufort's 1792 map and survey of Ireland.[27]

Indeed, James Hardiman would argue in 1820 that all of Connaught, 'its natural advantages, former transactions, numerous and respectable families, &c., ... remained almost unnoticed and unknown'.[28]

The more detailed county surveys conducted under the direction of the Dublin Society (re-named Royal Dublin Society in 1820) went some way to addressing this shortfall. It was Robert Fraser who first suggested that a survey be taken of the agriculture and other resources of each county in Ireland, with suggestions proposed as to how these could be improved. The government granted £3,000 to the Dublin Society to cover the expense of making and publishing the surveys.[29] James McParlan's survey of Mayo was published by the society in 1802. However, Hely Dutton's survey of Galway was still in the making and was not published until 1824.

The authorities regarded the coastal districts of Mayo and Galway as remote, backward and lawless wastelands. Cut off from the rest of the country by expansive bog lands, mountains and sea, they were, for the most part, the property of absentee or impoverished landlords and were totally neglected by them and by the inept local authority, the Grand Juries. Efforts were made to identify the underlying problems, to supply remedies and to apportion blame and responsibility, but little was done to improve the situation.

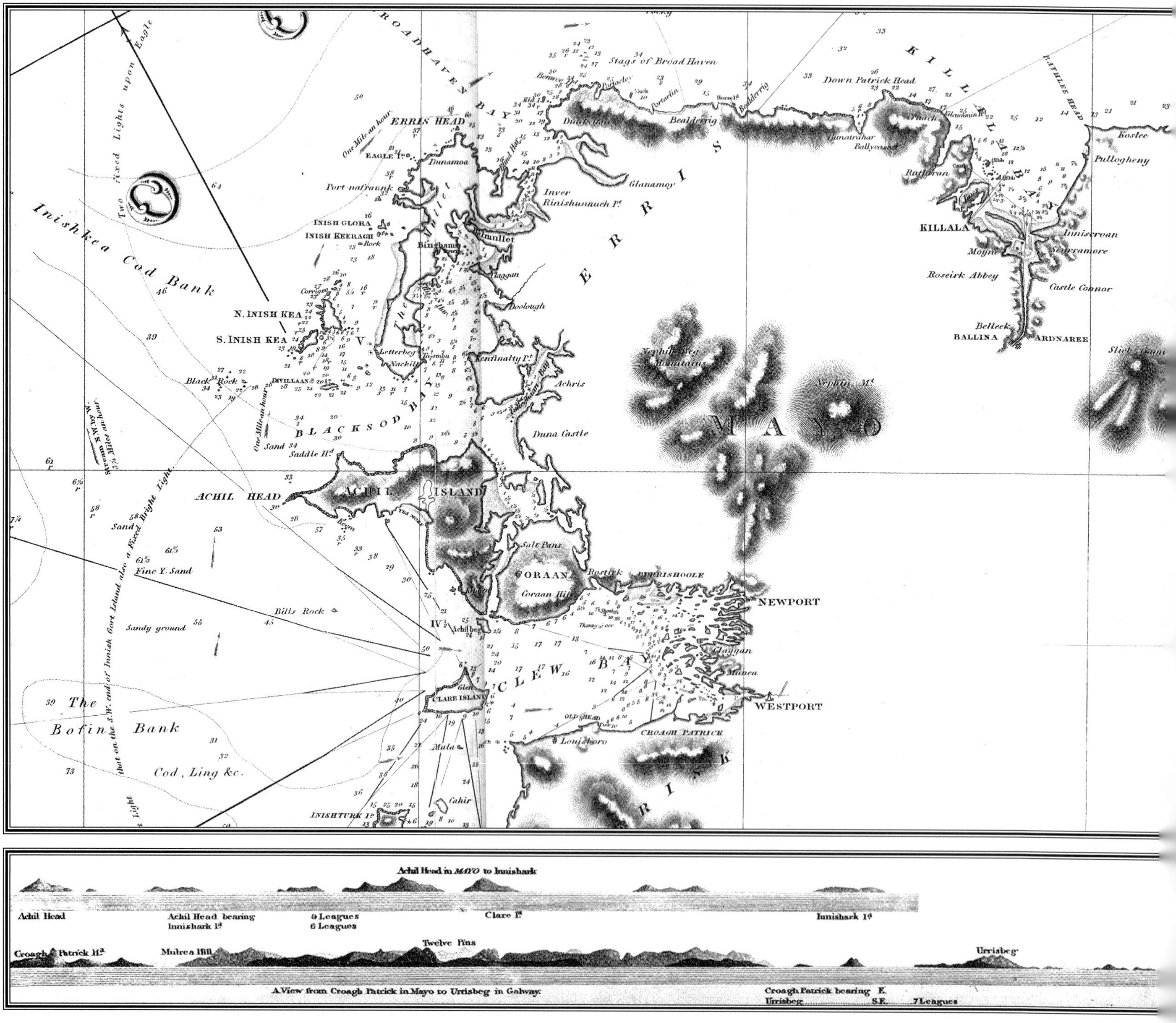

Details from Nimmo's chart of Mayo coast, circa 1831
(*Chart of St George's Channel and the coast of Ireland*, by A. Nimmo)

Independent commentators and government appointed committees reported time and again on the appalling conditions to be found in the west. They told of the poverty of the tenant class, the poor housing, the disease, the backward state of the agriculture and fisheries, the periodic distress brought on by crop failure, the unlawful behaviour of the people and the total lack of government control. In the early years of the nineteenth century the Napoleonic wars had brought an increase in agricultural prices and in the quantity of corn and young calves raised in the west. It had also boosted the kelp trade by preventing the import of barilla, a cheaper alternative. However, this improvement in income for both tenant and landlord would see a reversal when peace was declared. The agricultural depression that followed the end of the Napoleonic wars affected both classes and contributed to the fall in the numbers employed throughout the country. This was more acutely felt in the west where the only employment to be had was that of farm labourer. The lack of employment, a recent increase in population figures and the total dependency of so many on the land, had contributed to high rents, subdivision and a dependency on the potato, which was the only crop that could produce a large return from a small portion of land. The conclusion reached by the various bodies was usually the same; the poor conditions of the people were attributed to bad estate management, unemployment and the underdeveloped agriculture and fishery industries. The landlords were accused of allowing overcrowding, subletting and subdivision on their estates and of doing little to encourage improvements by way of expenditure or example.[30]

The rundale system

The population of the remote regions tended to occupy the few arable acres to be found in the interior, but mostly they lived on the coast. Land close to the shore was more fertile than that of the interior and here the dual occupations of fishing and farming could be attended to quite easily. The smallholdings occupied by the poor tenants were capable of supplying only enough to sustain the family and the fishing paid the rent. In many instances the rent was paid in day labour. The system of land tenure permitted by many landlords in the west, known as 'rundale', drew particular criticism. This was a system of collective farming, usually by extended families, that encouraged overcrowding and subdivision, both of which reduced the size of holdings and pushed up rents. Under the rundale system, several tenants came together to rent a portion of land, which usually included some arable, mountain and bog. They built their houses in clusters, forming a rundale village or *baile,* and divided the land so that each house received a portion of garden, on which to grow the potato crop, with a few acres of mountain and wasteland to raise their stock. Any attempt by the tenant to reclaim a portion of the wasteland was met with an increase in rent or eviction, 'to make way for some other tenant who was willing to pay an increased rent'.[31] They could, however, increase their holding by renting conacre; this was land taken on a yearly basis to raise an extra crop of potatoes or oats for

***Clochan* or *baile* settlement circa 1862**
(*The West of Ireland*, by Henry Coulter)

sale in the market place. But rent on conacre was high and had to be paid in advance, with the result that, should the crop fail, the tenant lost everything. The *bailes*, described as being 'almost as numerous as the tenanted farms', give a false impression on early maps of a larger population inhabiting the region than was in fact the case.[32]

Some saw the rundale system as the most effective use of a region that contained very little arable land and was made up principally of mountains, high moors, bog and wasteland. Others felt it left the tenants very poor and vulnerable, with little possibility of increasing their holding or of advancing their circumstances. In the west, as in the rest of the country, it was customary for the tenant to be six months or a year in arrears. This suited the landlord as it gave him power over the tenant and insured that he adhered to his will in all things, whither it be in accepting rent increases, or casting his vote in times of election. The practice of leasing land on verbal rather than written agreements further added to the vulnerability of the tenant.

Fishing

Valuable fish banks were known to exist off the west coast but the local boats were too small to venture that far out. They preferred instead to fish close to the shore and, 'on the first appearance of a smart breezes or sudden change of weather, immediately ran for shelter into the next creek or harbour'.[33] The most common boat used by the fishermen was the currach, a light, wooden-framed boat, covered with horse or bullock hide. The currach had been in use on the west coast for centuries. Although considered unsuitable for the open sea, they had, however, many advantages. They were extremely buoyant, with the result that few accidents happened to them, they could be run on to a beach and were

West of Ireland Currach, circa 1862
A light, wooden-framed boat, covered with horse or bullock hide, usually with a higher bow than shown here. (*The West of Ireland*, by Henry Coulter)

light enough for a fisherman to carry home on his shoulders.[34] The fishermen fished the herring that twice yearly, in autumn and winter, flooded the bays, making it unnecessary for them to venture out into the open sea. The herrings were sold locally to larger vessels, for resale in Galway and Westport, or to ships from as far away as England, Scotland and the south and east coast of Ireland.[35]

As well as currachs, the fishermen of Galway Bay had also larger boats. In 1762, it was reported that two hundred boats were employed in Galway Bay, of these 'about 160 belonged to the town', and the rest to the Clare side of the harbour.[36] Arthur Young, writing in 1776, tells us that the boats used by the Claddagh fishermen in Galway Bay, the only exclusively fishing village on the west coast, were 'from 4 to 6 tons, some 9 tons'. The cost of building a boat was £20, 'and the nets and tackle, £15'; the nets were made of 'hemp, tanned with bark'. Usually there were five or six men to a boat; 'they fish by shares, dividing into sixty'.[37] In some cases the boat and nets belonged to one person and were hired out; the owner got a half share and the remainder was divided amongst the crew.

Housing

The houses of the poor in the first half of the nineteenth century were described as being, 'the merest hovels, unfit for swine to inhabit. They were constructed for the most part of mud, and were sunk in the ground so that the walls of the upper part could be constructed' from the excavated mud.[38] A step up from that were the small low thatched cottages with earth floors, usually occupied by large families in the company of their animals. The method of building cottages was similar in both Galway and Mayo. They were

Cottages circa 1862
(*The West of Ireland,* by Henry Coulter)

usually made of stone without mortar. In some places the foundation was of stone, and the remainder of the wall was 'either of sods or tempered clay'.[39] They tended to be 'badly roofed and thatched, and full of smoke and dirt, as they had neither chimneys nor offices [outhouses]'. But there were indications that improvements were under way and many houses were being built with lime and stone. Some still shared their accommodation with the cows, pigs and other farm animals, but more frequently a separate building was erected for the animals.[40]

Food, clothing and education

The people made their own clothing, manufactured from the wool and flax produced locally. Their fuel was turf, cut from a nearby bog and their food was principally potatoes, with a little oat bread, milk, butter and herrings; meat was eaten 'on two or three grand festival days in the year'.[41] The potato was a favourite with all classes and large quantities were required to sustain the population. The quantity used by a family of six was reckoned to be about twenty-two stone per week. Added to this was the amount given as feed to the animals and that reserved for seed potato for the following year. The *Lumper* was the most common potato used, because it was the most productive. However, according to Hely Dutton, it was 'a wretched kind for any human creature; even pigs, I am informed, will not eat them if they can get any other kind'.[42]

The national school system had not yet been introduced, but both counties did have a number of private and charitable schools and these were well attended. Although English reading and writing was taught in the schools, it was usually only spoken by the children. Irish, however, was understood by the landlords and used in their dealings with their tenants: 'In Connaught, the gentry understood Irish, which facilitates their intercourse with the peasantry; therefore they are, consequently, enabled to become acquainted with their wants, to assist them with advice, and restrain them by admonition'.[43]

Landed estates and their owners

Almost all of the landed estates in the west were held under the rule of primogeniture and were therefore inherited by the eldest son. In the past, these estates were not the prime source of income for many of their owners; some had more profitable lands in other parts of the country. But more recent years had seen a change in the fortunes of many of the families and the proprietors at this time tended to be younger men who had inherited greatly reduced estates and these usually carried heavy debts. Attached to many of the estates were 'fixed payments', these related to long-term loans, acquired either through personal expenditure or inherited debts and the proprietors were struggling to meet their payments.[44] Previous generations had viewed their western estates as inaccessible and backward. However, for this generation the west coast estates were recognized as potential earners, possessing unexplored possibilities for mining, unlimited wasteland for cultivation and underdeveloped fisheries off their coast.

To develop their estates more fully the proprietors needed to attract a more prosperous tenant than heretofore, men who were prepared to invest in their holding, rather than simply establishing themselves as middlemen by subdivision. They also needed to attract artisans and merchants to help create viable market centres on their estates. To succeed they sought to improve communications and to develop new towns. Major Dennis Bingham had recently established a town north of Elly Bay, on the Mullet peninsula, which he named Binghamstown, and John D'Arcy was attempting to develop his town, Clifden, on his estate in Connemara. The government appointed commissions continuously pointed out the necessity for some form of government intervention by way of grants and loans, to assist in the development of these poorer regions. But the government paid little heed to its own commissions. Reluctant to take over what it considered to be the responsibilities of the landlords and Grand Juries, the government instead restricted itself to providing limited financial assistance only in times of severe distress. Such thinking did little to

Ballynahinch Castle circa 1862
(*The West of Ireland,* by Henry Coulter)

Clifden Castle circa 1862
(*The West of Ireland,* by Henry Coulter)

advance conditions in the west and was frequently denounced by enlightened commentators who called on the government to display the same interest in Ireland as they had done in Scotland and to invest public money in road communications and in the fisheries.[45] The information gathered by the engineers for the Bog Commission would later be used to strengthen this argument.

The principal landlord resident in Erris when the bog survey got under way was Major Denis Bingham. Mrs Carter, the proprietor of the largest estate on the Mullet peninsula, had never even visited the place. In Galway there were very few absentee landlords, but having a resident landlord was sometimes not enough. Wakefield in 1812, the year Nimmo arrived in Connemara, is critical of the lack of enterprise displayed by Colonel Richard Martin, the proprietor of the largest estate in Connemara. The Martin estate, Wakefield felt, would benefit from an increase in tenants: 'the property of Mr Martin, remains uncultivated and nearly in the state of nature; nor is it likely to exhibit a more pleasing aspect, till from some fortunate change in the agricultural system of Ireland, it becomes divided among a great number of occupiers, and is improved by the hand of industry. At present there are scarcely any roads through it'.[46] He was appalled by the total neglect and bad management exhibited by a resident landlord: 'the land is let without measurement, miserably grazed by half-starved stock, and yet this gentleman is a constant resident on his estate'.[47] McParlan complained of the backward state of agriculture in Mayo, where the farming implements were described as being 'of the most inferior kind'. The ploughs were inadequate and their work 'miserably executed'. Erris had its own peculiar spade, described as, 'a two-bladed or double spade, which the natives call a *gowl-gob*'.[48]

Nimmo's 1812 survey of Connemara

Nimmo began his survey of the bogs west of Lough Corrib in late 1812. He arrived in the district accompanied by a number of surveyors, chainmen and labourers. The people of Connemara at first observed the actions of Nimmo and his team with suspicion, until eventually word went out that a war was imminent, 'for the officers of the army were fortifying the hills'. This news was greeted with approval, 'as they expected the price of corn would rise directly'.[49] The district contained roughly 350,000 Irish acres, which Nimmo broke down as follows: about 25,000 arable land, 120,000 bog, 200,000 mountain and upland pasture and 5,000 rock, mostly limestone.[50] He was issued with a copy of William Larkin's map of Galway, which he complained did not provide a 'correct picture' of the district and proved inadequate for the purposes of the survey. Larkin (1788-1819) was producing Grand Jury maps for several counties in Ireland at this time.[51]

Nimmo approached this survey with the earnest enthusiasm that would be the hallmark of his future career. He began an exact survey of the bogs between the bays of Cashla [Costello] and Ardbear and 'along the foot of the mountain range to Lough Corrib', and proceeded to draw up a map based on the survey.[52] His progress, however, was interrupted in November when he was instructed by the Commission to continue with a general survey only and his map of Connemara was left unfinished. Before handing in his report, Nimmo returned to Connemara in the following year (1813) and spent the summer studying the geology of the region in the company of George Bellas Greenough, President of the Geological Society of London. Greenough had produced a geological map of England and Wales and was a recognized authority on the subject. The two men, accompanied by another, travelled on to Erris in Mayo and may have met with William Bald, who was then conducting his survey of Mayo for the Grand Jury. The knowledge gained on this trip would be a tremendous asset to Nimmo in later years. Greenough confirmed Nimmo's findings in relation to the geology of Connemara, and indeed of Ireland. And Nimmo included his findings in his report to the Bog Commission.

Although this was to be only a general report, Nimmo continued to work to the original guidelines outlined by the commission and went on to produce a more detailed account on Connemara than that produced on Erris by Griffith. In it he gave a full account of the geology and topography of the region. He named the principal proprietors of each district. He gave details of the geographical spread of the population and discussed the soil, manures and mode of cultivation adopted by the people, including their method of kelp making and lime making. He identified the principal bogs and suggested ways of draining them, using trenches and canals to draw the water to the many lakes. He pointed out that to reclaim the vast bogs of central and southern Connemara and convert them to fertile land, large quantities of fertilizer would be needed. He recommended that seaweed, shells and coral from the coast, and limestone from the east, be brought into the interior for use as fertilizer. To facilitate the transport of the fertilizer he suggested that the lakes

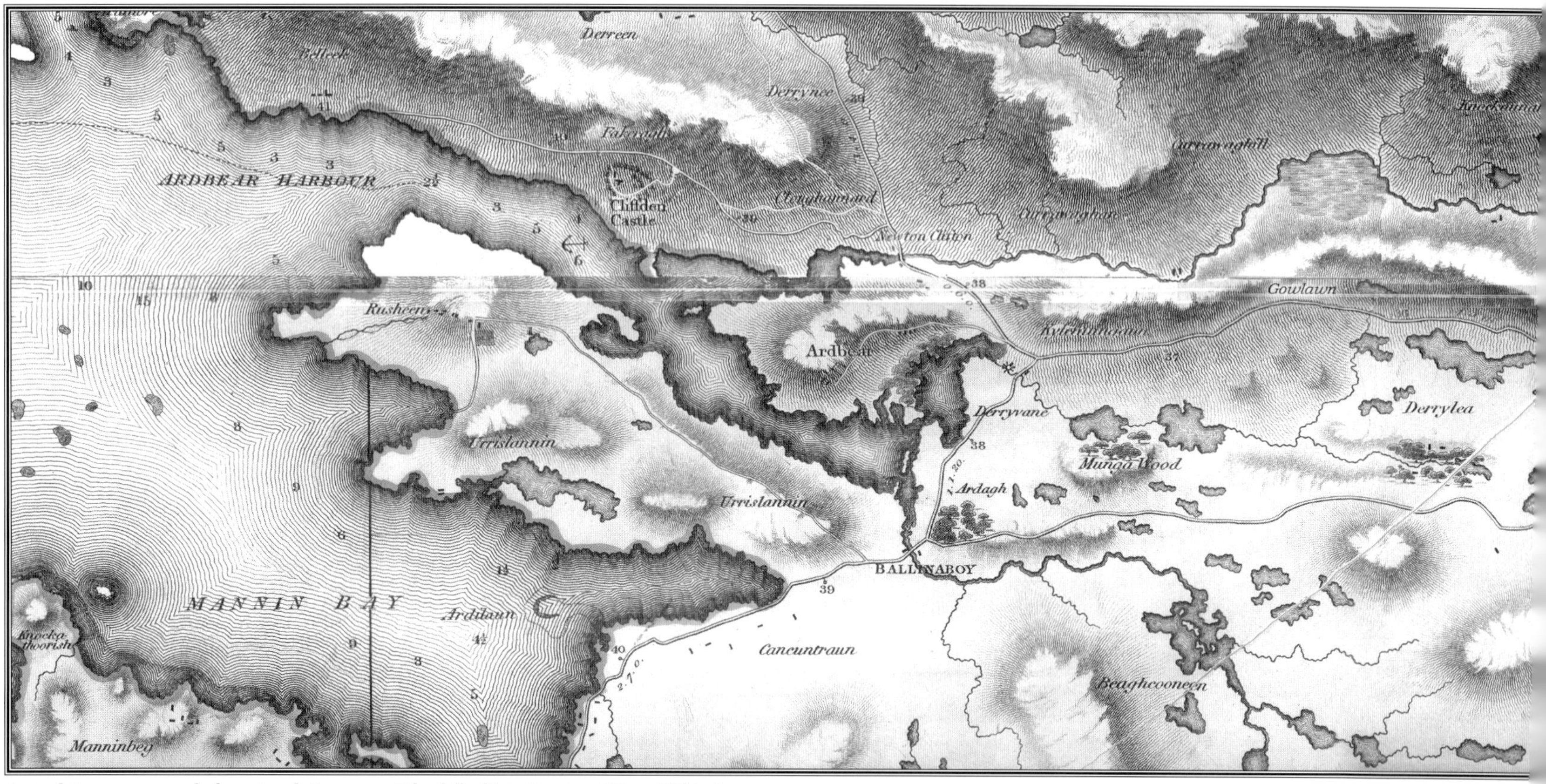

Central Connemara roads from Lough Garroman (Glendollagh) to Newton Clifton (Clifden).
Showing the old county road from Ballynahinch to Ballinaboy and the new D'Arcy road from Emlaghdauroe through Gowlan and Killymongaun to Ardbear and Clifden. (William Larkin's 1819 Map of County Galway)

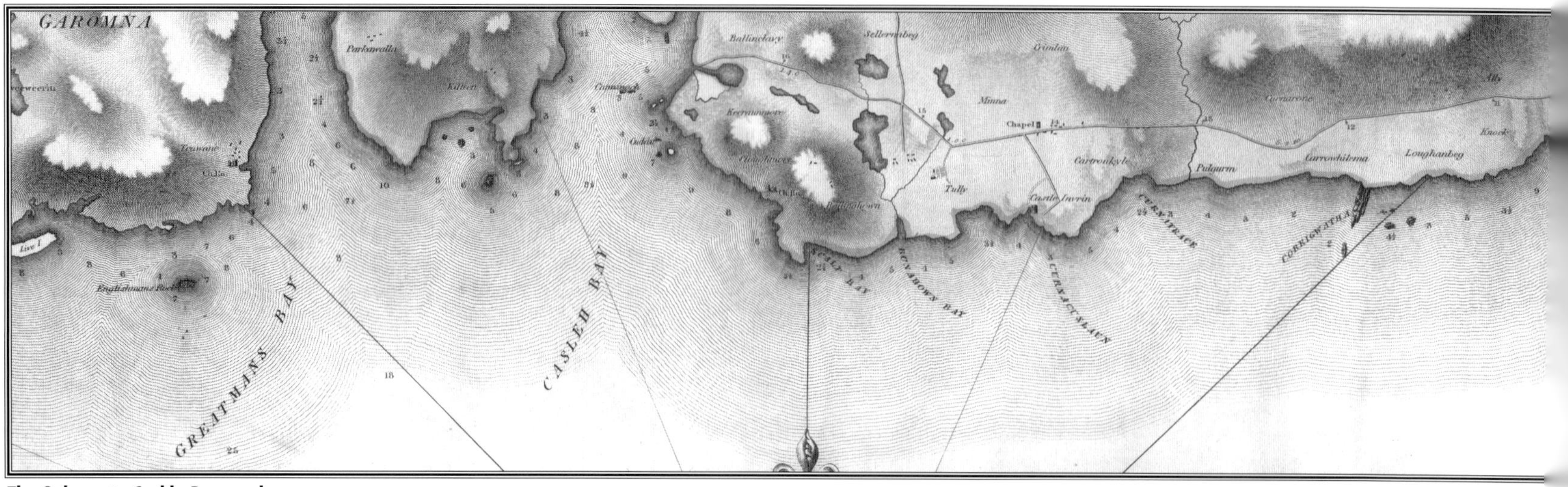

The Galway to Cashla Bay road
This was the only road in Connemara Nimmo considered suitable for a carriage in 1812. (William Larkin's 1819 Map of County Galway)

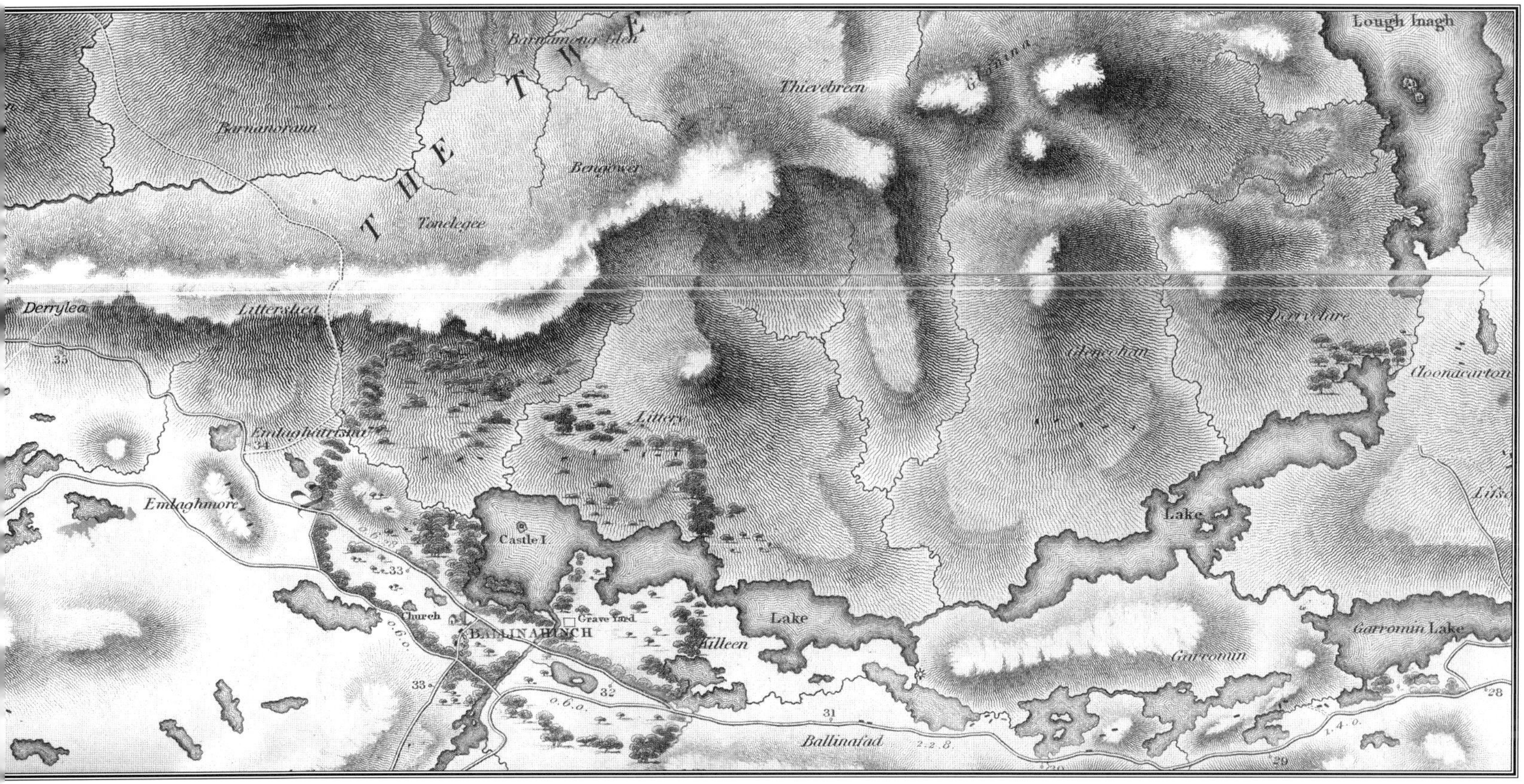
Lough Inagh
Thievebreen
Bengower
Tonelegee
Derrylea
Littershea
Emlaghmore
Castle I
Church
Grave Yard
BALLINAHINCH
Killeen
Lake
Lake
Garromin Lake
Garromin
Ballinafad

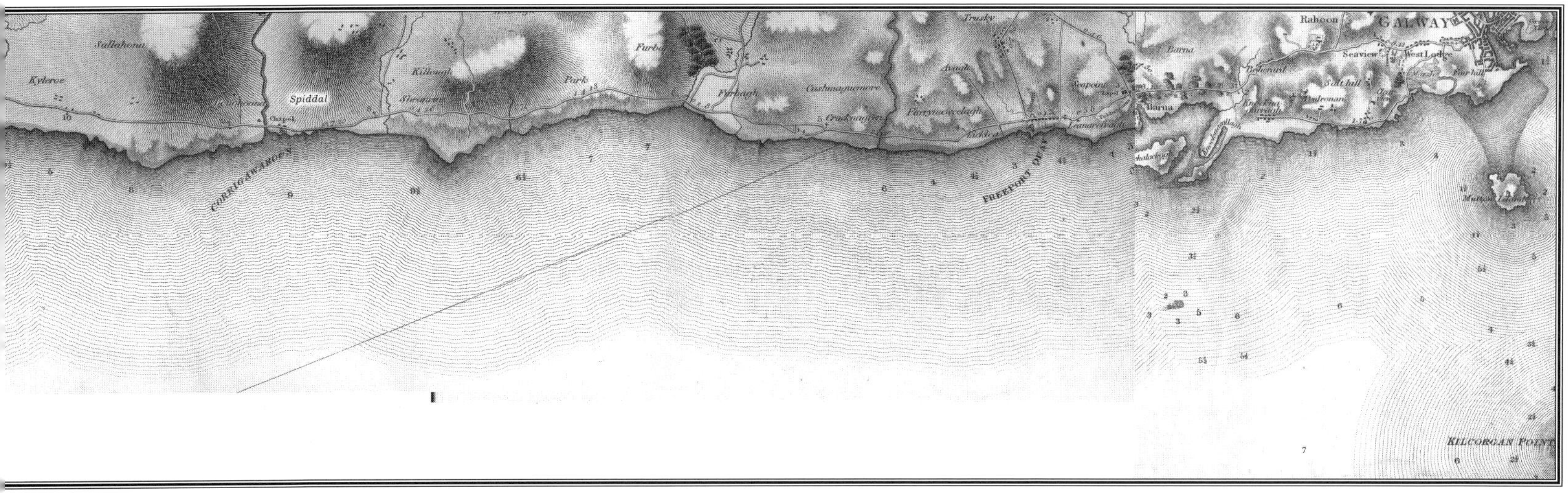
Spiddal
GALWAY
Rahoon
Barna
Seaview
West Lodge
Salt hill
FREEPORT QUAY
KILCORGAN POINT

and rivers of Connemara be converted into inland waterways and that a new network of roads be laid down. But these, he warned, 'must be designed with more attention than the present attempts at bridleroads, which seem only calculated to perpetuate the present sterility'.[53] Bridleroad was the name given to a road which could only be travelled on horseback, or on foot, and which was never intended to accommodate a carriage.

Connemara roads in 1812

In 1812 roads in Connemara had deteriorated to such an extent that the transport of people and goods was done almost entirely by sea. The few roads that did exist tended to be narrow tracks running from the *bailes* to the shore, or to the peat bogs and hill pasture of the interior. These roads were commonly referred to as boreens or bridleroads. Some of these had been in existence for centuries, others had evolved from well-used footpaths and more were of recent formation, constructed by local landowners under the presentment or Grand Jury system. The latter were poorly constructed; they were laid out so as to avoid the construction of bridges and tended to meander over long distances in an effort to retain the high ground.

In the whole of Connemara Nimmo found just one road, 'scarcely passable for carriages', and it ran along the shore from Galway to Cashla Bay.[54] There was a very poor road running through the centre from Oughterard to Ballynahinch and Ballinaboy. From Ballynahinch it ran north of Lough Fadda and Loughanilaun, then south of Munga to Ballinaboy. A little east of Ballinaboy, a branch line ran north at Ardagh to Clifden. Although a later map by William Larkin (printed in 1819) would give the impression that this was a substantial road, it was, in fact, badly laid out over rocks and bog and was sometimes no more than a track.[55] The stretch from Oughterard to Ballynahinch had been improved some years earlier with funding from the Grand Jury, but it came to an end at Emlaghdauroe, situated west of Ballynahinch Lake at the edge of the Martin estate. Sometime after 1810, under a Grand Jury presentment, John D'Arcy continued the road to his new town, Clifden, at the head of Ardbear Bay [later re-named Clifden Bay].[56]

Work on this stretch of road was under way while Nimmo was in Connemara and he may well have influenced the route chosen by D'Arcy.[57] D'Arcy's road ran north of the old road, through Derrylea, Gowlan and Killymongaun, to enter Clifden from the south, crossing the Owenglen River by Ardbear Bridge at the waterfall. The branch line from Ballinaboy and Ardagh met D'Arcy's road just south of Lough Phreaghaun. Following the construction of D'Arcy's road, the old road fell into disrepair and was abandoned. However, the Martin – D'Arcy route from Oughterard to Clifden, referred to hereafter as the old centre road, was never satisfactory and was made too narrow to take a carriage. It traversed high ground and bog, making it difficult to travel even on horseback, and was frequently flooded in winter.[58] From this old centre road, bridleroads ran north and south, through mountain passes and along the foothills into more remote regions. A bridleroad ran from Lough Oorid [Lough Úraid], along the foothills of the Maumturk Mountains to Killary Harbour [Western Way], one from Oughterard to Cashla Bay was under construction and several more were to be found to the north of the district.

Landowners of Connemara

Nimmo began his report by defining the very name of the district and the area it implied. The term Connemara was frequently attributed to that tract of land between Lough Corrib and the Atlantic Ocean, but this was, in fact, subdivided into three separate districts, Connemara Proper, Iar Connaught and Joyce Country. Each of these districts was enclosed almost exactly within the baronial divisions of Ballynahinch, Moycullen and the Half Barony of Ross. Connemara Proper, (Barony of Ballynahinch), was the most westerly, lying along the Atlantic coast. Its eastern boundary could be defined by a line drawn from the head of Kilkieran Bay to Shannaunnafeola mountain and along the ridge of the Maumturk mountains to Killary Harbour. The two other districts, Joyce Country (Half Barony of Ross) and Iar Connaught (Barony of Moycullen), lay north and south, respectively, of a line from Shannaunnafeola to the upper part of Lough Corrib. Joyce Country took its name, he explained, 'from the prevailing family name of the natives'.[59]

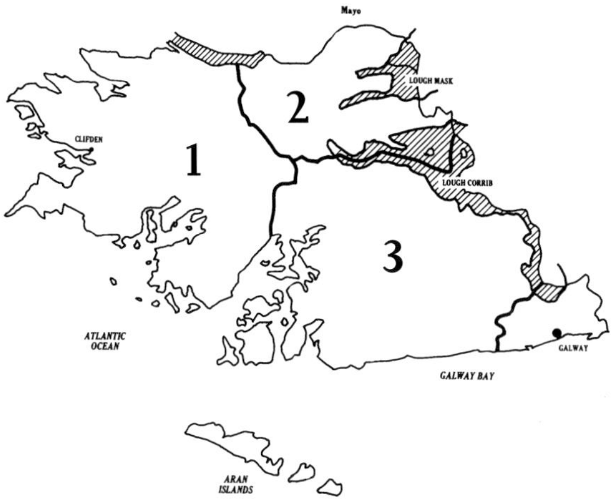

Baronies of West Galway
1 Ballynahinch
2 Half Barony of Ross
3 Moycullen
The barony was an important administrative division within each county, made up of varying numbers of townlands.

The region had been in the hands of the same families since the middle of the seventeenth century, when the land was confiscated from the O'Flaherty clan following the civil war and the Cromwellian land settlements. Almost all were descendants of the Galway Tribes, the old merchant families of Norman descent. The principal proprietors of the bogs in the Barony of Moycullen were Sir John Blake, Colonel Richard Martin M.P., Lord French, Mr Martyn of Ross, along with the Browns, Blakes, O'Flahertys and Lynches. The chief proprietor in the Half Barony of Ross was Trinity College. The land around Killary and down through Joyce Country was held by the Provost of the college. Sir Neal O'Donnell held a large estate around Cong and General Millar held over eight thousand acres at Salruck or Little Killary. William O'Flaherty, the last of the O'Flaherty to hold sway in this part of Connemara, held Renvyle on a long lease from the Blake family. The Blakes had purchased the 12,000 acres estate in 1678 and had leased it out to the O'Flahertys, who acted as middlemen. William O'Flaherty was, at the time, an old man in his late eighties and the Justice of the Peace for the district.

Colonel Richard Martin M.P., held the largest estate in Connemara, a little under 200,000 acres, and his lands stretched from Oughterard almost to the shore of Ardbear Bay. In 1810 he signed over the greater part of his Connemara estate to his son Thomas. Richard had other lands in east Galway, but throughout his life he showed more interest in politics than agriculture and he periodically occupied a seat at Westminster. The Martins resided at Ballynahinch Castle. Arthur St George of Tyrone House, in Clarinbridge, held

Thomas Martin (1786 – 1847)
© Martin Family

John D'Arcy (1785 – 1839)
© Adrian Lead

lands south of the Martin estate, along the coast at Gorumna Island [Garumna]. John D'Arcy was the proprietor of over 20,000 acres on the bay of Ardbear. He also held estates at Kiltullagh in east Galway and was referred to by Nimmo as 'John Darcy of Killolla'. D'Arcy had inherited his Connemara estate in 1804 from his cousin, Patrick D'Arcy, brother-in-law of Richard Martin. John Geoghegan O'Neill held 13,000 acres at Bunowen. The Cromwellian Commissioners granted the Bunowen Estate to the Geoghegans in 1656, in lieu of their forfeited estates in Westmeath. John had been a member of the Irish Parliament, but was then an absentee landlord living in England. He had changed his name to O'Neill by Royal Licence and when in Connemara he resided at Bunowen Castle.[60]

Thomas Martin and John D'Arcy were the principal resident landlords in the Barony of Ballynahinch. They were both young, energetic men and were making a determined effort to improve Connemara by reclaiming lands and building roads. D'Arcy was making every effort to establish a town, Clifden, at the mouth of the Owenglen River on his estate. The two men were cousins and were close in age to Nimmo. Nimmo was the eldest at twenty-nine; D'Arcy was aged twenty-seven and Martin twenty-six. Nimmo and Martin seem to have struck up a friendship from the beginning, while with D'Arcy the relationship would appear to have been on a more professional footing.

On his visit with Greenough in the summer of 1813, Nimmo may well have stayed at Ballynahinch Castle. Richard Martin and his family were spending the summer at Ballynahinch. Thomas was recuperating from a war wound received while serving with the 88th Foot Regiment in the Peninsular Wars and had only arrived home in the April. Another visitor to Ballynahinch that summer was Julia Kirwan, who would later become Thomas's wife. Greenough, like Richard, had been a Member of the recently dissolved parliament at Westminster and the two might already have been acquainted. Both Richard and Thomas failed in their bids to be nominated as candidates in the upcoming election, Richard for Galway county and Thomas for Galway city.[61]

In the following years, Nimmo became a frequent visitor, if not a guest, to the homes of many of the landowners of Connemara. He became acquainted with the families and familiar with their aspirations for their own estates and for the region in general. He offered advice and made recommendations on engineering matters, on agriculture and on forestry, and these were frequently acted upon. Maria Edgeworth, writing in 1833, tells us that Thomas Martin admired Nimmo's engineering ability: 'I believe Mr Martin had some knowledge of engineering and so could estimate Nimmo's abilities'. His wife, Julia Kirwan, however, 'admired his scientific knowledge and writings the more, because she knew nothing of science'. He was also credited with later giving their daughter, Mary (born 1815), a course in engineering. Nimmo was acquainted with Maria's father, Richard, and her brother, William, both of whom worked with the Bog Commission, and he became a frequent visitor to their family home in Edgeworthstown.[62]

Although Connemara was mountainous, Nimmo explained, it was not an upland country; three-fourths of Connemara Proper was lower than 100 feet above sea level. A great part of Iar Connaught rose 'from the shore of Galway Bay, in a gently sloping plain, to about 300 feet, at the upper edge of which there are some hills of about 700 feet, and beyond them a low limestone country extends to the edge of Lough Corrib'. The lake itself was only fourteen feet above sea level. Joyce's Country, on the other hand, was 'an elevated tract with flat-topped hills of 1,300 feet to 2,000, interspersed with deep and narrow valleys'.[63] The entire district was nearly surrounded by water, with the Atlantic to the north, south and west, and Loughs Mask and Corrib to the east. This vast expanse of seashore produced large quantities of seaweed for use as manure or in the manufacture of kelp. The seacoast and all of the lakes abounded with fish and the district seemed to offer many natural advantages. There were 'upwards of twenty safe and capacious harbours, fit for vessels of any burthen; about twenty-five navigable lakes in the interior, of a mile or more in length'. Lough Corrib was navigable into the town of Galway and 'could easily be made so to the sea'. However, despite these advantages, this was among the most uncultivated parts of Ireland, 'a continued tract of bog and mountain,' with the arable land amounting to one tenth, 'perhaps not one-twentieth' of the whole.[64]

Nimmo estimated that the rental of the whole district amounted to about £50,000 per annum, of which the kelp produced about £6,000. The principal part of the rent from shore estates in Galway was said to result from kelp.[65] Almost the entire population of about thirty thousand lived along the coast, half of these in Connemara Proper, leaving the interior almost deserted. The three hundred or so families to be found in the interior lived mostly on the bridleroads or on the shores of Lough Corrib. The south shore of the lake was the most cultivated section of the region. Here the arable or 'dry land' was interspersed with extensive tracts of naked limestone and it was 'only after incredible labour, that a few patches of soil' were won from the 'general waste'.[66]

Along the south coast the shore and islands were well inhabited and each man held a parcel of land as well as a boat. Land close to the shore was reclaimed using seaweed and coralline deposits. The work was hard but once reclaimed the land proved very fertile. But farming and fishing did not go well together and Nimmo was of the opinion that one was usually advanced to the detriment of the other. In this case the people seemed to have a preference for fishing over farming. The usual size of boats here was about nine tons, and cost £40, exclusive of tackle, and almost every farm had 'one or two of these, besides smaller'. This convinced Nimmo that 'a considerable capital [was] applied to navigation, though very little in agriculture; for the only implement of husbandry is the spade'.[67] When not fishing, the boats were used to transport manure and turf. All along the south coast, and on the large islands, large quantities of turf were cut for export to Galway, Clare and the Aran Islands. The boats returned with limestone as ballast, which was usually used for building but could equally be used as fertilizer.[68]

Connemara place names

Reading Nimmo's report on the bogs in Connemara it is interesting to note the place names that were in use in 1812. Some of these would change over the years; others would lose their prominence to neighbouring districts elevated by the erection of a pier or the construction of a new road. There was at the time no standardized spelling for townlands, villages, rivers or mountains. These varied from one individual to the next, depending on education, nationality and familiarity. As in Scotland, Nimmo may well have employed the services of an interpreter while in Connemara, but given his aptitude for languages, he perhaps made a better attempt than most at formulating the names as given to him by the natives. But even he was not consistent. The principal mountain range in Connemara, he refers to as the 'Group of Burabola, commonly called the Twelve Pins'. In contrast, the Maumturk mountains had no name in one reference: 'a chain of hills runs along the boundary of the barony of Ross or Joice's (sic) Country, and terminates about one mile and a half down the south side of the lake ... This chain has no particular name; the passes over it are known by the name of Ma'am, a term also used in the highlands of Scotland'.[69] However, he later refers to this mountain range as 'Mamturk', and he goes on to use Maam Ina, for Inagh Valley. In his survey of the boundaries of Inverness-shire, Nimmo describes Mam or Maum as 'usually applied to those high and elevated paths which run across the summits that divide two parallel glens'.[70] The Owenglen River is 'Clifton Stream'. Roundstone and Cleggan Bays are mentioned, as is Letterfrack; these names would later be given to new villages that would develop out of the erection of piers and the construction of new roads in their locality.

Seaweed, kelp and fertilizer

Nimmo described the wealth and variety of fertilizers available to the people throughout the region. The east side was well supplied with limestone, situated as it was on the edge of the great limestone field of Ireland. Numerous veins of good limestone were also to be found running through the centre of the region. Banks of shell, coral sand and calcareous sand were available along the coast; these were not as frequently used as seaweed, although their value was generally acknowledged. The most common manure used along the coast was the red seaweed. It was cast ashore by the tide and was 'frequently cut in the deep water by people in boats; two or three boat loads, of about six tons each, are usually applied as manure over an acre of potato ground'. The right to the seaweed belonged to the proprietor of the land that ran down to the shore and it was jealously guarded. Because of this, land close to the shore attracted a higher rent than that further inland. All along the coast the numerous patches of cultivated ground bore testimony to the value of seaweed to the tenants of smallholdings.[70]

Kelp Stacking at Carna circa 1940s
© Failte Ireland

The black seaweed was used in the manufacture of kelp and provided employment for a great many people. The method of manufacturing kelp was introduced into Ireland in

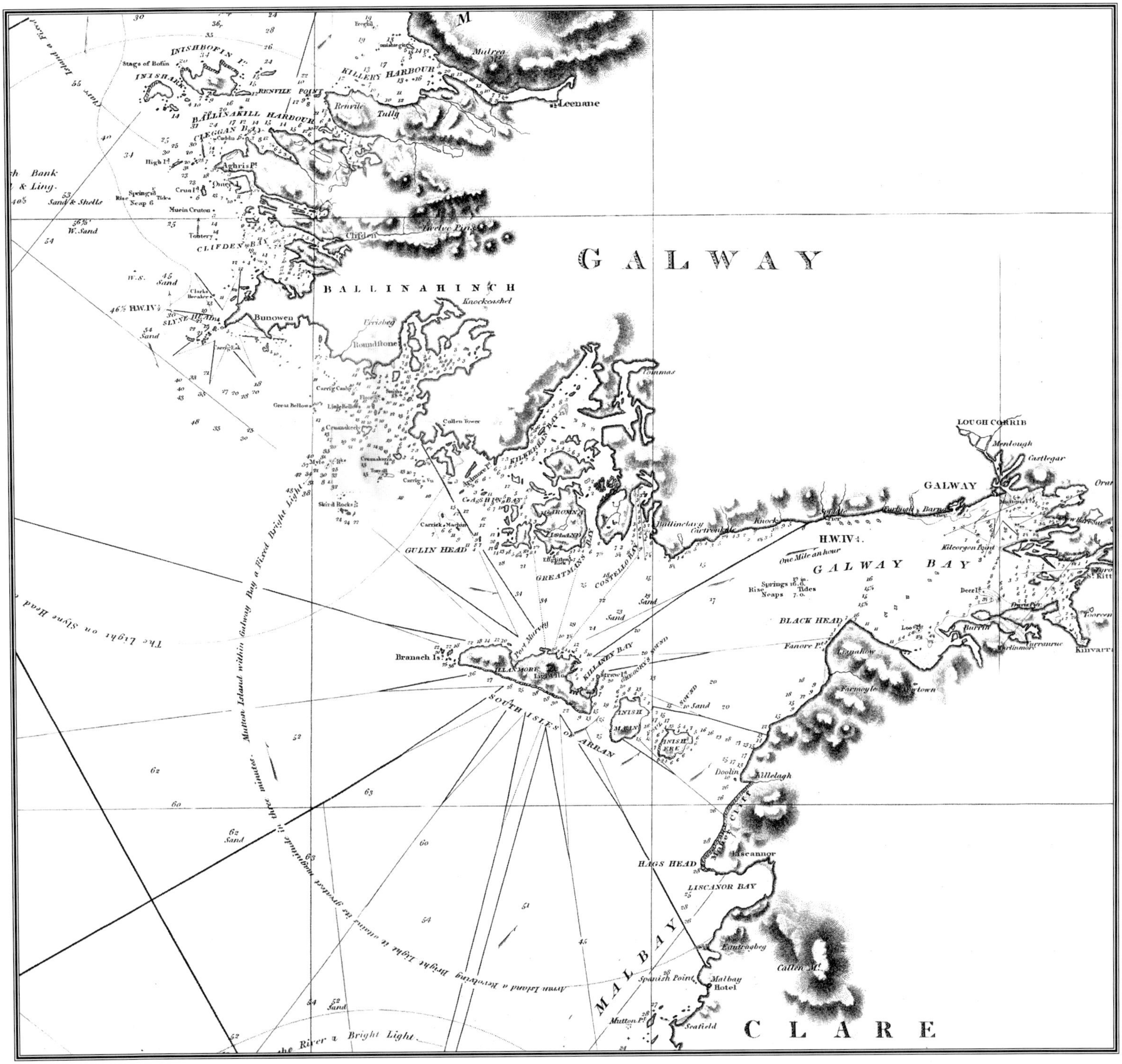

Details from Nimmo's chart of Galway coast, circa 1831
(*Chart of St George's Channel and the coast of Ireland*, by A. Nimmo)

about 1720, and was carried out on the west coast from Clare to Donegal.[72] The tenants were reported to begin cutting the weed in May 'after finishing the tillage, and employ on it all the time that can be spared from turf and potatoes until Michaelmas [29 September]'.[73] The weed was cut from the rocks with a hook and gathered on the shore to dry. Then it was burnt in a kiln, 'rudely constructed of stones in a circular form, four or five feet in diameter, and about one in depth'. The ash condensed into kelp, 'a dark blue or whitish coloured mass, nearly of the hardness and solidity of a fragment of rock'. This was purchased by agents and taken out of the district for use in the bleaching of linen cloth and in the manufacture of glass.[74] Nimmo estimated that two thousand people were employed in the manufacture of kelp along the Connemara coast. On average one kelper made two tons in a season, an able man could produce three, and 'three men in one house may make from seven to eight in a season'. In the past this had been a lucrative business but since the end of the Napoleonic wars barilla was again imported and the price of kelp had fallen: the annual production of kelp a few years prior to 1813 was valued at £50,000, but more recently this had been reduced to £16,000. Nimmo questioned its continuance, when much more long-term benefit could be derived from applying the weed to the land:

> The quantity manufactured in the whole of Conamara is about four thousand tons, which may require about fifty thousand tons of the sea-weed, and it is probable might suffice for manuring four thousand acres of land; and this, after one course of cropping, would be let in permanent pasture, worth 5s to 10s an acre at least; while during the cropping the produce cannot be reckoned at less than £10 to £12, and the rent to the landlord £2 to £3 per annum.[75]

However, the transfer of manure and labour from the manufacture of kelp to reclaiming the land, might, he contended, prove too expensive to be considered.

In the north, in the neighbourhood of Killary, there was no limestone, calcareous sand or shell sand to be found. Sand, however, was imported from Ballynakill Bay and limestone was brought in from the east by way of Lough Corrib and Lough Mask. The greatest disadvantage in this northern district was 'the want of roads, of which there [was] not one in the whole district fit for a wheel carriage, nor [was] there a single bridge'. William O'Flaherty had reclaimed about a thousand acres of bog at Renvyle: 'he removed the cottagers from their old stations, and settled them on the bog; this they reclaimed with potatoes and sea-weed, treating it afterwards with the sand of the shore, which [contained] no calcareous matter; the effect has been very great'. The reclaimed bog was on rising ground, interspersed with rock, and was well suited for use as commonage.[76]

To fertilize the lands north of Killary, as far as Clew Bay, limestone was imported at Westport and transported to the interior on horseback. The development of a market at Westport had 'given a spur to the agriculture of its neighbourhood' and Nimmo suggests that establishing similar towns and laying down good roads would bring similar benefits to Connemara.[77]

Forestry

Nimmo was of the opinion that, although the general improvement and cultivation of Connemara 'would seem an undertaking of the most arduous description', the region did have many advantages which might make it 'more worthy of attention than many other of the wet lands of the kingdom'. The climate was mild, the cattle were never housed, and the mountains and hills offered considerable shelter. The summers, however, were wet and it was exposed to heavy westerly winds, but with the extent of bog in the district, the supply of fuel was 'evidently inexhaustible'.[78] The region also possessed many advantages for the raising of timber. And although Connemara was 'destitute of wood', in some sheltered spots oak, beech and hazel appeared in abundance and with a little care could be cultivated more widely. Most of the old timber had been consumed by several iron works established a century before, and he was surprised that no attempt had been made to replace them. For Nimmo, planting was the obvious solution for the 'extensive moors, which [were] so far from manure or limestone', and the advantages of 'sheltered vales, navigations, and abundant water-power', all pointed towards the possibility of developing a timber industry in the future.

Between Galway and Oughterard several of the proprietors had planted trees, principally for ornamentation, and they seemed unaware of their potential commercial value. In contrast, nurseries and copses, already thrived along the coast:

> Mr John Darcy (sic) of Killolla (sic), has succeeded well in raising most kinds of timber at Clifton (sic) and Ardbear Bay, though immediately exposed to the Atlantic. His nursery there is in good order, and he is extending his plantations. The copses of Ballinaboy, Munga, &c. are also very near the ocean; and the wood of Cloonill (sic), in Birterbuy Bay (sic), grows down to the water edge.
>
> Mr Martin had planted extensively at Ballinahinch, but for want of enclosures, much of it has been ruined. Mr Thomas Martin, his son, has formed an extensive nursery, and begun to plant upon a very great scale; there could not be a finer field than what his demesnes afford. He is also rapidly fencing in his numerous copses.[79]

Nimmo's vision for Connemara

However, by far the greatest necessity for the future development of the region was the construction of roads, for which he gave a rough estimate of £41,270. These would open up the interior and develop a link between the deep bays in the south and Killary and Westport in the north. Working his way through the country, Nimmo ascertained the general route the roads should take and outlined these in his report. He proposed a 'great road' from Galway to Clifden, running close to the southern shore and then cutting through the interior bogs to approach Clifden from the south. The proposed line would leave Galway by Rahoon, cross the Spiddle River about four miles from its mouth, then travel north to Screeb, where it should turn west to Cashel Bay and on to cross the

Ballynahinch River at the salmon weir. From there it would be taken across the moor to Derrigimlagh, Mannin Bay, where it would divide, leaving one branch running off to Ballyconneely and a second branch running north to Clifden. Several branch lines would run off this road to penetrate the various peninsulas to the south and the interior of central Connemara to the north. He recommended that existing bridleroads be improved and extended to link Spiddle with Moycullen and Oughterard, and Cashla with Oughterard. To develop a substantial south north route, he suggested that the bridleroad running through Rosmuck be taken across the bog to Boheeshal and proceed up the valley towards Fionasclainn, to the east of Lough Inagh, and on to Killary. This road would become the 'great road' to Killary Harbour and Westport.

Thomas Martin's road from Ballynahinch to Cloonisle
This road intersected at Gleann Toircin with the old bridleroad running from Loch na Cúige Rua to Glinsk on Bertraghboy Bay. (William Larkin's 1819 Map of Galway)

Nimmo emphasized the necessity to develop towns, villages and ports to advance the region. Galway and Westport were the nearest towns and west of Oughterard there was nothing until one reached the struggling town of Clifden. He pointed out several sites, which he felt were suitable for new villages and ports. One such site was at Bealandangan, opposite the south point of Rosmuck, which had enough depth of water for a small port. The mouth of the Ballynahinch River is suggested as another possible site for a port. Here the lakes of Ballynahinch and Derryclare could easily be 'converted into one piece of navigable still water', which, if not made navigable to the sea, could be linked by a good road of about two miles. With no regular carriage road linking the 'central vale' of Connemara with the sea, this could become an important port and centre for improvements in the area. Nimmo pointed this out to Thomas Martin and suggested that the construction of such a road was 'one of the greatest improvements that the country required'. By the time he submitted his report the road was already finished and Nimmo praised Martin for the quality of the work, which was conducted 'in a style which shows that the roughness of Cunnemara may be no obstacle to its possessing a land-carriage of the most effectual description'.[80] John D'Arcy's 'village … on the stream of Clifden', he felt, would 'have a considerable effect in promoting the improvement of the neighbourhood'.[81]

Nimmo conducted an exact survey of the flat moor between the Ballynahinch River and Mannin Bay [Roundstone Bog], which contained 14,000 Irish acres of red bog and the proprietors of which were Colonel Richard Martin of Ballynahinch and John Geoghegan O'Neill of Bunowen. Nimmo estimated that the bog held about 143 lakes, of different sizes, many of them having numerous and intricate arms: 'When viewed from an elevation, this appears to be a complete labyrinth, in which it is difficult to perceive the direction of the drainage'. Nimmo suggested how some lakes could be made navigable, 'so that limestone or calcareous sand loaded at Ballinaboy, might be transported over most of a tract of about two miles long'. A similar navigation could be achieved on the lakes north of Errisbeg and brought close to the shore at Roundstone Bay. Here again more roads were proposed and it was suggested that the existing bridleroad south of Errisbeg should be improved by bringing it to lower ground.[82]

From Ardbear northwards to Killary an entire new set of roads was needed to improve communications with Westport. Nimmo laid out a line for a large section of this road and handed it over to the landlords. Before he had left the district this road was already in progress, but he held out little hope of sufficient funds being allocated by the Galway Grand Jury to have it completed to the required standard. He proposed Creggauns as the most suitable site for a village on Ballynakill Bay and suggested either the mouth of the Bunowen River, which had deep water, or the mouth of Béal na Coille River, which seemed 'best calculated for communication with the interior', as suitable sites on the southern shore of Killary Harbour.[83] A village was also needed at the edge of Lough Corrib, which should have a ferry to improve communications between Galway, the Joyce Country and County Mayo. Letterfrack, Leenaun and Maum were later established close to the proposed sites.

Nimmo outlined a new route for the old centre road running through central Connemara, to bring it to lower and more level ground. Improvements were needed on the existing road from Oughterard to Lough Bofin, carrying it through Glengowla and along the south side of Loughs Bofin and Shindilla. At Lough Bofin he suggested that the existing bridleroad to Lough Corrib [Western Way] be improved and taken along by the lakeshore to Joyce Country, from where a branch could be taken west to Killary at Béal na Coille and another east to Lough Mask, to join the Mayo road (this road was later built a little further west, running through Maam Valley). West of Lough Shindilla, he proposed that the old centre road should be brought down from the high ground to the shore of Lough Oorid, on to Boheeshal and from there to Garroman, taken across Derryclare Lake to proceed along the north side of Ballynahinch Lake to Emlaghdauroe and from there by Gowlan and Clifden. As previously mentioned, a section of this improved line was already in progress under D'Arcy. However, this stretch of road was destined to cause many a headache for Nimmo in the years to follow. Lack of funding, disagreement with the contractor and public criticism over its slow progress did much to sully his reputation in later years and in the end he would not live long enough to see its completion.

Nimmo left Connemara with clear directions as to the best route for future road construction in the region. Other roads would no doubt become necessary 'as the country

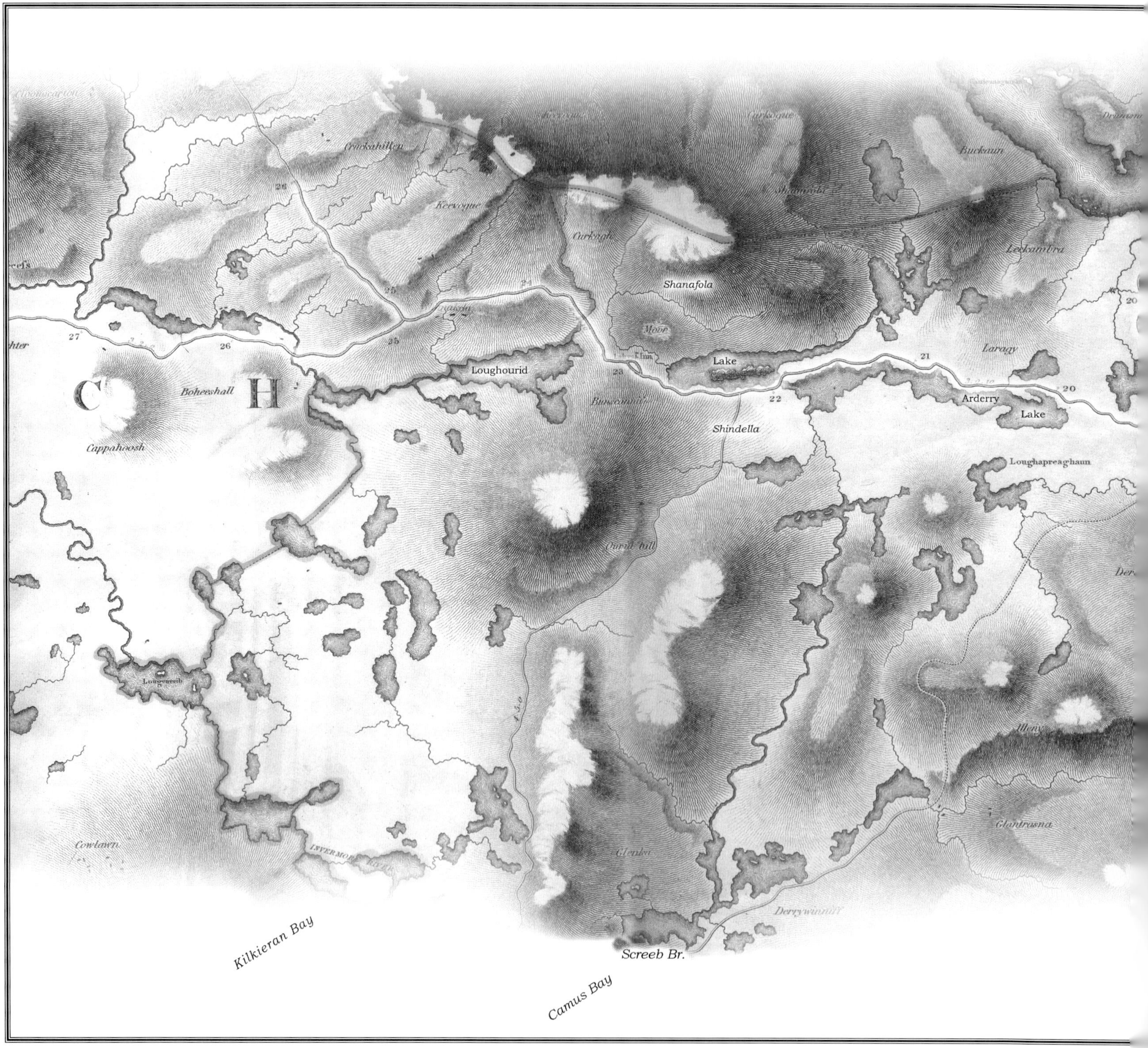
Cruckahillen
Keevogue
Carkogh
Carkogue
Buckaun
Leckambra
Shanafola
Moor
Lake
Loughourid
Laragy
Arderry
Lake
Shindella
Loughapreaghaun
C
H
Boheeshall
Cappahoosh
Cowlawn
Glenka
Glantrasna
Derrywinnay
Kilkieran Bay
Screeb Br.
Camus Bay

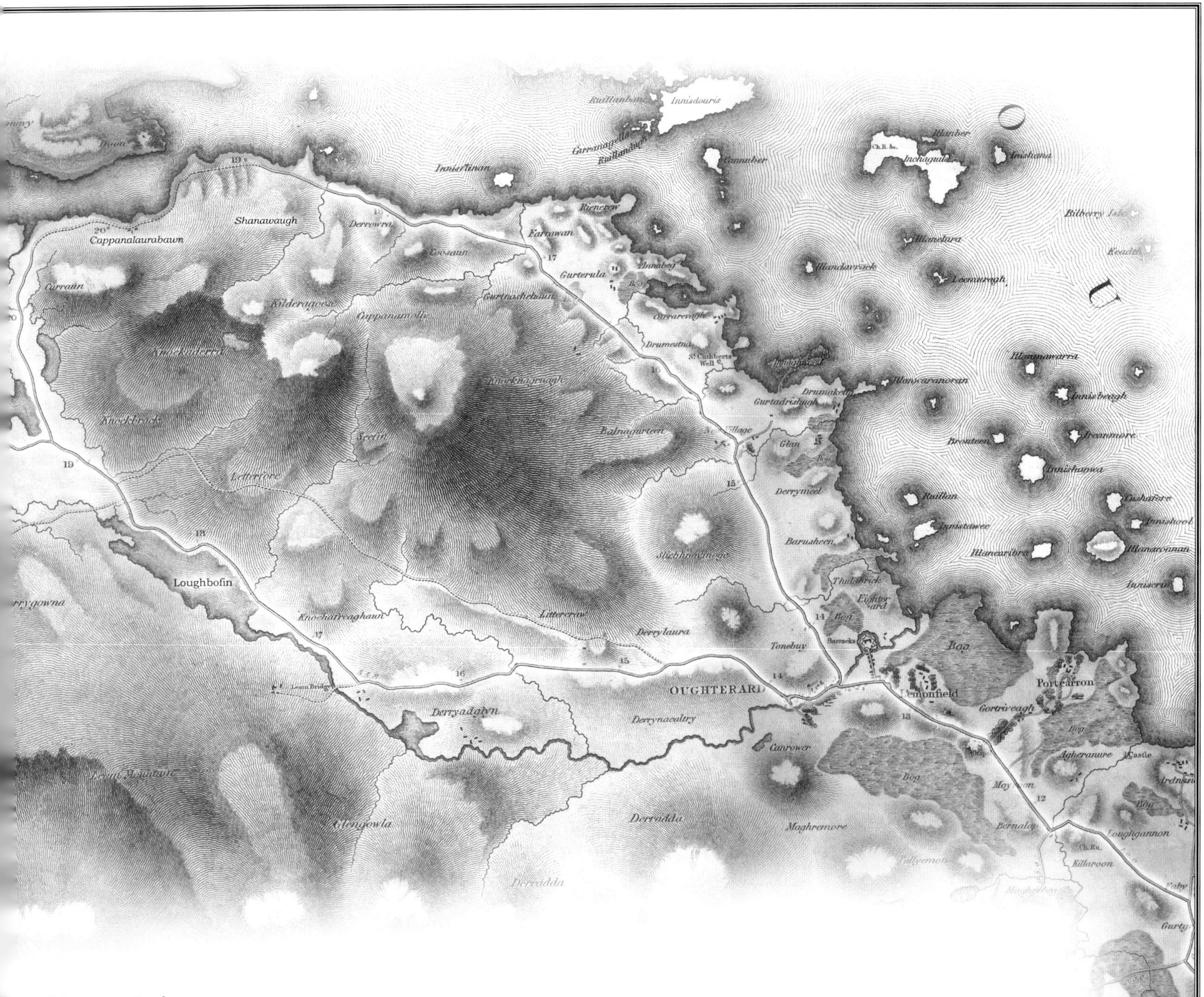

entral Connemara Road

howing the bridleroads from Lough Shindilla to Kilkieran Bay and from Lough Bofin to Screeb Bridge at the head of Camus Bay. Also own is the ring road from Oughterard to Lough Corrib and Lough Bofin. From markings on Larkin's map this would appear to be tle more than a track between Shanawaugh (Shannawagh) and Curraun (Derryherbert). (William Larkin's 1819 Map of Galway) advanced in cultivation'. However, he warned that the construction of the roads was 'far beyond the ability of the present possessors', but worthy of 'the attention of a munificent Government'.[84]

To further improve communications between Galway and Mayo, he pointed out that Lough Corrib was underutilized and should be made navigable to the sea. This was a topic that Nimmo would return to a number of times in the years to come and one that would employ the talents of several engineers before the work was eventually completed in 1854 under the direction of the County Surveyor, Samuel Usher Roberts.[85] But here Nimmo put forward his suggestions for opening up a channel between the lake and the sea for the first time:

> A good chart, with soundings and sailing directions, should be published; the shoals and rocks cleared or beaconed, and a communication opened with the sea.
>
> There will be only two locks required, which (exclusive of the purchase of property of no great value) would cost about £6,000. I should propose them to be of large dimensions, so as to admit vessels of burden, as this would give to Galway all the advantages of wet docks; two weirs in the river, at the upper level of these locks, will secure a much better supply and greater fall to the different mills, and give every one of them a water-carriage up and down; neither of which is enjoyed by any of them at present. The river and lake being deepened across the shoals, small vessels may run up into the lake and carry up sea manure as a back freight for the turf, which would greatly benefit the agriculture of the interior.[86]

In his conclusion, Nimmo was very positive regarding the future development of the Connemara region: 'On the whole, it appears to me that the improvement of this district, so far from being difficult or hopeless, is a thing highly feasible, and if vigorously but steadily pursued, is likely to meet with fewer obstructions and greater ultimate success, than perhaps in any other part of Ireland'.[87] With regard to converting the bogs into productive land, he stated his conviction that 'any species of bog is by tillage and manure capable of being converted into a soil fit for the support of plants of every description; and with due management, perhaps the most fertile that can be submitted to the operations of the farmer'.[88]

Griffith's and Bald's surveys of the Mayo bogs

Richard Griffith was also in no doubt that the bogs surveyed by him could be reclaimed and cultivated, but considered his district in Mayo too poor to take on the considerable costs associated with such work. He conducted a survey of the bogs in the Baronies of Tirawley, Erris, and Burrishoole, in Mayo, and of Tireragh, in Sligo. His report was completed by May 1812 and included a map, which he himself said was not to be considered 'perfectly accurate', but which showed all the prominent features of the country. The sea coast had been copied from MacKenzie's charts; 'and the more minute details of the map [had] been filled in from sketches taken from the summits of the mountains, corrected by angles; also taken from the summits of the mountains to every object which could be seen from two places'.[89]

Having been instructed to produce a general report only, Griffith confined himself to observations on the topography, geology and general condition of the district, and on making recommendations for improvements through drainage and road construction. On the subject of improvements he had high praise for the late Marquess of Sligo and for the works carried out by him in reclaiming wasteland and developing Westport town. The Marquess had encouraged his tenants to reclaim wasteland by giving them long leases at low rents; this allowed them, and their children, 'to enjoy the fruits of their industry'.[90] Thousands of acres of mountain land around Westport had been reclaimed in this way. This had brought about an increase in the production and export of corn and oats from Westport harbour. Griffith was of the opinion that the example given by the Marquess should force the owners of mountain estates, 'to pay more attention to them than they have hitherto done, by clearly proving that in so doing they will not only benefit their country, but enrich themselves'. After Westport the next harbours of importance were Newport and Killala. The Barony of Erris had two excellent harbours, Broad Haven and Blacksod, but these were underutilized: Broad Haven was exposed to the northern winds and Blacksod was little known and seldom used, but was considered by Griffith to be the better of the two.[91]

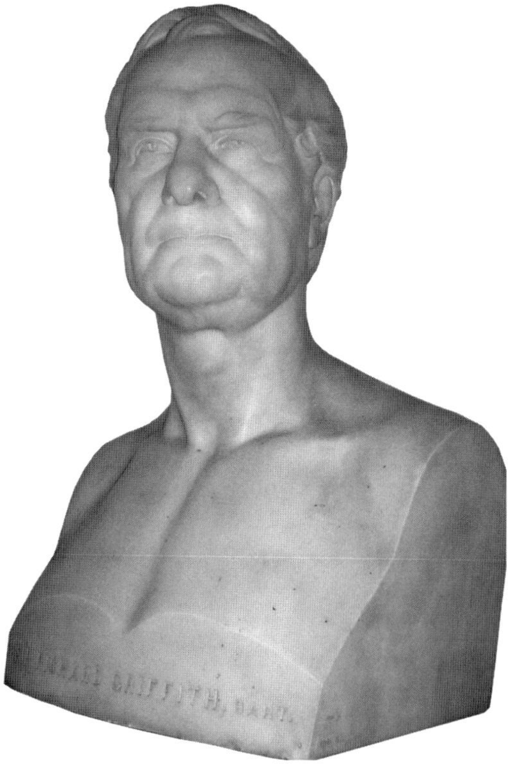

Richard Griffith (1784 – 1878)
Marble bust on display in the Library of the Royal Dublin Society.

A map in McParlan's survey of Mayo, dated 1802, gives the impression that the county was well served with roads. And indeed he declares as much, stating that the roads and bridges were in 'so very good a state that one can hardly complain of the few that are bad'.[92] Griffith, however, was severely critical of the method in which roads were constructed in Mayo: 'immense sums of money are annually levied on the county for the making of new roads, laid out without system, or reference to general communication, and for the repair and rebuilding of bridges and other public edifices, fallen into decay on account of bad foundations and ill construction'. Griffith's explanation as to how this had come about could be equally attributed to Galway:

> A landlord gets a presentment from the county, and his steward, or country agent (however ignorant) lays out the road, without reference to the other roads, either in that county or the next, and without the smallest consideration of the height or levels of the country. In the same inconsiderate manner, bridges are built without proper foundations, which, as might be expected, are torn down by floods, and the county is put to the expense of rebuilding them a second, sometimes a third, and even a fourth time: at length it is discovered that the line of road is bad, and the sites of the bridges are worse; a second presentment is then laid before the Grand Jury for a new line of road; and I have known instances where the second was conducted with as little judgment as the first.[93]

There were only two roads in the barony of Erris and both of these had been recently constructed; one ran from Crossmolina to the isthmus of Belmullet and the other from

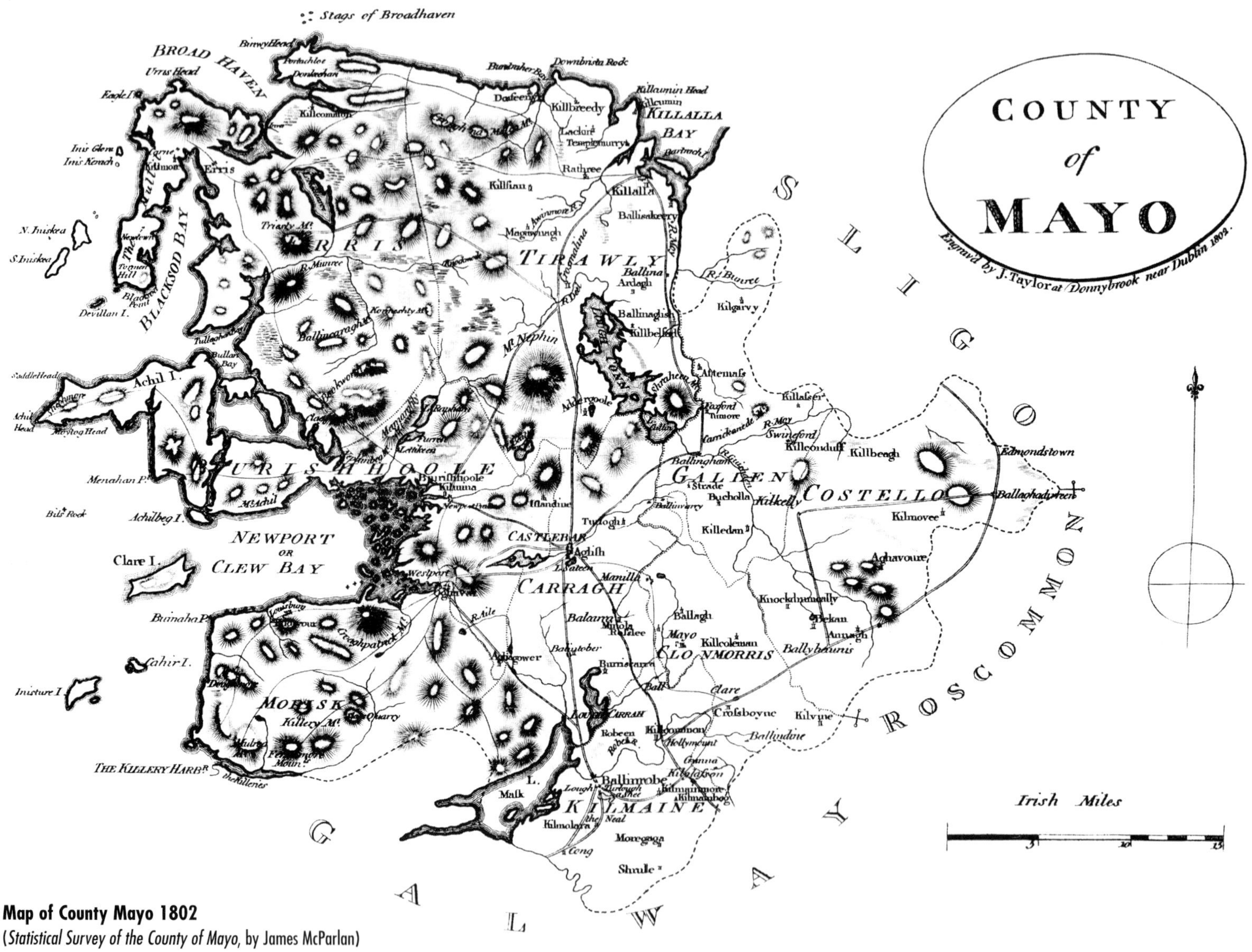

Map of County Mayo 1802
(*Statistical Survey of the County of Mayo*, by James McParlan)

Newport to Tullaghan Bay at Aughness. The Crossmolina to Belmullet road was very badly laid out, with 'no attention having been paid to avoid hills, deep gullies, or other obstacles'. The road to Aughness was an extension of an old road running from Newport Pratt [Newport] to the entrance to the Corraun peninsula at Rosturk. This road had been gravelled for some time, but horses were unable to pass 'owing to the want of bridges'. Griffith suggested two new roads, pointing out that without roads there was no inducement 'held out to the industrious man to cultivate more land than [was] absolutely

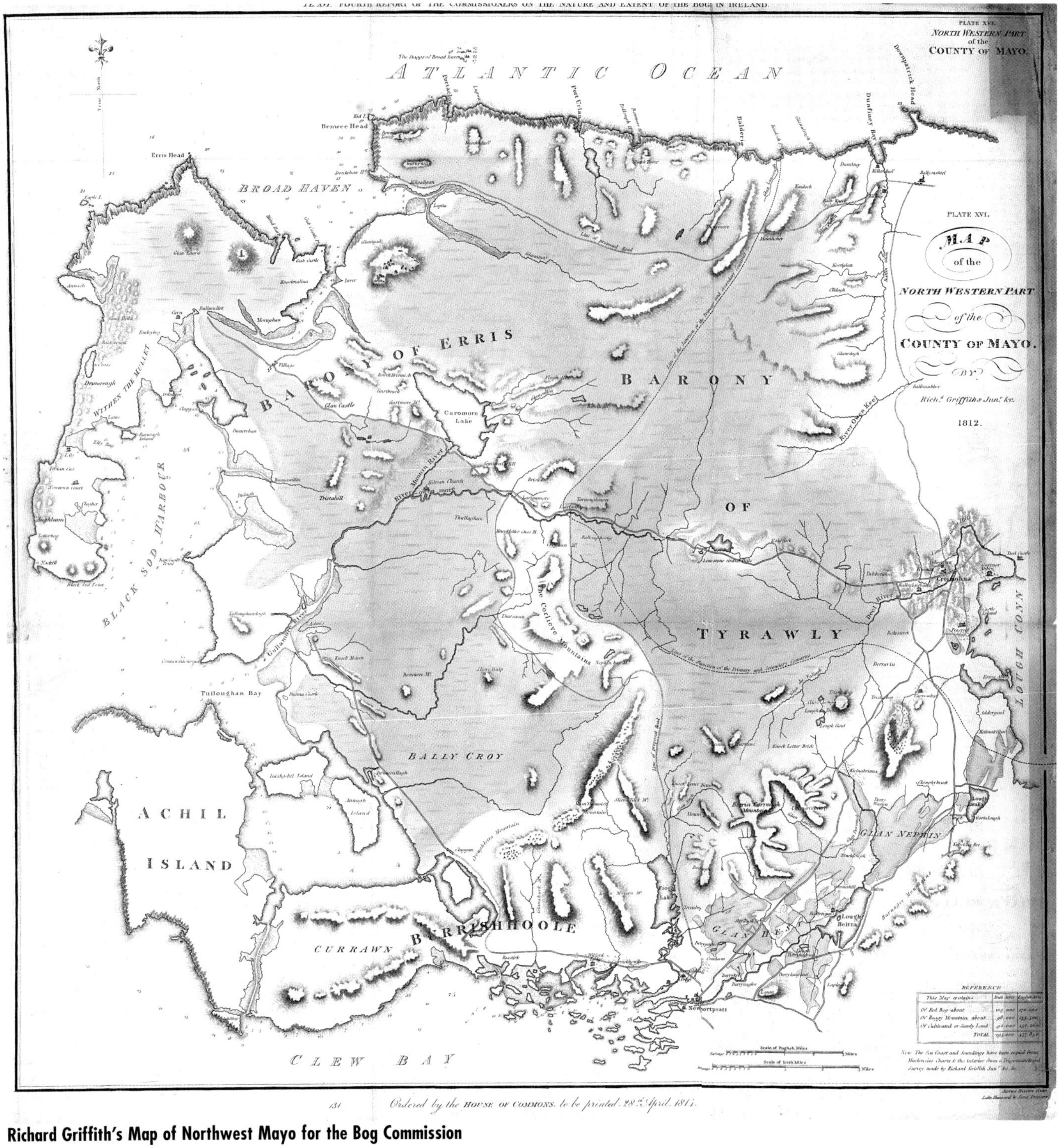

Richard Griffith's Map of Northwest Mayo for the Bog Commission
(Fourth Report of the Bog Commissioners 1813-14)

necessary for his own immediate subsistence, as he possesses no means of bringing his produce to market'. He outlined the proposed routes on his map, but stated that he had not spent sufficient time in the county to determine if these were the best routes. The first would connect Ballycastle to Rinroe Point on Broad Haven Bay and the second would run from Newport to the interior of the Barony of Erris.

The road from the village of Ballycastle to Rinroe Point would give the people of that district the opportunity to convey their corn to the markets at Ballina and Killala, 'and would take away all excuse for illicit distillation, which is now carried on there to a considerable extent'.[94] At the time almost all of the corn produced in the mountain districts along the west coast was converted into poitín. It was a lucrative business and was one of the few success stories coming out of the west, although not considered so by the government, which was anxious to put an end to this illegal activity, 'with its consequent depravation of morals and loss to the revenue'.[95] But not all of the poitín produced was consumed locally, 'the greater part of [it was] taken out of the country on horse backs, over moors and mountains, upon which even man find it difficult to walk'. Griffith warned that if 'this illicit distillation were now stopped, the growth of corn in the country would be put [to an end]', because without roads the inhabitants had 'no possible means of taking their bulky produce to market'. The second road he proposed, between Newport and the interior of Erris, should commence at the edge of Lough Feeagh, as a good road already existing between the lake and Newport, and from there run north to meet the existing Crossmolina to Belmullet road. From the proposed new roads branch roads could be added, 'as population and improvement increased'. But, he warned, 'whenever it shall be deemed expedient to carry them into execution, it will be necessary to appoint a skilful person to conduct them, for it requires both patients and judgment to lay out level lines of road in hilly countries'.[96]

William Bald's survey of the bogs in the interior of Mayo was conducted over a longer period of time and produced a number of detailed maps. Prior to his appointment to the Bog Commission on 22 May 1810, Bald was already working on a survey and map of County Mayo. The map was commissioned by the Mayo Grand Jury in 1809 and turned over to them in 1813, but it was not published until 1816.[97] Some years later Bald made four models of parts of Mayo. One 'containing 200 square miles; the perpendicular and horizontal scales [were] four inches to an English mile', showing 'every road, rivulet, lake, wood, town etc', this was given to the museum in Trinity College Dublin. He also made a model of Clare Island, which was given to the Royal Irish Academy, and another of Achill Island, 'the largest on the coast of Ireland'.[98] For the Bog Commission he surveyed three large districts in Co Mayo; (1) the bogs between Clew Bay and Castlebar, (2) those lying east of Castlebar and running into Counties Sligo and Roscommon and (3) the bogs between Clew Bay, Killary Harbour and Lough Mask. For all of these Bald outlined the individual bogs, suggested ways of draining them and proposed new roads to facilitate their cultivation.

The maps accompanying Bald's reports also show a number of roads running through the centre of Mayo. Many of these had been constructed over the previous thirty years and, bearing in mind what Griffith had to say regarding the construction of roads in Mayo, it seems safe to assume that almost all of these would benefit from the attention of a professional engineer. In the southwest of the county, south of Clew Bay, there were roads connecting Westport to Louisburgh and to the head of Killary Harbour. From Louisburgh a road ran south to the village of Cross at Cross Lough on the coast, and there were several branch roads running from it to the east and west. The Westport to Killary road came along the south bank of the Erriff River to enter Killary at Glenanane. There were no roads on the north or south shores of Killary Harbour, although the presence of settlements or *bailes* would indicate that there might well have been tracks: the village of Bundorragha would appear to be accessible only by sea. Bald suggested new roads and improvements to existing roads in all his districts and he outlined the direction these should take. He also suggested that efforts be made to improve the fishery along the coast, where there was a 'remarkable fine fishing banks of cod and ling'. Killary Harbour, he pointed out, was known to be one of the best fisheries for herrings on the west coast and was considered deep enough to accommodate 'a first-rate man of war'. The construction of new roads here would greatly improve the area and 'enable the inhabitants to send their produce to the market towns of Westport, Castlebar and Ballinrobe'.

Bald also held the opinion that the bog could be easily reclaimed and cultivated. The rise in population and the practice of subdivision had created a demand for additional land and he saw no difficultly in attracting tenants to occupy and cultivate the reclaimed bogs. He compared the bogs to new colonies: 'The reclaimed bogs in Ireland would have all the advantages of new colonies, both as to room and food in abundance, and none of their difficulties in distance and transport'. And in areas where the bog was on high ground, too high for the cultivation of grain, he suggested that it be fenced off and planted with trees.[99]

The reports submitted to the Bog Commission by all three engineers were positive in their findings and practical in their recommendations for future developments. Staying true to their brief they gave a clear picture of the geography of the counties and the infrastructure that existed at the time. However, with the exception of Nimmo, they said little about the people, except to point out that the tenants were extremely poor and that the landowners lacked the capital to finance the improvement works as outlined in their reports. While conducting their surveys and drawing up their maps, the engineers would have informed the landowners of their intended recommendations and this may have raised expectations among them that some form of government aid would be made available for bog reclamation. But any hope of the introduction of state funds to aid the development of the west was soon crushed with the commission's recommendation, in its final report in 1814, that the development of the Irish bogs should be left to private enterprise.[100] This resulted in no action being taken by the government to implement any of the suggestions

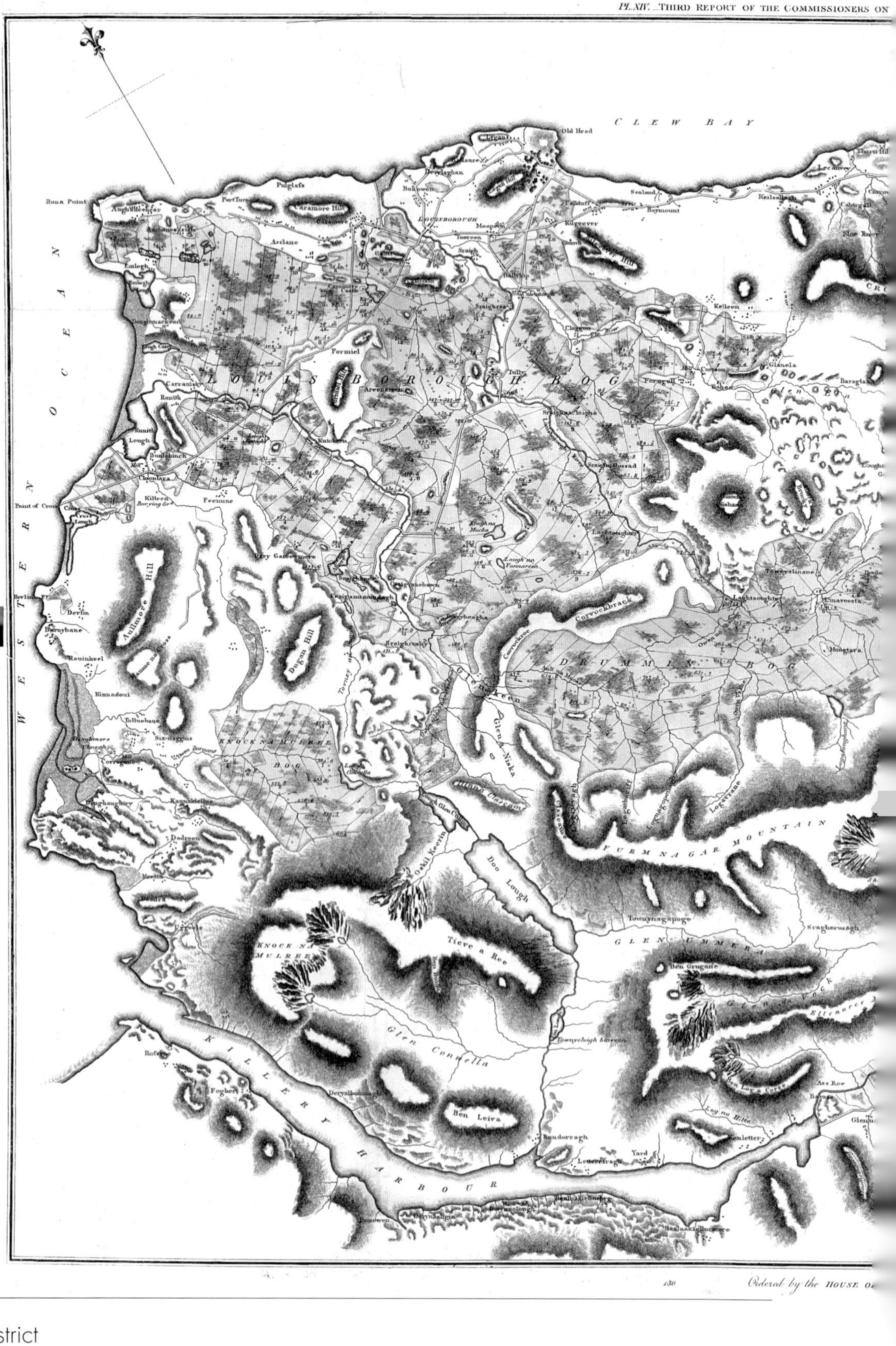

William Bald's Map of Southwest Mayo for the Bog Commission
(Third Report of the Bog Commissioners 1813-14)

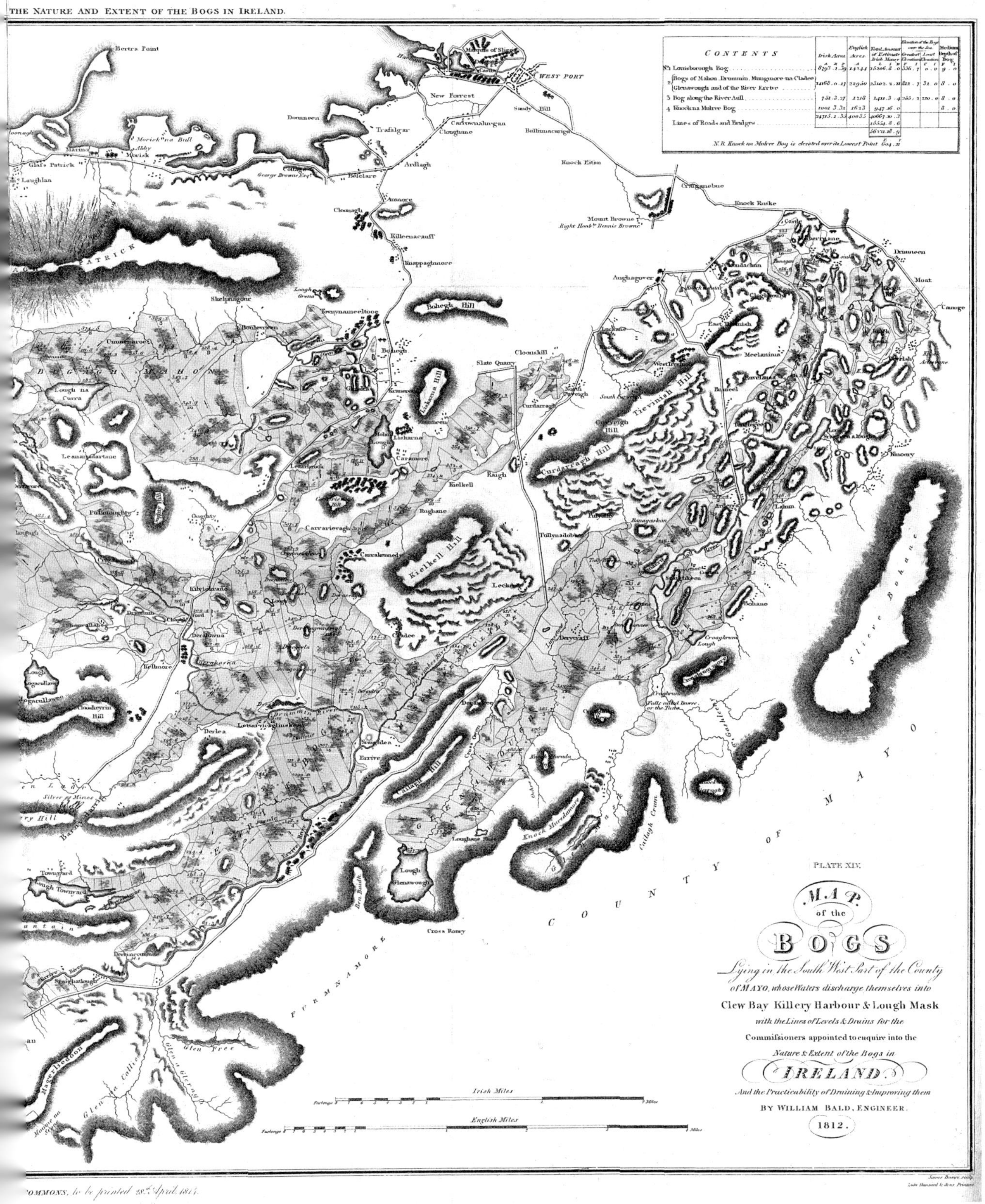

CONTENTS
Irish Acres
English Acres
Total Amount of Estimate Irish Money
Elevation of the Bogs over the Sea
Greatest Elevation
Least Elevation
Medium Depth of Bog
No. 1 Louisborough Bog
2 Bogs of Mahon, Drummin, Mungmore na Cladee Glenawough and of the River Errive
3 Bog along the River Aull
4 Knock na Mulree Bog
Lines of Roads and Bridges
N.B. Knock na Mulree Bog is elevated over its Lowest Point 604 . 11
Bertra Point
West Port
New Forrest
Trafalgar
Ardlagh
Knock Ruske
Bohegh Hill
Slate Quarry
Kielkell Hill
Tievinish Hill
Curdareagh Hill
Slieve Bohane
Lough Glenawough
Cross Roney
COUNTY OF MAYO
Irish Miles
English Miles
PLATE XIV.
MAP of the BOGS
Lying in the South West Part of the County of MAYO, whose Waters discharge themselves into
Clew Bay Killery Harbour & Lough Mask
with the Lines of Levels & Drains for the
Commifsioners appointed to enquire into the
Nature & Extent of the Bogs in
IRELAND.
And the Practicability of Draining & Improving them
BY WILLIAM BALD, ENGINEER.
1812.

put forward in the engineers' reports and their findings were in danger of being pushed aside only to gather dust in some back room in Dublin Castle. Some of the reports, however, were independently published, including one of Nimmo's Kerry reports in 1818, and these later become blueprints for future development in these regions.[101]

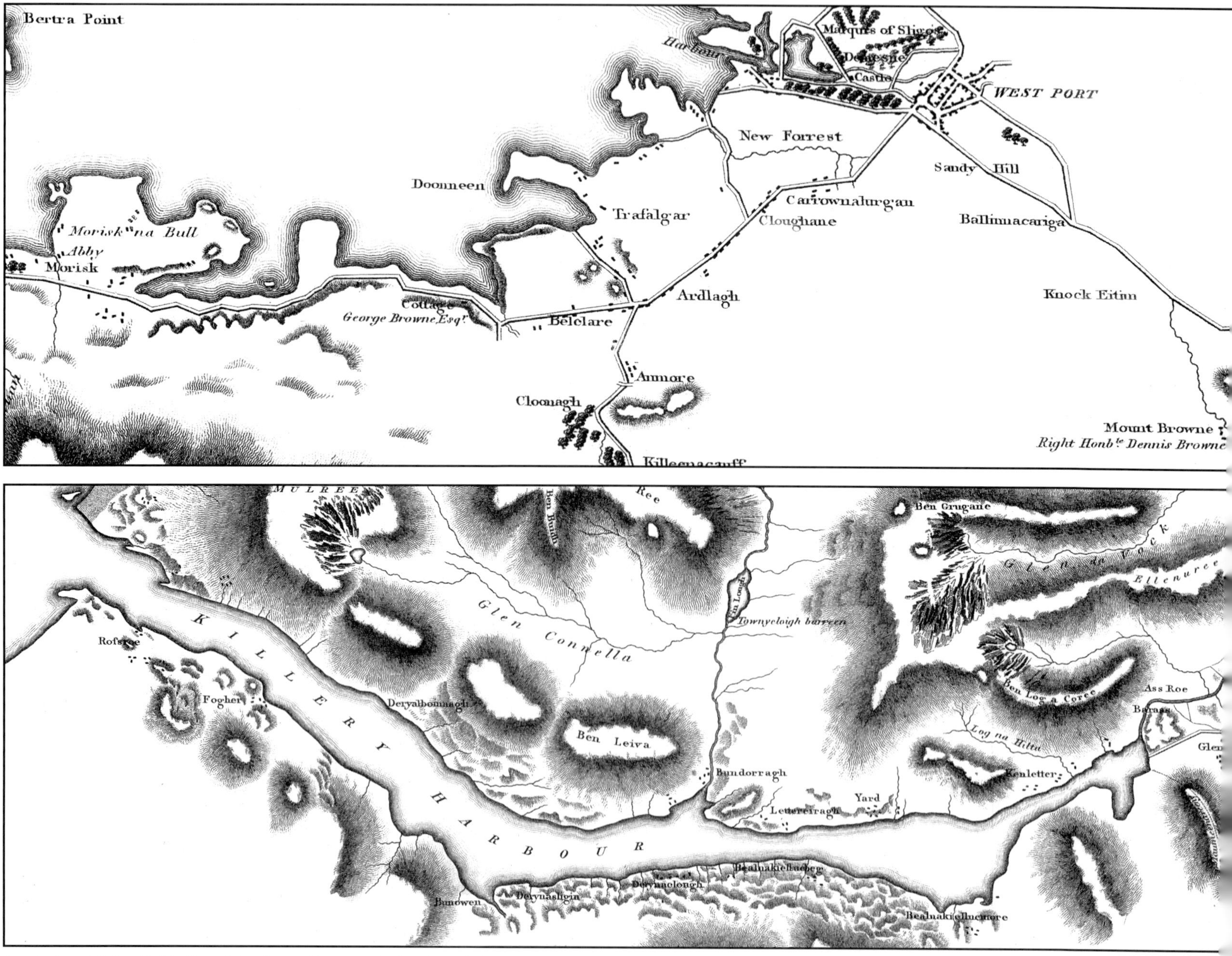

Details from William Bald's Map of Southwest Mayo for the Bog Commission
(Third Report of the Bog Commissioners 1813-14)

Chapter 2

Recommending developments for coastal regions 1815–1821

On completing his work with the Bog Commission, Nimmo travelled to Europe in 1815, visiting civil engineering works in France, the Low Countries and Germany. In Holland, the British ambassador introduced him to the minister with responsibility for public works and he toured the principal drainage works there, taking away plans and ideas for future use.[1] When his travels were over, he returned to Ireland and set about establishing a private practice. Over the next few years he surveyed several small harbours, as well as some of the principal harbours in the country. His employers were usually harbour commissioners and improvement committees, eager to take advantage of government loans made available through the Loan Commission to develop the fishery and industry in their towns. The Loan Commission issued funds from the Consolidated Fund to county Grand Juries, and individuals, for public works. These were to be paid back with interest over a specified term.[2]

Nimmo surveyed the harbours of Sligo and Cork for their local harbour commissioners.[3] He surveyed Waterford harbour for the postmaster-general and the bay of Tramore for the treasury. Killough in County Down was surveyed for the proprietor of that district.[4] As well as surveying the harbours, he frequently assisted the local committees in preparing Bills to be brought before parliament to solicit funds for the works.[5] He constructed Portumna Bridge in Galway for the Corporation of Portumna in 1818: the Corporation received £3,000 from the Loan Commission towards its construction.[6] Nimmo's bridge was replaced in the 1840s by a swivel bridge built by Thomas Rhodes.[7]

Nimmo's chart of Cork Harbour circa 1831
(*Chart of St George's Channel and the coast of Ireland*, by A. Nimmo)

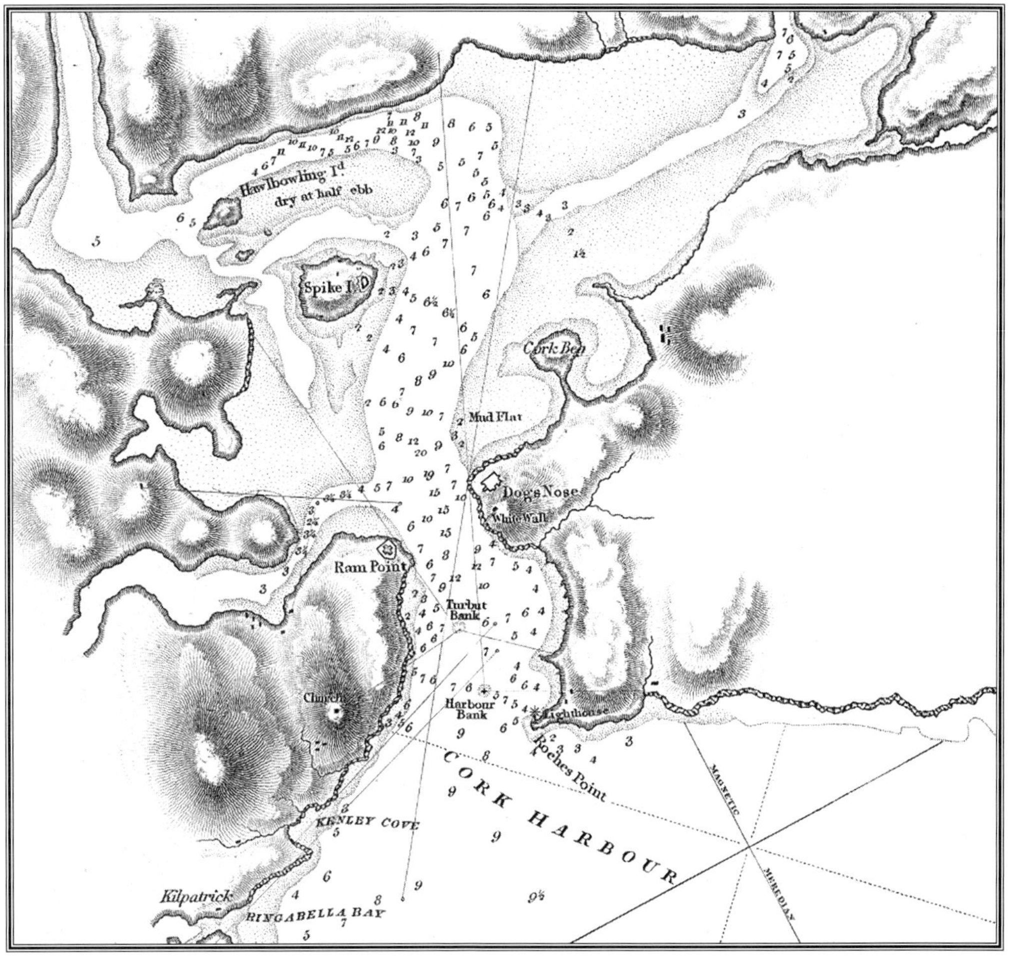

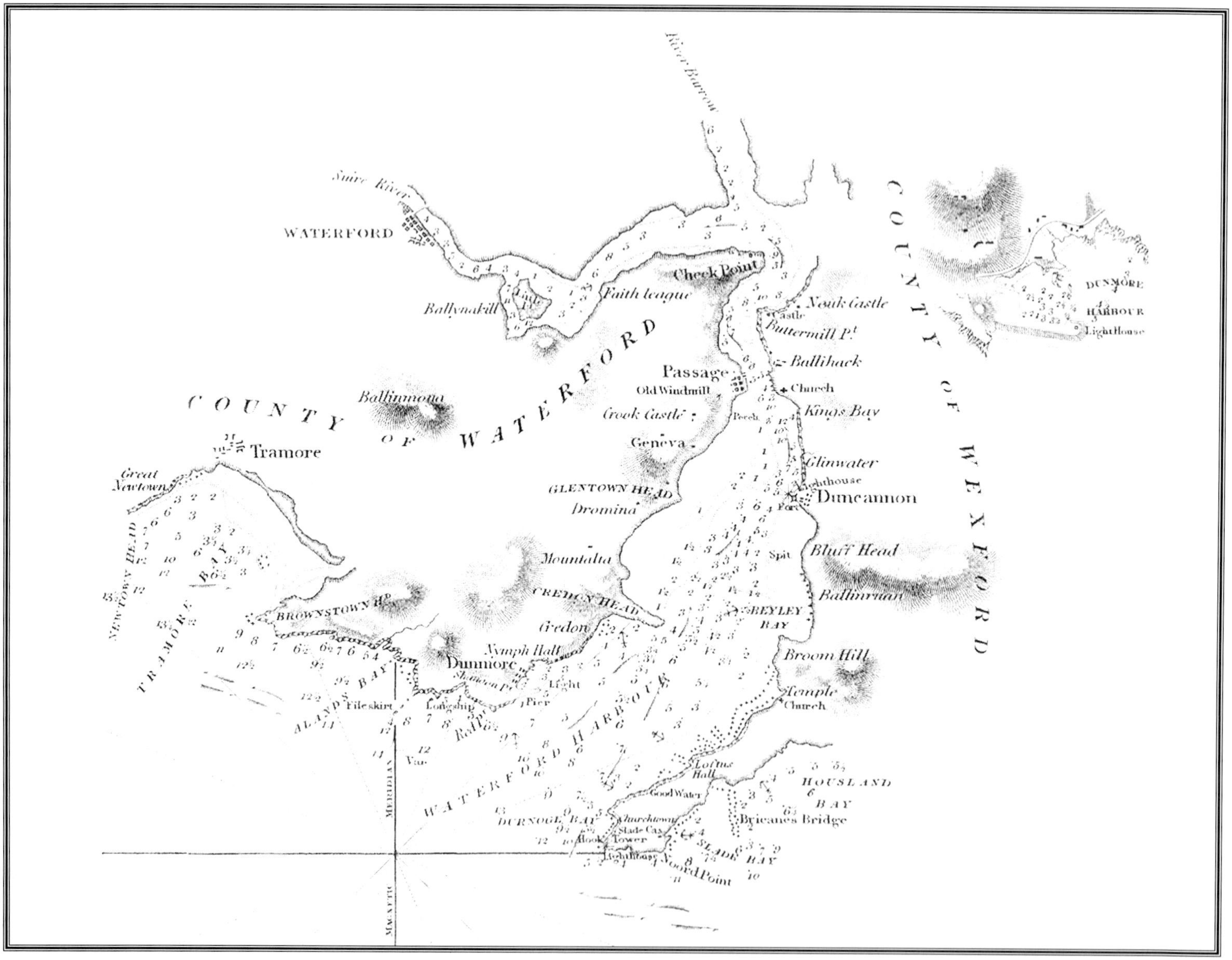

Nimmo's chart of Waterford Harbour circa 1831
(*Chart of St George's Channel and the coast of Ireland*, by A. Nimmo)

The General Post Office adopted his plan for Dunmore Harbour, County Waterford, and the site was considered suitable as a station for the mail packets from Milford Haven in south Wales. Under Nimmo's direction, work began on deepening the harbour and constructing a pier to his plan. The management of the works was undertaken by Alexander McGill, a man Nimmo would later bring to assist him in the construction of piers on the west coast. During the course of the work Nimmo designed an improved diving machine to remove underwater rocks, which Thomas Telford later endorsed. Dunmore harbour was eventually opened in 1821.[8]

Nimmo's early road building

Between 3 November 1820 and 20 February 1827, Nimmo was periodically employed by the Loan Commission, at a rate of three guineas per day, examining plans and estimates of proposed works and reporting on their progress while under construction. With the Commission he worked alongside the canal engineer, John Killaly. Killaly's career would overlap with Nimmo's on several occasions in the years to follow, until the death of both men in 1832.[9] During this time Nimmo was employed by the Kerry Grand Jury to superintend the construction of the mail coach road between Tralee and Limerick. He was also appointed, along with Killaly, joint overseer on a stretch of road from Bantry to Bear Haven. The road was to replace a mountain bridleroad. It was begun prior to 1822 and ran for about seventeen Irish miles along the western shore of Bantry Bay.[10]

While working with the Loan Commission, Nimmo was able to observe the method of constructing roads under the Grand Jury system and to identify the possible opportunities for fraud and jobbery. Lessons learned here were later put into practice in the west. He noted that, when proposals for works were brought before the Grand Jury, it was a common practice to overestimate the cost and at the same time to swear an oath as to the validity of the figures, without regard for the solemnity of the oath and with no sense of obligation. Nimmo found he frequently had to alter the plans and revise the estimated. The practice of swearing false oaths was so widespread that, in Nimmo's opinion, it deterred honest men from getting involved in road making and resulted in the poor quality roads that were evident throughout the country. Despite this, he considered that 'an immensity of good [had] been done by the Grand Jury presentments of Ireland' and that all that was required to ensure that works proposed by Grand Juries were successfully and inexpensively carried out was 'a good system of audit'.[11]

Nimmo considered the wages offered on public works to be low at 8d per day, although that was the average wage for labourers at the time. He was critical of how the labourers were paid. The method of paying labourers in food or having their labour set against their rent, left them unsure of the value of their labour and denied them 'an immediate reward for their industry'. It also left them suspicious and uncertain that they were being dealt with fairly.[12]

Prior to this, it was customary to employ adults only on the construction of roads. At first, Nimmo found that the labourers refused to 'work by measure', or 'piece-work'. This system was preferred by Nimmo as it enabled an entire family to take on a stretch of road and the wife, as well as the children, could assist the father in completing it. But the people did not understand this system and preferred to be paid in wages. In order to achieve the highest wages possible, they would 'combine', demand a high wage and refuse to work until this was agreed. Nimmo was offering the customary wage of 8d per day and was usually prepared to wait it out until the workers accepted his offer. If an agreement was not reached, he would simply move on to another sector and would only return when requested to do so by the workers themselves. However, in cases where time was of the essence, Nimmo went to the landlord, who usually exerted pressure on the men to set to work. While agreeing that some form of pressure was needed, Nimmo did not always approve of the method adopted by the landlord. On one occasion the landlord threatened to distrain the tenants' cows to force them to start work; this, Nimmo felt, although it was to his advantage, was an illegal use of the landlord's power.[13]

The suggestion that government assistance be extended to the poorer regions in the west did not end with the Bog Commission. In the years that followed, various government committees proposed government aid as the only possible solution for the future development of the west. In 1819, a government committee set up to investigate the state of disease and condition of the labouring poor in Ireland recommended that some of the suggestions put forward by the engineers be implemented. They recommended that measures be taken by the government to improve the mountainous districts along the coast by reclaiming the bogs and developing the fisheries. The committee was advised by Nimmo, described as 'one of the most eminent of the engineers employed under [the Bog] Commission', Robert Fraser, an expert on the fisheries in Britain, and the author of a recent publication highlighting the need for government encouragement of the fisheries of Ireland, and Thomas Telford, identified as the engineer responsible for constructing the road and bridges in Scotland, along with a 'considerable number of harbours, landing-piers and ferry-piers'.[14] On the advice of these three men, the committee made recommendations for improvements and developments that would, in time, be implemented in the west under Nimmo.

The committee returned to the recommendations outlined by the engineers and suggested that the bogs be reclaimed, so as to provide employment for the poor, that financial encouragement be made available to the fishing industry, similar to that granted to the industry in Scotland, and that roads be laid out to facilitate the fisheries in districts that had not benefited under the Grand Jury system.[15] They examined the methods adopted in Scotland in the construction of roads and piers and stated how these had contributed to the development of the fisheries and the country in general. The government paid half of the construction costs for the Scottish roads and the remainder came from the proprietors of the county through which they passed, or from the private subscriptions of interested individuals. The salaries for the engineering and superintendence were 'defrayed entirely by government', Nimmo told the committee, and the 'habits of industry and skill which those works introduced' had greatly benefited the districts in which they were carried out.[16] Telford held the opinion that public works ought to be considered 'in the light of a working academy', where men were taught how to work and how to use tools.[17]

Nimmo informed the enquiry of the findings of the Bog Commission and of reclamation works taking place in Holland, Flanders and Westphalia, which he had visited during his tour of Europe.[18] Before the enquiry he discussed the advantages to be gained by reclaiming the bog, the employment it would give and the land it would bring into cultivation. Nimmo was of the opinion that the improvement of the bogs and wasteland throughout Ireland, 'would provide for an additional agricultural population of two millions'. When questioned about the isolated fishery stations and the need to construct roads to open up communications with the coast, he admitted that he was not well acquainted with the fisheries, but knew that there was an abundance of fish to be found on the coast of Galway and Mayo. The boats of the fishermen were, he told them, totally inadequate; 'the boats of Mayo and Galway are half-decked hookers, those of Clare are chiefly made of wicker and hide, and in Kerry they had only yawls'.[19]

Robert Fraser explained that the biggest drawback to the development of the fisheries in Ireland was the heavy duty on the importation of rock salt, which cut down the quantity of fish cured for local consumption and for export. The best means of encouraging the fisheries and improving the sea-coasts of Ireland, he maintained, would be to enable the proprietors of land to raise funds, as had been done in Scotland, and to encourage them 'to build towns, villages and fishing stations' and to make and maintain 'harbours, roads and bridges'.[20] A Bill to assist the Irish fisheries had been rejected in the previous session of the House of Commons. On the request of the chief secretary, Fraser had made recommendations for changes, which, it was hoped, would be sufficient to see it pass through the House in the autumn.[21]

The three men came before the commissioners on different days in May 1819. However, given that Nimmo and Telford were already well acquainted, it seems safe to assume that the three would have met and that the topics raised before the commissioners would have been discussed in more detail, perhaps in a more private venue. The conversations would have gone on well into the night as the three discussed and debated how past experience and tested methods of development could be applied to the poorer regions of Ireland. The erudite Nimmo would have learned a great deal from the experience of both men; Telford, with his knowledge of road construction and fishery development in the Scottish Highlands; Fraser, with his proposals for establishing fishing villages along the Irish coast and for the government to encourage English capitalists to invest in Irish fisheries and manufacturing industries. All of this would have helped prepare him for the task that would occupy so much of his time and energy in the decade that followed.

Irish Fishery Board

The Irish Fisheries Act was eventually passed and resulted in the setting up of the Irish Fishery Board in 1819, to promote the development of the Irish fishery industry. The board received an annual grant of £5,000, from which it paid bounties for boats and on catches and issued loans for boat building and net mending. The board employed Nimmo on 17 August 1820 to undertake a general survey of the coast of Ireland and he remained in their employment until 1824.[22] The survey, it was hoped, would point out places of shelter along the coast which 'would form the most advantageous sites for Fishing Stations', and would indicate suitable routes from these, through the mountain passes, to the interior of the country.[23] It would also contribute to the construction of 'a more correct map of the coast of Ireland', which would prove useful until a general map would arise from a trigonometrical survey of Ireland which, it was anticipated, would soon be undertaken by the government.[24]

At various times over the next four years, Nimmo worked his way around the coast surveying the coastline, the hinterland and the sea, in so far as it related to the fisheries. The coast from Sligo northwards to Dublin was surveyed between 1820 and 1821, and from Leitrim southwards to Clare, between 1822 and 1824. The rate of pay for this work is unknown. However, the published reports of the Fishery Board show Nimmo to have received several sums for his work on the survey over this period, totalling £4,576.18.8d. The funds received were to cover all expenses incurred while conducting the survey, that of his own time and professional services, along with his attendants and assistants.[25]

From the outset Nimmo advised the board that, in order to save costs, he would draw information from all existing maps and surveys and would only survey those sections of the coast where existing surveys proved impossible to obtain or where they were shown to be defective. He used Murdoch MacKenzie's 1776 chart of the coast as his principal guide. This was the chart generally used by the admiralty at the time and was considered by Nimmo to be on the whole fairly accurate, but at times dangerously inaccurate. Where defects were found he drew up his own charts of the bays, and the most dangerous stretches of water.[26]

In the end the only county maps to be 'sufficiently accurate' for use were those of Louth, Dublin and Mayo; the remaining coastal counties had to be surveyed.[27] Nimmo made no comment on William Larkin's map of Galway, published in 1819, but it was already attracting criticism from other quarters as not being entirely accurate.[28] William Bald's map of Mayo was published by the Grand Jury in 1816 and it attracted high praise. This was the first full map of Mayo and, according to Patrick Knight, Bald's assistant on the project, it was 'a perfect general topography of the county'. It gave details on the seacoast, 'rivers, and streams, to their very sources, - lakes, roads, villages, mountains, with their principal elevations, were absolutely measured, and laid down'.[29]

To Bald's map, Nimmo added his own findings and those from recent surveys, which he knew to be accurate. For Clew Bay he used the findings from a survey commissioned by the Marquess of Sligo for the improvement of Westport harbour, and added these to his own survey of 'all the rest of the creeks' in the bay. Under his supervision, his assistant, William Grant, took the soundings and a detailed survey of Killala Bay. To Grant's findings were added details from Nimmo's own survey of the river Moy as far as the bridge of Ballina. Inishboffin Island and some of the other harbours of Mayo were also attributed to Grant.[30] Grant was one of the surveyors employed by the Bog Commission and he later worked for Nimmo on the public works in Sligo.[31] Patrick Knight, another surveyor from the Bogs Commission, surveyed the Inishkea islands. Knight had assisted Bald on the Mayo map and he too would work with Nimmo on the public works in that county. The coasts of Galway and Clare were 'minutely surveyed from Cleggan to Blackhead (sic)', for the Fishery Board, by Nimmo himself, with the exception of Kilkieran Bay, which he had previously surveyed for the Bog Commission.[32]

The findings from Nimmo's survey were sent to the board in a number of comprehensive reports, which were later published. The reports were far more detailed than anything Nimmo had produced heretofore and appear to have drawn on material from a wide range of sources. Many of the county surveys were already published by this time and Wakefield had provided a good deal of information on the fisheries of Ireland in his survey published in 1812. There is no doubt that much of the background information on the coastline was obtained locally while conducting the survey, but for a man who declared just a year before taking up this appointment that he was not well acquainted with the fisheries, it is obvious that he had certainly done his homework in the intervening period.

In his reports Nimmo gave details of each bay, harbour and headland along the coast, communications between them and the interior, the existence or non-existence of coastal roads, the geography and geology of the region, the agriculture and industry conducted in the hinterland, the conditions of the people and the general state of the country. He listed existing piers and pointed out their advantage, or otherwise, to the fishery industry. In accordance with his brief, he pointed out suitable harbours in need of piers and breakwaters. In many cases he provided plans and estimates for these and gave a full account of the landscape surrounding the proposed site. He wrote about the fishing grounds, listing the type and quantity of fish caught. He also wrote about the fishermen, their vessels, their fishing practices and the obstacles preventing them from exploiting the fishery to its full potential.

The east coast was generally fertile and well provided with good roads made under the Grand Jury system. The west coast was generally mountainous and boggy, with deep inlets that had very poor communications with the interior. On the west coast, where the best fishing grounds were to be found, the principal obstacles were poverty, the lack of piers, suitable vessels and insufficient curing houses. The men employed in the fishery tended to be fishermen-farmers. They divided their time between the land and the sea, to the detriment of both, he emphasized once again. The east coast fishermen were better off. They benefited from trading with Scotland and England and had a ready market for their fish in Dublin, Liverpool and the many other towns on both coasts. They were better equipped with boats and tackle and the coast was dotted with long established piers.

In contrast, the lack of piers had greatly hampered the development of the fisheries on the west coast. Without piers the fishermen in the west were confined to small boats. There were no decked vessels, and very few half-decked, 'between Killala and Newport, or between Westport and Galway'. However, there were large numbers of open sailboats and rowboats.[33]

Skerries Hooker circa 1848
(*The Deep Sea and Coast Fisheries of Ireland,* by Wallop Brabazon)

One of the Irish herring fleet (Galway Hooker) circa 1848
(*The Deep Sea and Coast Fisheries of Ireland,* by Wallop Brabazon)

These were more suited to the coast, but less suited to the boisterous Atlantic. The local open boats were easier adapted to trade in turf and seaweed, which was their principal occupation when not employed in the fishery. They were also able to land in small safe harbours without quays and were easily pulled inland to protect them from the storms. Small boats, however, restricted the fishermen to fishing the herring close to the shore, leaving the lucrative deep-water white fish banks to the larger, better-equipped fishing vessels from other parts of Ireland.

The principal white fish bank on the west coast was situated between Inishboffin and Achill Island and had been fished for centuries by some local fishermen.[34] The bank was considered by Nimmo to be 'the most productive on the Irish coast'. It was well stocked with 'all kinds of white fish, cod, ling, glassen, mackerel and gurnet, especially in spring, as also a few herrings'. Large decked vessels came from as far away as Rush and Skerries on the east coast, and Kinsale in the south, to fish this bank.[35] Twenty-four vessels from the east were found fishing there in 1822.[36] These boats were far superior to the local boats and were more suited to the heavy sea of the Atlantic. The boats landed their catch on the shore, where a member of the crew was left to attend to the drying of the fish, and when a full cargo was made up it was removed to the Dublin market.

Early Mayo boat circa 1832
(*Wild Sports in the West,* by W.H. Maxwell)

Another lucrative bank found off this coast was the sunfish bank, located about thirty miles west of Achill. Here the sunfish, or basking shark, were fished at the end of April or early May. The sharks 'were pursued and struck with harpoons for the sake of their oil' and, when successful, the results were very rewarding. One fish alone was capable of yielding seven to ten barrels of oil, 'worth about £40 sterling'.

***'Spearing the basking shark or sun fish off the west coast of Ireland'*, circa 1848**
(*The Deep Sea and Coast Fisheries of Ireland,* by Wallop Brabazon)

However, Nimmo believed that the high cost of outfitting the large number of boats engaged in the fishing made the business unprofitable and kept the boats away from the more dependable white fishery.[37] Time spent on the west coast convinced Nimmo that a more detailed examination of the white fish and sunfish banks was called for and he recommended that this be done under the direction of the Admiralty; 'more appropriately entrusted to nautical men than to an engineer'.[38]

There were almost no local boats fishing these banks in the 1820s. The poverty of the fishermen of the west made it impossible to aspire to such vessels without financial assistance from some government or charitable body. Nimmo advised the Fishery Board of the necessity of providing such funding: 'The superiority of the wherries and hookers over the county boats is equally conspicuous along all the west coast of Ireland and Scotland, and as far as the Isles of Shetland; and as these crafts are equal not only to stand the sea in the fishery, but also to be employed on the coasting trade, and to make considerable voyages, they seem to be deserving of encouragement in this maritime nation'.[39]

Wherry, hooker and yawl were terms used frequently by Nimmo to describe light coastal vessels used in the fishery and in the transport of goods. These had a single mast and included decked, half decked and open boats. They varied in size and in rigging, but common features were a forward mast, a bowsprit, a mainsail and jib, with some boats having a foresail. Nimmo used the term wherry to describe the largest of the three, the

Statement of the Number of Boats within the Westport District, extending from the Western point of Killala Bay, namely Lacken Point to Sline Head, which is the Northern point of the entrance of the Bay of Galway, with their description, Tonnage, Number of Men employed and their general occupation when not fishing, for 1820 and 1821.

Places belonging to.	1820. Decked Boats.	1820. Half Decked.	1820. Open Sail-Boats.	1820. Row-o ats.	1820. Tonnage of Sail-Boats.	1820. Number of Men employed.	1821. Decked Boats.	1821. Half Decked.	1821. Open Sail-Boats.	1821. Row-Boats.	1821. Tonnage of Sail-Boats.	1821. Number of Men employed.	
Lacken Creek	...	...	7	61	21	115	...	..	7	16	21	115	
Bally Castle Bay	...	...	1	12	3	72	...	...	1	...	3	6	
Killuduff Creek	...	...	...	10	...	50	...	...	.	12	...	60	Occupied
Balderick do.	...	...	...	6	...	30	..	...	..	6	...	30	as stated in
Porturlin do.	...	...	...	3	...	15	...	...	...	3	...	15	last years
Broad-haven Harbour	...	...	...	77	...	385	...	...	...	88	...	440	Report.
Portnafrank Creek, in B.F.	...	...	...	3	...	15	...	...	...	3	...	15	
Black Sod	...	...	2	10	7	65	...	...	7	16	29	115	
Clew Bay and Islands	5	14	16	235	378	1034	5	18	24	334	461	1460	
Killiney Harbour	...	...	1	9	8	42	...	...	1	9	8	42	
Ballinakil Harbour	...	1	11	50	80	261	...	1	10	48	76	281	
Claggan Bay	...	1	3	13	20	62	...	1	3	13	28	68	
Achus	...	...	1	10	6	46	...	...	2	13	10	75	
Isle of Boffin	...	2	12	46	91	190	...	3	14	47	101	256	
Isle of Omey	...	2	...	8	15	33	1	2	...	11	30	56	
Isle of Innistenk	...	...	3	3	23	30	...	..	1	...	2	4	
Isle of Clare	...	...	2	4	10	26	...	...	4	11	20	60	
Isle of Achile	...	...	7	1	33	32	...	...	21	5	90	104	
Isle of Inniskea	...	...	1	...	2	4	...	...	1	6	7	24	
Isle of Inisshark	...	...	...	1	...	4	...	..	1	2	6	14	
Streamstown	...	...	...	...	...	...	...	...	...	2	...	12	
Kingstown Bay	...	...	...	...	...	...	...	...	2	29	13	150	
Ard-Bear Bay	...	...	...	...	...	...	1	...	3	7	35	50	
Manning Bay	...	...	...	...	...	...	...	...	2	6	15	40	
Total for 1820	5	20	67	517	607	2516	7	25	104	687	955	3492	**Total 1821.**
							5	20	67	517	697	2516	
Increase for 1821							2	5	37	170	258	976	

Statement of the Number of Boats within the Galway District, extending from Sline Head exclusive, to Black-Head inclusive, with their description, Tonnage, Number of Men employed, and their general occupation when not fishing, for 1820 and 1821

Places belonging to.	1820. Decked Boats.	1820. Half Decked.	1820. Open Sail Boats.	1820. Row Boats.	1820. Tonnage of Sail Boats.	1820. Number of Men employed.	1821. Decked Boats.	1821. Half Decked.	1821. Open Sail-Boats.	1821. Row-Boats.	1821. Tonnage of Sail-Boats.	1821. Number of Men employed.	
Islands of Arran	...	...	36	8	288	212	...	...	41	17	328	273	
Bunnown Bay	...	..	10	40	90	210	...	...	10	40	90	210	
Roundstone Bay	...	...	26	53	208	342	...	...	32	53	256	372	Occupied
Burtrabue Bay	...	...	5	72	40	313	...	..	5	72	40	313	as stated in
Ard Bay	...	1	23	34	184	251	...	1	26	34	208	266	last years
Kilkern Bay	...	...	2	88	16	362	...	...	2	88	16	362	Report.
Cosheen Bay	...	...	62	...	449	310	...	...	65	2	473	325	
Greatman's Bay	...	...	53	2	424	265	...	...	54	5	432	282	
Costla Bay	...	...	40	...	320	200	...	...	40	3	320	215	
Galway Bay	1	2	136	399	1088	2076	3	4	142	38	1256	2296	
Tyrone Bay	...	...	8	29	64	156	...	...	11	29	88	171	
Kenvara Bay	...	...	6	22	48	118	...	...	7	25	56	123	
Burren Bay	...	...	21	20	168	185	...	...	21	20	168	185	
Malbay	...	...	...	1	...	5	...	...	...	...	...	5	
Total 1820	1	3	428	768	3387	5005	3	5	456	826	3731	5398	To tal 182.
							1	3	421	768	3387	5005	
Increase for 1821 in this District							2	2	28	58	344	393	

Table giving the number, type and tonnage of boats working out of each bay and island on the Mayo and Galway coast for 1820 and 1821
(*Account of the Fisheries … from Killala Bay to Kerry Head* by William H. King)

smallest being the yawl. The wherry, like the other two, was light enough to negotiate the harbours of the west coast, but strong enough to fish in deep waters and it was able to stay out for a number of days without having to return to shore at night. The wherries on the east coast were clinker built and were capable of twenty to fifty tons burden. They were frequently to be found fishing off the west coast. West coast wherries and hookers were carvel built and capable of carrying from about six to twelve tons. In latter years the term hooker was used, almost exclusively, to describe those boats working out of Galway Bay. However, in the early 1820s, they were also to be found working out of Kinsale and other ports in the south and Nimmo puts them on par with the wherries.[40] James Hardiman tells us that the Galway boats had improved greatly by 1820 and that the common sailing boat was then 'from 8 to 10 tons, and some of the largest from 12 to 14' tons. The normal cost of these boats was from £40 to £50, 'exclusive of nets and fishing implements'.[41] Yawls were carvel-built, typically of two and a half tons burden, with four oars and some had a lugsail. They usually had an eighteen-foot keel, six-foot beam and were two feet ten inches deep. Some had a round stern others were sharp at both ends.[42]

Apart from the cost, large vessels had one big disadvantage; they required piers to load and unload their cargo and very few piers existed along the entire west coast. William King, the Inspector-General with the Fishery Board for the west coast, noted in his report to the board in 1820 that, apart from a temporary pier at Clare Island, built by the late Sir Samuel O'Malley, there was no safe harbour on the Mayo coast between Broad Haven and Killary. He pointed out the difficulty this presented to the fishermen: 'the vessels fishing on this part of the coast run considerable risk, while out on the Fishery, for want of an harbour, in case of being blown from the Isle of Boffin, or coast of Connamarra (sic), there being no place of refuge for them to run to, from the Kilerries (sic) to Broad-haven (sic): and if they put into either of these, they may be detained there for weeks without being able to get out'.[43]

Nimmo's report on the fishing stations on the north coast of Ireland

Nimmo began his coastal survey on the Sligo coast on 16 September 1820. A surveying cutter was hired by the Board to take him around the coast, but bad weather prevented him boarding. Instead, he was forced to conduct the survey from the land until more clement weather and a suitable boarding place was eventually found at the port of Derry. This gave him the opportunity to observe road communications along the north west coast and the agriculture of the neighbourhood. He found the roads in county Sligo to be in good order 'and tolerably level'; however, those in Leitrim and especially Donegal were 'universally hilly and bad; the lines injudiciously chosen'.[45]

Within Sligo Bay there were seemingly very few fishermen, no decked boats or wherries and not more than twenty sailing boats in the whole district. Between the northern point of the bay, at Raghly Head, and the southern point, Aughris Head, ran a

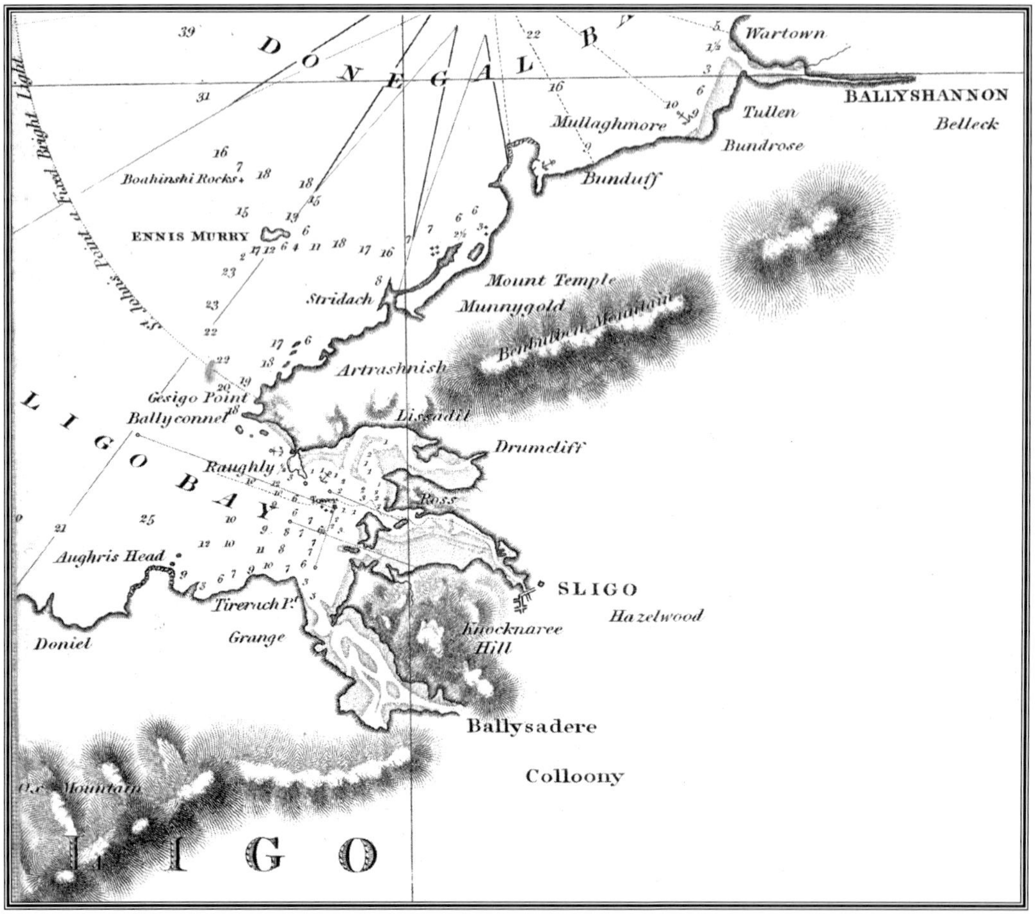

Nimmo's chart of Sligo coast circa 1831
(*Chart of St George's Channel and the coast of Ireland*, by A. Nimmo)

reef of rocks called the Ledge. The fishermen of the bay rarely ventured outside the reef, preferring instead to fish the inner bay in small open boats.[46] Nimmo recommended Raghly Head as the most suitable site for a pier as it could be approached on either side, depending on the winds, and was in the vicinity of the fishing grounds. Good fishing was to be had off the coast at Mullaghmore and from here the counties of Fermanagh, Leitrim and the neighbouring parts of Sligo and Donegal were supplied with white fish. In the previous year the herrings were so plentiful off this coast that they were sold for five shillings a thousand. However, although Mullaghmore was the only place between Sligo and Donegal where the fishing boats could find shelter, a proper pier here would prove too expensive and so, at this time, Nimmo was recommending improvements to the cove and landing place already used by the fishermen.

Donegal town had a recently constructed quay, but this was too far up the river to act as a rendezvous for fishery boats. Instead they collected at Salthill, below Mountcharles. The numerous bays on the southern coast of Donegal offered shelter to the rowboats and yawls fishing the herring during the season. But the land was poorly cultivated and the little corn that was produced was used in the distilling of poitín. In general the agriculture here was backward and the people were dependent on supplies from Connaught:

> The whole agriculture of Donegal is late; the crops in Boylagh and Banagh are in many places yet (29 September 1820) uncut, the hay unsaved, and the late heavy rains have made much of it unserviceable. The chief cause of this is, probably, late sowing; the inhabitants are well satisfied if they hear of a good crop in Connaught.[47]

Nimmo approached Killybegs by a hilly road, 'through a poor mountainous country.' He found the place in decline, although it was the safest harbour on the coast and 'several hundred sail of vessels' could be seen there in the past, 'purchasing or curing herrings'. Less than a dozen vessels were visible to Nimmo and 'none of them employed in the fishery'. Its decline he attributed to its location; it was some distance from the fishing

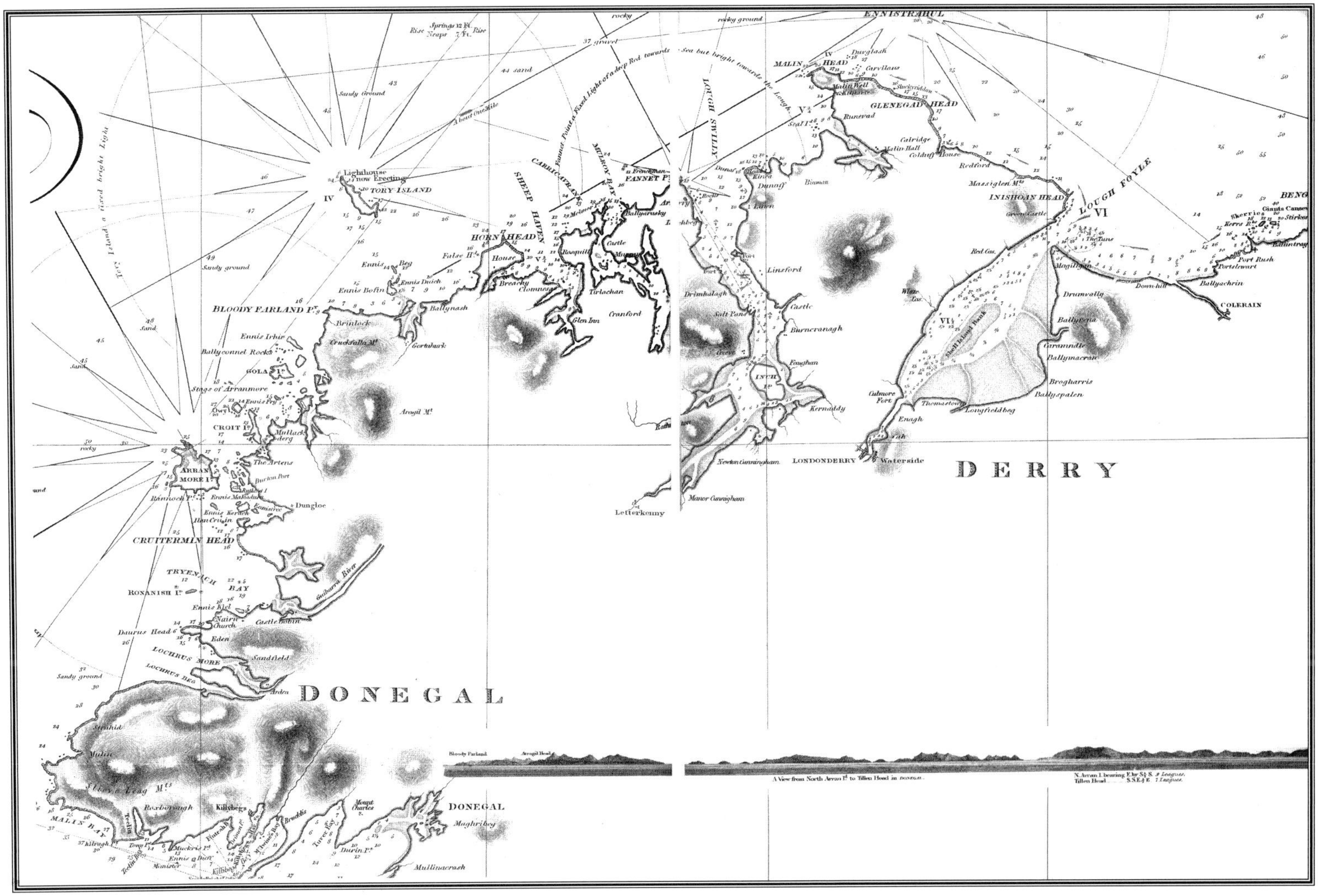

grounds and no vessel could put to sea from there in west or south-west winds. A good carriage road ran from Killybegs to Ardra, but the coast road was 'scarcely practicable on horesback'.

Nimmo's chart of Donegal and Derry coast circa 1831
(Chart of St George's Channel and the coast of Ireland, by A. Nimmo)

A natural increase in the population along the west coast at Dungloe and The Rosses had resulted in a good deal of the bog land being reclaimed and cultivated. This population increase was expected to benefit the fishermen where provisions were expensive and bad roads meant that 'the traffic of the country [was] entirely done by horses and sacks or baskets'. Among several piers recommended by Nimmo was one for Arranmore Island where the people pursued the fishery, 'with more zeal than any persons on this coast, but they have no boats fit for going outside of the sound'. Poor roads prevented Nimmo from making the journey along the north cost of Donegal and he was forced instead to return to Burtonport and journey to Letterkenny by way of Fintown. The neighbourhood around Letterkenny was well cultivated but the town was 'too far up the Swilly to be of any importance to the fishery except as a market'.

At Derry, Nimmo was able to board his cutter and continue his survey from the sea. However, he considered Lough Foyle to be unsafe for small craft and suggested that one or two small piers, in particular one at Greencastle, would be of great service to the fishermen. On the north coast, from Derry to Belfast, he found the inhabitants 'industrious, persevering people'. But the coast itself was so 'totally devoid of protection for even the smallest vessels, that no progress' could be expected in the fishery without the erection of piers. The exception was Portrush, where plans were under way to build a pier in deep water. This, in Nimmo's opinion, would be of tremendous benefit to trade vessels of the North Channel, which were without shelter from Lough Swilly to Belfast.

There were piers west and east of the Giant's Causeway at Portballintrae and Ballycastle. However, both could be improved with financial assistance from the board. Villages along the east coast of Antrim traded with Scotland, 'limestone for coal', and would benefit greatly from piers. There were a good many salt works in the district; the salt was mostly sent to Belfast for export. Larne had 'an excellent natural harbour' and Carrickfergus had a good pier. Hardly any fishing craft worked out of Belfast, but it proved a valuable market to the fisherman. It was now October and Nimmo had succeeded in surveying the coast from Sligo to Belfast Lough in just two months. He decided to conclude the first section of the survey at this point and drew up his report, which he handed in to the Fishery Board on 30 January 1821.[48]

Early in 1821 Nimmo was back on the sea continuing his survey along the coast of County Down. This coast he found to be well cultivated and 'thickly populated by an industrious and intelligent race, a great proportion of whom are occupied in the linen manufacture'. The pier at Bangor was built in 1757 and was frequented by

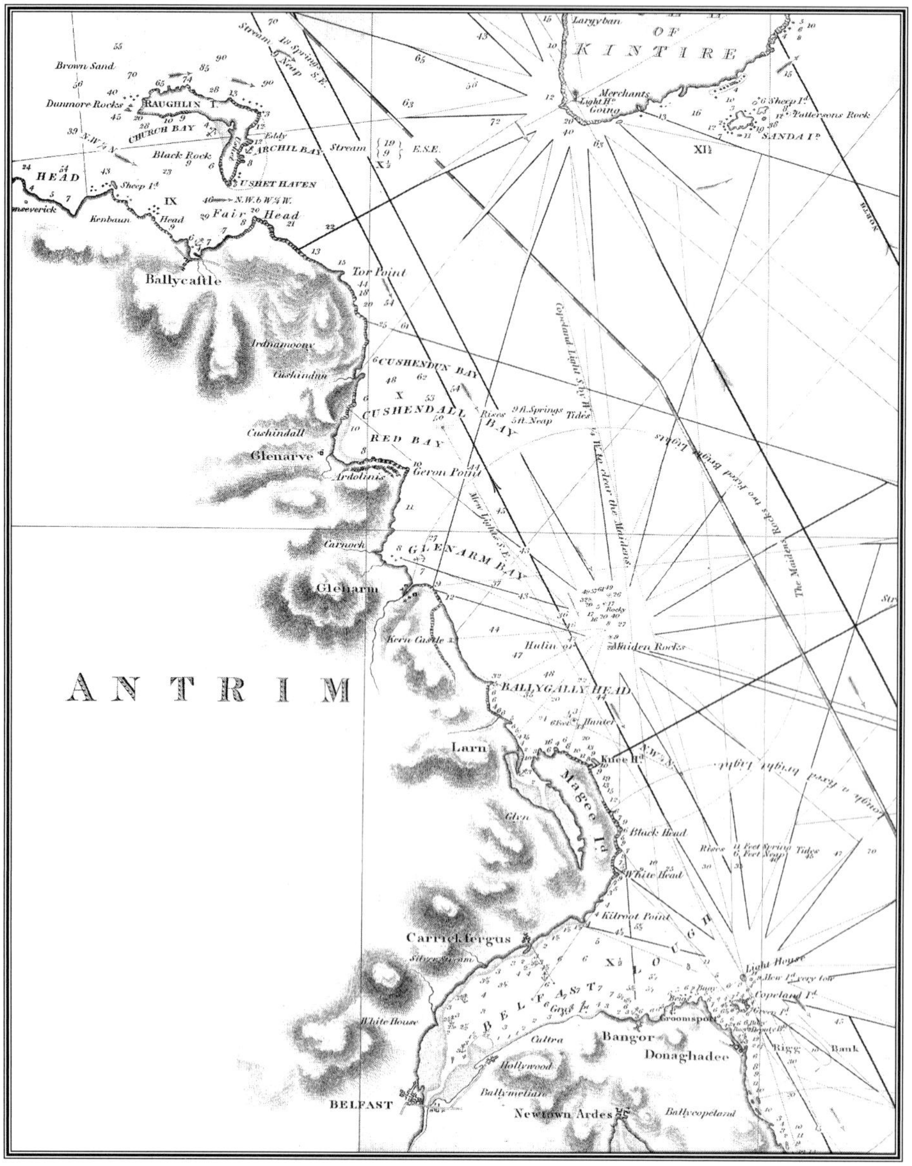

Nimmo's chart of Antrim coast circa 1831
(Chart of St George's Channel and the coast of Ireland, by A. Nimmo)

boats crossing the channel from Portpatrick in Scotland. As well as the fishery, a good live cattle export trade was conducted through here, but the harbour was small and improvement works were ongoing for a number of years. Nimmo offered suggestions on ways of further advancing the works, but held out little hope of these being completed to the standard required.

From Bangor to St John's Point a number of villages were actively engaged in the fishery in half-deck boats of twenty to thirty tons. From April to July they trawled for flat fish and from July until late winter, they fished the herring. There were piers or sheltering places to be found and to many of these Nimmo recommended small improvements at reasonable costs. Good herring fishing was also to be had from St John's Point to Annalong. Newcastle, 'a beautiful and well-frequented bathing-place', had an old pier in need of repair. Nimmo was of the opinion that this should be extended into deep water, but seemed resigned to the fact that the board had insufficient funds to tackle expensive works: 'I presume it is out of the question to look for funds sufficient to carry this pier forward into deep water, and that we must be restricted to the simplest plan by which the present structure can be rendered valuable'.

Fifty years prior to this the herring fishery along this coast was 'only a shore-fishery, pursued by small boats', but now it was in deep water 'from twenty-eight to fifty fathoms, pursued

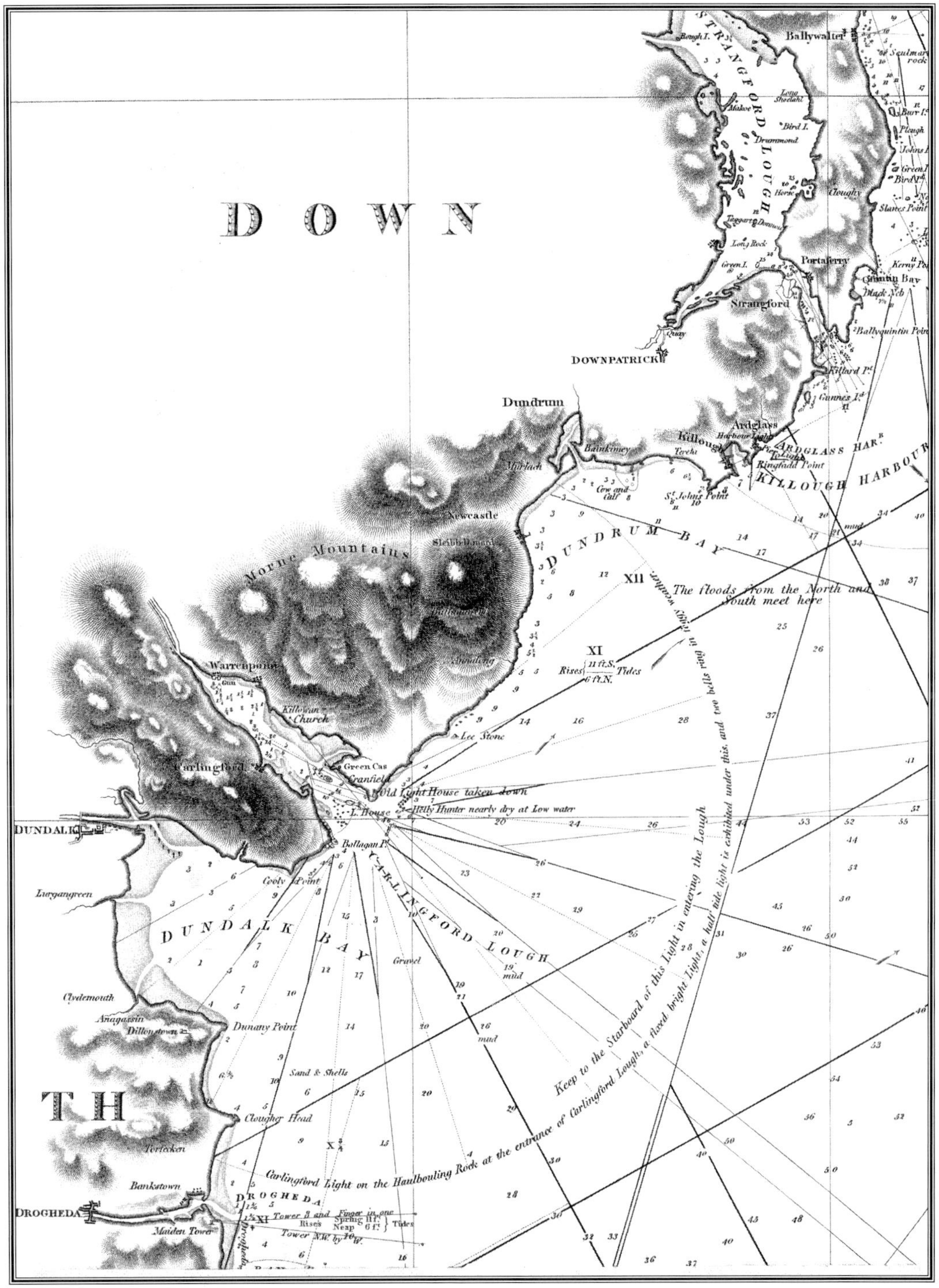

Nimmo's chart of Down and Louth coast circa 1831
(*Chart of St George's Channel and the coast of Ireland,* by A. Nimmo)

in large smacks and wherries'. However, English fishermen from St Ives, using superior nets, were succeeding better than the native fisherman. The fleet working this coast was accompanied by lighter vessels, which purchased large quantities of the fresh herrings and ran with them to Dublin, Liverpool and 'other towns on both coasts'. Fresh herrings were also sold to the neighbouring counties and the remainder was cured.

Strangford Lough, because of the 'supposed intricacy of its navigation', was frequently avoided by strangers, but was considered by Nimmo to be the safest haven on the coast. Here Nimmo found MacKenzie's chart to be inaccurate. He decided to 'make a new survey and chart of the entrance into this haven' and recommended that the board publish it immediately. Carlingford Lough, on the other hand, he considered not entirely safe, 'being much exposed to squalls from the mountains'. While surveying the Lough, Nimmo's own cutter was forced on several occasions to run to Warrenpoint for shelter. He made 'a very exact survey' of the Lough and drew up a chart, 'where the soundings and all known dangers are carefully expressed'. This he hoped would be of assistance to the navigator, as he considered this harbour, 'a more dangerous navigation than has been imagined'. The fishermen lived chiefly on the southern or Louth side, and were, in Nimmo's opinion, 'a different race from those of the county of Down, their language chiefly Irish' and their boats, 'those long half-decked smacks, with falling masts', were similar to those used in Dublin, 'above Carlisle Bridge' [O'Connell Bridge].

Shipwreck was a regular occurrence off the County Down coast. In the few months that the survey was under way, six vessels, with a value of from £40,000 to £50,000, were wrecked between Donaghadee and Strangford, and many more were seriously damaged.[49] During the course of the survey a large vessel from Liverpool went aground on some sunken rocks off the County Down coast that were not marked on MacKenzie's chart. Without visiting the spot and relying on information received, Nimmo included them on his chart, only to be afterwards advised by the Admiralty that they were laid down in the wrong position. He returned later and after surveying the area, he laid down the rocks in their correct position on the engraving.[50]

The counties of Louth and Meath were 'almost entirely destitute of shelter', and Nimmo pointed out several sites between Dundalk and Drogheda where piers could be constructed or existing piers repaired. South of Drogheda the coast of Meath was a broad shallow strand, 'having no natural advantages for constructing piers'. The obvious need to develop the harbour of Drogheda for commercial use he decided not to dwell on, it not being associated with the fisheries. Along the northeast coast of Dublin he noted 'a variety of towns and villages actively and extensively engaged in the fishery', and piers had been built at several places for the protection of craft. Here the land was rich and fertile. The most common vessel used was the wherry, capable of between twenty to fifty tons burden. Shellfish was caught in pots inside the islands of Rockabill, Lambay and Ireland's Eye. Outside the islands was the trawling ground and 'beyond that the long-line fishery':

> [The wherries] are not only employed in trawling and hooking for white fish for the supply of Dublin market, but also with long lines in the Channel, and on the north-west coast for cod, ling, etc; their skill in catching, curing and drying these is well known. Many of these wherries are also engaged in the herring-fishing, not only on the coast of Down and Leinster, but also of Donegal and Connaught. The wherry-men are rarely to be met along the Munster coast.[51]

Safe harbours were to be found at Rogerstown near Rush, Malahide and Baldoyle. There was a pier at Balbriggan, Skerries and Rogerstown. Howth was the rendezvous of all those engaged in the supply of the Dublin market. Fish landed at Howth was carried over land to Dublin; this enabled the boats to put to sea again without delay. A thorough examination of Lambay Island convinced Nimmo that reefs on the north side, called the Taylors, were the most suitable place for a fishing village and pier. He drew up the plan for a pier that would offer shelter to between a dozen to twenty wherries, but the place was capable of 'considerable enlargement' if future needs demanded it. This work would, Nimmo contended, 'certainly be of very great use in promoting the coast-fishery' in the east. Nimmo was anxious to provide charts of the entire coast as far as Dublin, but was prevented by bad weather from doing so and planned to return early in the next season to complete the task.[52]

The board was highly satisfied with Nimmo's work. On receipt of his second report, they concluded that the benefits from the survey, once published, would extend beyond the fisheries 'to the commercial interests of the country', and contribute to the development of uncultivated wildernesses. The

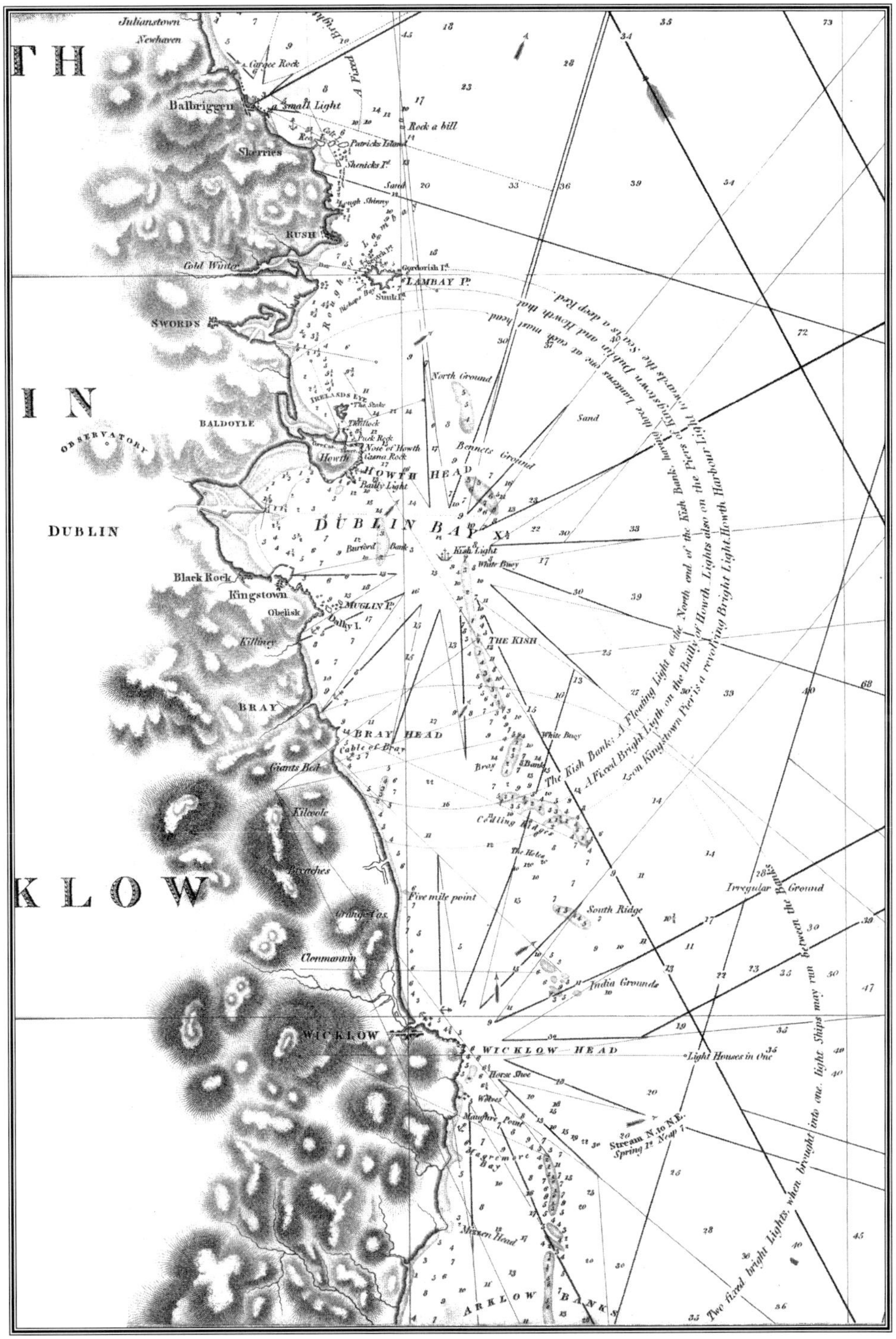

Nimmo's chart of Louth, Dublin and Wicklow coast circa 1831
(*Chart of St George's Channel and the coast of Ireland,* by A. Nimmo)

report indicated numerous creeks and bays where each season shoals 'of the finest herrings' were to be found, 'but from the want of places of shelter, either naturally or artificially constructed', these were not being fished to their capacity. The board decided that the best way to promote the coastal fisheries of the country was to build small piers to facilitate the fishermen, at the same time specifying that, as these piers would also benefit the neighbourhood in general, half of the costs must be contributed by local landlords and other interested groups.[53] However, in the districts where the piers were most needed, along the west coast, the people were too poor to pay the excess, resulting in very few applications emerging from the poorer regions.

Nimmo's first and second reports on the fishing stations on the north coast, along with a scaled down version of the plans and charts, were published with the board's fourth annual report in 1823. His charts later went on sale and were described by William Bald as, 'among the finest engraved specimens of our hydrographic surveys yet published'.[54] Satisfied with the progress of the survey the board decided to purchase the cutter used by Nimmo and agreed a price of £1,203.7.4d with its owner. The cutter was refitted early in 1822 and was proceeding with the examination of the east coast when Nimmo received instructions from the board to visit the coast of Clare. The potato crop had failed in the previous year and the government was receiving reports of distress among the poor. Nimmo was being sent to the west to ascertain what steps could be undertaken by the Fishery Board to alleviate the distressed conditions of the people and, at the same time, benefit the local fisheries.[55]

Chapter 3

Fishery piers and the Western District 1822–1824

In the spring of 1822 requests seeking aid for the destitute poor were coming in from the west and south of the country. The potato crop, the staple food of the poor, had failed. Over the previous three years the people had endured bad weather resulting in poor harvests that had left them in a state of 'unparalleled misery' and in need of immediate assistance.[1] The weather had deteriorated throughout the summer and autumn of 1821. After a cold and frosty June, heavy rain began to fall towards the end of August and continued 'almost without intermission through September'. Serious flooding was experienced in the west, especially Mayo and Galway. The floods damaged the wheat and 'not only carried off the hay, but literally swept away whole crops of potatoes. Towards the middle of the month all outdoor work was suspended', and the people grew anxious for the harvest. The rains continued through October and November and the press reported that the country was in 'a wretched state'. Very few potatoes were dug and those that were proved to be small and deficient. December began with a violent storm and the Shannon overflowed its banks and flooded the country. However, the full extent of the crises only became known in 1822 when the farmers opened their store of potatoes and discovered that most had rotted in their pits.[2]

By April 1822 reports were reaching the press that cattle in the west were dying 'by dozens' of starvation and in some places they were being bled 'for the sake of the food which the blood afforded'. People were reduced to living on 'salt leaf' [Dillisk], nettles, and wild plants, which brought on dysentery and other bowel complaints. By the end of the month typhus had broken out in Mayo, Galway and Clare. In Achill several people were found dead on the roadside. As well as dysentery and fever, many were suffering from swollen limbs and were unable to walk to their neighbours for help. Others made their way in large numbers to the nearest towns and villages seeking employment or charity. But employment was not to be had and as the streets filled with the wretched poor, typhus began to spread, endangering all who came into contact with them. Galway was overrun with the Connemara poor: 'Such misery was never witnessed before in the town of Galway, whole clans pouring in from the mountains of Connemara in quest of food'.[3]

The counties and towns worst affected were, for the most part, situated along the coast; Sligo, Leitrim, Roscommon, Mayo, Galway, Clare, Limerick, Kerry, Tipperary and Cork, along with the cities of Limerick and Galway. Over half the population in the affected districts were said to be dependant on charitable assistance for their support. The nature of the distress was described as 'peculiar', in that there was plenty of food to be found in these counties, but the poor were without the funds to purchase it. There was no shortage of corn or oatmeal, prices were reported to be 'very moderate' and export trade was good, with the result that, 'those districts in the south and west presented the remarkable example of possessing a surplus of food, whiles the inhabitants were suffering from actual want'.[4]

The vulnerability of the poor was explained by their total dependency on what they themselves grew on their own small plot of ground, which was mainly the potato, and by the lack of employment available in those areas. Without employment, the poor were unable to purchase food in the market place regardless of price: 'When the produce of the peasant's potato ground fails, they are unaccustomed to have recourse to markets, and indeed they seem rarely to have the means of purchasing'.[5]

Dublin Castle Relief Commission and the London Committee

As the appalling conditions of the people became widely known, a quick response was called for. Efforts to assist the poor began within their own communities with the setting up of local committees to raise funds and to solicit help from the government and charitable bodies. Collections were made in the churches in Dublin. In London a public meeting was held in May, a committee was formed to raise subscriptions and to oversee their distribution. Contributions were forwarded to the committee from all over Great Britain and from as far away as Calcutta, Madras, Bombay and St Helena. A similar committee was established at the Mansion House in Dublin. The lord lieutenant, the Marquess of Wellesley, set up the Commissioners for Relief of the Distressed Poor, under the chairmanship of the under-secretary, William Gregory, at Dublin Castle. The Dublin Castle commission was to oversee the distribution of funds already in hand for such an emergency and also those received from charities at home and abroad. The commission drew up the principles on which relief was to be administered. It was agreed by all that the best means of assisting the people was to provide employment on public works and to issue cheap food to the poor. They purchased one thousand tons of potatoes from Scotland and four hundred tons of seed potatoes from Liverpool and issued instructions that every 'inducement should be held out for the cultivation of potatoes' in the distressed regions for the coming season. This was to try and prevent a similar shortage occurring in the following year.[6]

The London Committee succeeded in raising over £300,000. The funds were to be distributed to individuals, local committees and public bodies, to finance suitable projects in the distressed regions. The local committees proposed projects for adoption as public works and the London committee, and Dublin Castle commission, assessed their value to the poor and to the district in which they were to be carried out. Once approved, the government would contribute a portion of the cost of the project or pay for its supervision. Small local projects were to be chosen over large ones, the reasoning being that, works 'on a great scale [had] a tendency to invite an accumulation of numbers to a given spot, and probably to disappoint the greater part of them'. However, in assessing proposals, the employment of the poor, rather than the importance of the work, was to be the objective. But low wages were to be given at all times, so as not to attract large numbers, or interfere with the local demand for labour.[7] In conjunction with the projects put forward by the local committees, the lord lieutenant brought pressure on the Fishery Board to provide employment for the poor in the construction of piers, and it was to this end that Nimmo was dispatched to the Clare coast.

Nimmo's fishery piers

At the behest of the Fishery Board, Nimmo sent a list of thirty-four sites where, for less than £500, piers could be laid out all along the west and south coast. This was the maximum cost of pier construction towards which the board was prepared to grant funds. It was still the board's intention to contribute no more than one half of the overall costs of the works. However, as the proposed sites were in districts where the people were too poor to furnish the required local contribution, the board sought financial assistance for the remainder from the government and charitable bodies. Copies of Nimmo's list were issued to the commissioners at Dublin Castle and the board of the Mansion House relief committee. The Mansion House committee was willing to contribute one fourth of the cost of some of the piers, as they were 'calculated to combine the twofold advantages of national and permanent improvement' with bringing immediate relief to the distressed people.[8] Funds for the remaining fourth would have to be sought elsewhere.

Nimmo wrote to the chief secretary's office at Dublin Castle suggesting that the board's grant be increased so that they might take on the full cost of erecting the piers. He pointed out the utility of the works by stating that 'with the destitution, risk and difficulties encountered' on

OF IRISH FISHERIES; FOR THE YEAR 1822 — 21

Appendix, N° 15.—*continued.*

LIST OF PIERS.

		£.	s.	£.	s.
County Clare	Kilbaha	500	–		
	Seafield	500	–		
	Liscannor	500	–		
— Kerry	Bana	250	–		
	Dingle	400	–		
	Valentia	495	–		
— Galway	Costello	350	–		
	Roundstone	400	–		
	Cleggan	500	–		
— Mayo	Old Head	500	–		
	Achil Sound	400	–		
	Ely Erris	300	–		
Estimate Amount of 12 Piers				5,095	–
Galway	Barna	300	–		
	Spiddle	150	–		
	New Harbour	420	–		
	Ballynacourty	300	–		
	St. Kitts	250	–		
— Clare	New Quay	200	–		
	Canygaholt	400	–		
	Dunbeg	400	–		
— Mayo	Killalla	150	–		
	Bunaghtrahee	300	–		
	Ennis Tink	400	–		
	Tarmon	400	–		
— Kerry	Brandon Bay	400	–		
	Ballinskelligs	400	–		
	Cahirswine	400	–		
— Sligo	Roughly Point	800	–		
	Fullagheeny	500	–		
— Cork	Cullagh	400	–		
	Beerhaven	200	–		
	Continasheny	500	–		
	Clonakilty	500	–		
Great Island of Arran		250	–		
Lambay Pier		900	–		
Durus Pier, County Galway		300	–		
Clifden Pier, - - ditto		495	–		
Claddagh Pier - ditto		928	–		
Estimate Amount of 26 Piers				10,643	–
Total Estimate of 38 Piers			£.	15,738	–

List of piers recommended by Nimmo 1822
(Fourth Report of the Commissioners of Irish Fisheries)

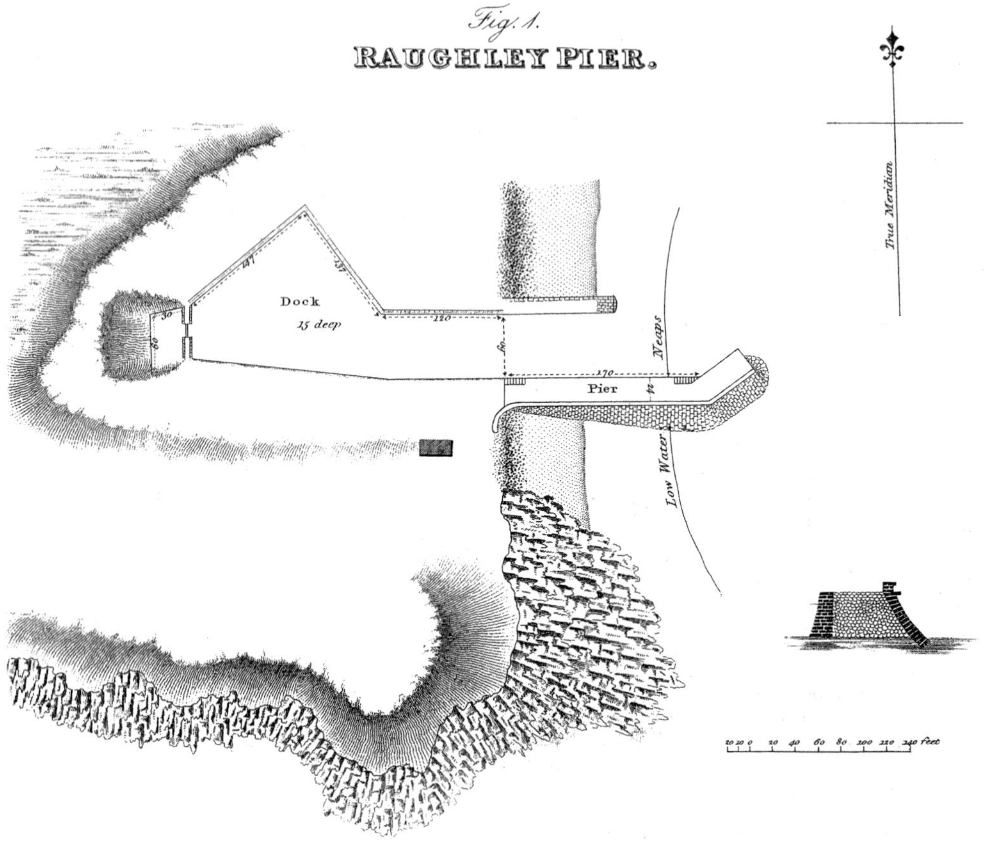

Nimmo's sketch of Raghly Pier, Sligo, circa 1824
(Nimmo's Report on the Fishery Stations of the Coast of Connaught)

coasts where no harbour existed, even works 'of a simple and economical description would be of the utmost benefit to the fishing craft'. In support of the fishermen he wrote that, in the course of his examination of the coasts, he had been 'strongly impressed by the appearance of industry and capital engaged in the fisheries in those stations where natural harbours and artificial piers and quays, even of a rude description, permit the fisherman to pursue his occupation with little interruption, and to preserve his property from accident'.[9] His proposal was brought before the Dublin Castle commission, which agreed to contribute the remaining one-fourth of the costs of twelve piers on the list. The piers were estimated to cost £5,095 and were located at the following sites: Co. Clare – Kilbaha, Seafield, Liscannor; Co. Kerry – Banna, Dingle and Valentia; Co. Galway – Costelloe [Rossaveel], Roundstone, Cleggan and in Co. Mayo – Old Head, Achill and Elly Bay [Saleen] in the Mullet peninsula.[10]

With this the Fishery Board decided to interrupt Nimmo's survey of the east coast and 'to transfer the whole establishment to the west of Ireland'. This would enable Nimmo to continue with his survey of the coast while at the same time superintending

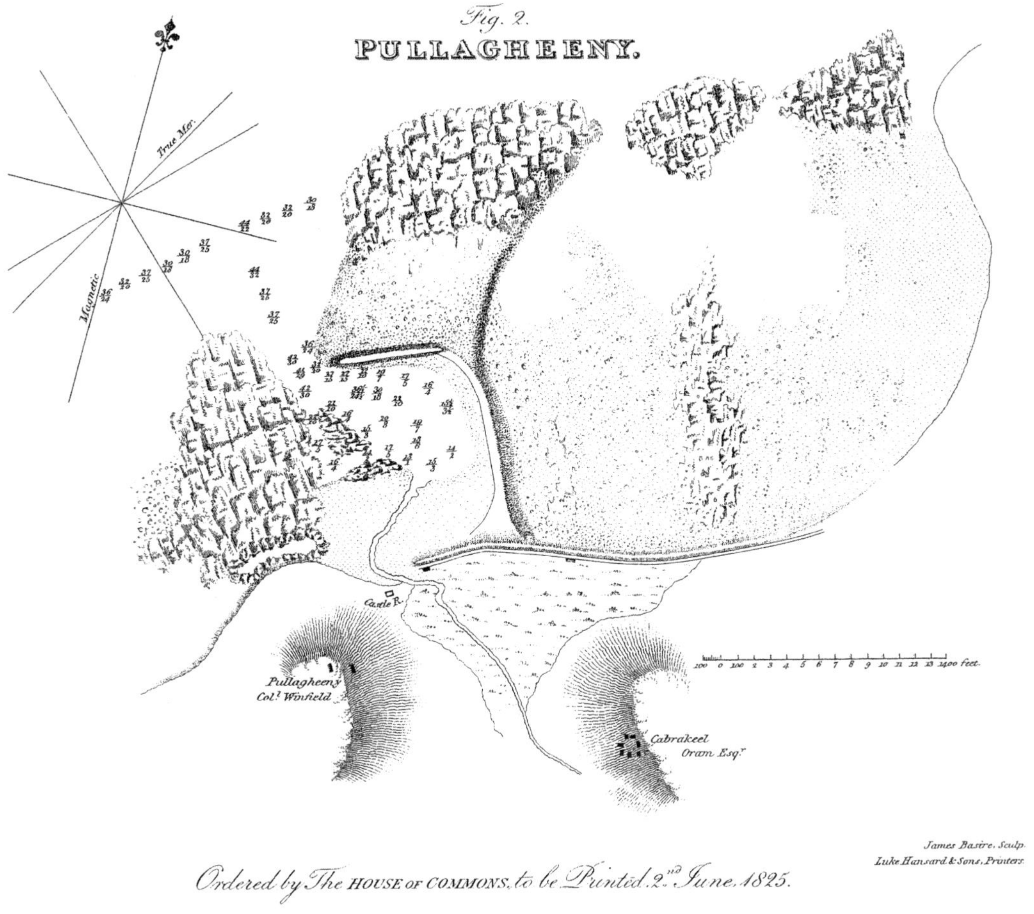

P57; Nimmo's sketch of Pullagheeny Pier, Sligo, circa 1824
(Nimmo's Report on the Fishery Stations of the Coast of Connaught)

the construction of the piers.[11] Nimmo was issued with instructions that work on the piers was to commence immediately, using his plans, and he was on no account to exceed his estimates. He was empowered to hire contractors, but the board had a preference for using direct day labour whenever possible as this would be more cost effective and leave the board in control. It is not clear exactly who paid Nimmo for his time spent overseeing the organization and commencement of the construction work of the piers. The Fishery Board Accounts indicate that Nimmo was paid for work associated with the coastal survey; as to whither this included his time spent on the piers or not, it is difficult to say.

In time, more funds became available and orders were issued to proceed with a further twenty-one piers, making up Nimmo's entire list. Again the Fishery Board contributed one half of the funding, with the London committee and the Dublin Castle commission contributing one-fourth each. The estimate for the twenty-one piers was a little over £8,000 and these were located as follows: Co. Sligo – Raghly Point and Pullagheeny [near Inishcrone]; Co. Mayo – Killala, Bunatragher [Dunfiny Bay], Inishturk, and Tarmon

[Blacksod Point]; Co. Galway – Barna, Spiddle, New Harbour [Rinville], Ballynacourty, and St. Kitts [Killeenaran]; Co. Clare – New Quay, Doonbeg and Carrigaholt; Co. Cork – Coolagh, Castletown Berehaven, Courtmacsherry and Clonakilty and Co. Kerry – Brandon Bay, Ballinskelligs and Cahirciveen. Máimín pier, on Gorumna Island in Co. Galway, was later added to this list.

The Fishery Board was convinced that the expenditure would be money well spent and that the benefits arising from the piers would be 'promptly felt, permanent in their effects, and embrace other points of national welfare besides that of the Fisheries'.[12] Piers were also ordered for Lambay in Dublin and for Inishmore, Durus [Dooras], Clifden and Claddagh in Galway. The extra funding needed for these piers would come from the districts themselves.

Nimmo travelled around the country securing sites and preparing plans. Agreement was reached with the proprietor of the site and the shore and adjacent land was usually handed over without charge. The only cases mentioned by Nimmo, in which disagreement arose, were at Raghly pier and Roundstone pier, and the difficulty was not with the proprietors but with the immediate occupier.[13] Wherever possible local men were hired as overseers and contractors, along with local stonemasons and stonecutters. The Fishery Board cutter arrived with the necessary tools and distributed these on site.[14] However, as the number of works increased, and on occasion when contractors proved inadequate, skilled men were brought in from the outside, such as Alexander McGill from Dunmore, a man Nimmo could rely on to do the job to his standard and his plan.

Engineers in charge of Public Works

In a short space of time the Dublin Castle commission had distributed £65,000 in the form of grants for various projects. But as the number of people seeking assistance began to rise and applications for public works increased, the Irish Administration sought extra government funding. In June and July funds totaling £300,000 were voted by parliament for relief works in Ireland.[15] To validate the necessity of any proposed work, and to protect against the misuse of public funds, the commissioners decided that they needed their own men on the ground. On 5 June 1822 the lord lieutenant appointed three engineers to take charge of all public works along the west coast from Sligo to Kerry. The men were to report to the commissioners at Dublin Castle and the lord lieutenant was to be kept informed of developments at all times. The coast was divided into three regions and placed under the supervision of the following men:

Northern – From Galway Bay to the Sligo Bay, Alexander Nimmo Esq.,
Middle – From the River Shannon to Galway Bay, John Killaly Esq.,
Southern – From South Coast to the River Shannon, Richard Griffith Esq.[16]

Nimmo's district covered counties Galway, Mayo, Sligo, Roscommon and Leitrim and in time became known as the Western District. At a time when few roads existed, the geographical spread of the district was quite daunting and would require long hours on horseback and at sea. Nimmo gives a glimpse of what he encountered in a typical day's journey when he described the difficulties met travelling through Erris, in Mayo, at this time: 'when a horseman attempted the journey it was necessary to have the assistance of six or eight guides on foot to cast the horse at every swamp, fasten his legs, and haul him over by ropes; an operation which is to me sufficiently familiar, being often resorted to from necessity in the course of our expeditions'.[17]

Nimmo, like Griffith and Killaly, worked as a consulting engineer on a per diem basis, this allowed his commissions to overlap and, while working as engineer for the Western District, he was able to retain his position as engineer with the Fishery Board. Over the next two years he continued his survey of the western coast and oversaw the construction of the piers, while at the same time constructing roads and bridges in the Western District. Time spent with the Bog Commission and the Fishery Board had demonstrated that Nimmo was an extremely capable and hard working individual, well suited to working long hours under difficult conditions. This new assignment, however, would be a good deal more demanding than anything he had previously undertaken. In his new position as engineer for the Western District, he would oversee a great many works spread over a very large area; he would employ a large staff and would be responsible for large sums of government money. He would come under pressure from Grand Juries, local proprietors and local committees to adopt their particular proposal for works and would be required by Dublin Castle to make decisions on the soundness of the project as well as attending to the obvious needs of the poor.

His term with the Bog Commission also meant that Nimmo was well acquainted with his allocated district and was conscious of the many difficulties it presented. He had already stated his conviction that the structural development of the west would only be achieved with government aid. Now the government was publicly committing itself to providing such aid. He must surely have felt excited at the prospect that lay before him. In his combined positions as engineer for the Western District and engineer with the Fishery Board, he was given the opportunity and funding to implement at least some of the recommendations that he and his colleagues had put before the select committee in 1819. He knew what needed to be done; the select committee had clearly indicated the steps necessary to develop the poor mountainous regions along this coast. He was also aware of all that Telford had achieved in Scotland and, perhaps modelling himself on the older man, he seized the opportunity to carry out cimilar infrastructural improvements in Ireland. It is certainly evident that, from the beginning, Nimmo had a definite agenda in mind. Whether he shared his thoughts with anyone else it is difficult to say, though hardly with his immediate employers, who would have viewed his appointment as a temporary response to a crisis situation. There was nothing long-term in their strategy and Nimmo would have known this.

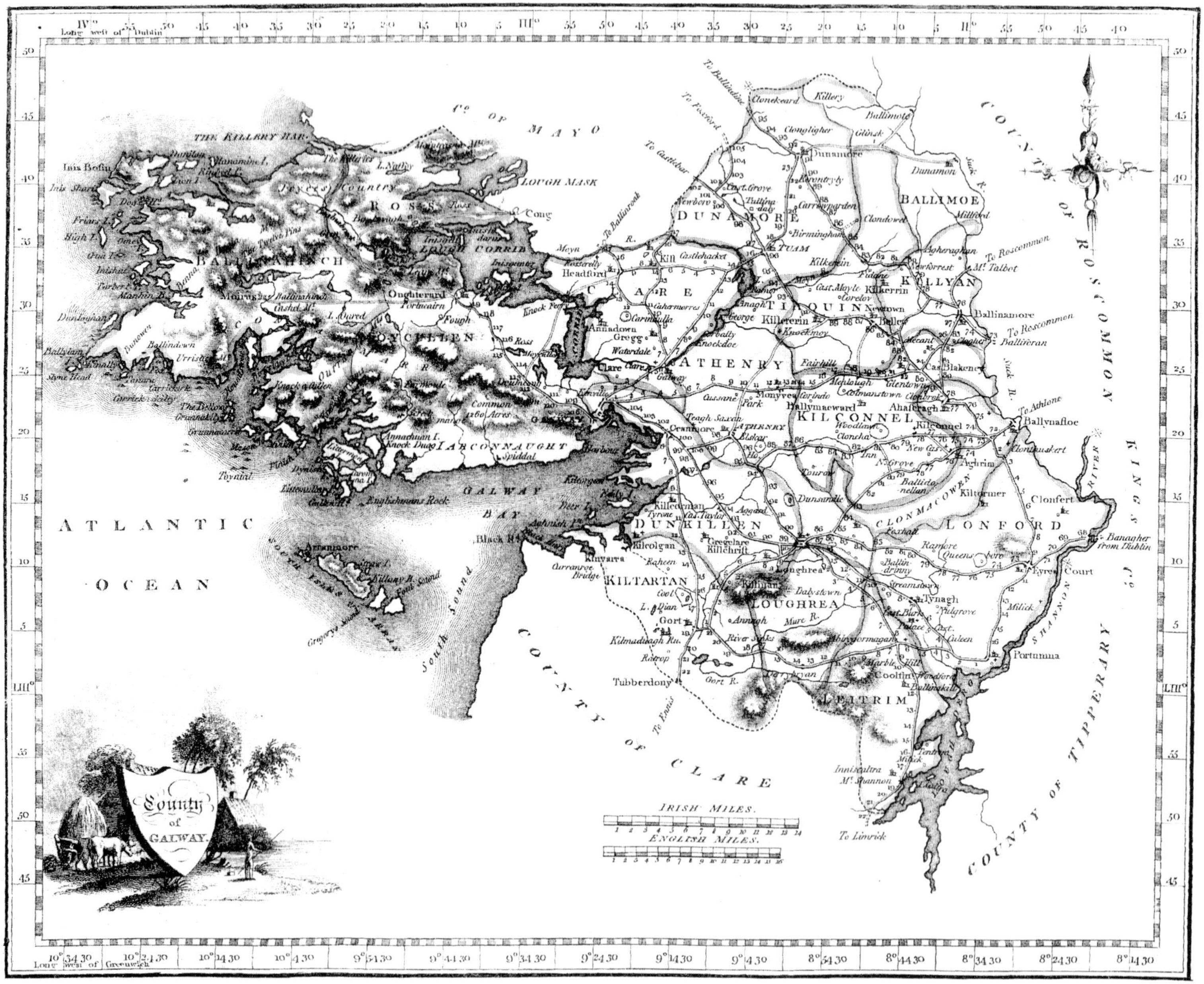

Map of County Galway circa pre-1825
(*New and Correct Irish Atlas*)

Proposing Public Works for Galway

Once appointed, Nimmo set to work immediately. Within a week of his appointment he was in Galway assessing conditions and determining works. Cases of famine and fever, both in the town and county, were rising daily. The citizens of the town were growing alarmed with reports of conditions in Connemara. The fever hospital in Galway was full and the 'principal inhabitants' were fleeing the town.[18] Nimmo arrived in Galway on 12 June 1822 and immediately reported back on conditions as witnessed on his journey across the country: 'The crop of corn and potatoes appeared in my way across the kingdom, to be very promising – near Galway there is much very fine wheat in the ear – no want of seed potatoes in the midland counties – nor much apparent distress until we arrive at Loughrea. The number of mendicant poor in Galway' was very great and the number of fever cases was higher than usual. Provisions had at last reached the town and five vessels carrying potatoes were at the quays. The potatoes were selling at 6d per stone in the town, and the charitable committees were selling them at 3d to the poor. One schooner from London had her cargo entirely spoiled and the best of these were given free to the poor for their pigs. The Aran Islands were 'much distressed' and boats had arrived from there seeking meal and herrings. To alleviate conditions somewhat, Nimmo sent money to the islands, along with instructions that the quarrying of Ashler stone was to begin immediately.[19] Ashler is a thin slab of stone particularly suited for use as facing material. The stone was to be used in the construction of the three Connemara piers at Cashla, Roundstone and Cleggan bays.[20]

The remainder of the month was an extremely busy one for Nimmo and was indicative of how the rest of his time in the west would be spent. On reading his many letters to Dublin Castle it becomes obvious that, throughout his time in the west, Nimmo moved about a great deal and seemed to spend no more than a day or two in each place; Sligo one day, Mayo the next, Galway the next, then back in Dublin, only to be found in London the following week. Throughout this time he wrote copious letters and reports to the chief and under-secretary, acquainting them with conditions on the ground, the state of the people, works in progress, works needed and so forth. Nimmo's many observations would provide the chief secretary's office with an insight into conditions in the west based on informed judgements. Both men would have welcomed such reports. The chief secretary, Henry Goulburn, had only arrived in the country in December 1821. Goulburn was initially slow to recommend government intervention in relieving the poor and was anxious that all funds be spent wisely. William Gregory had held the position of under-secretary for ten years and was a man who prided himself on being well informed on all things taking place in the country.[21] The lord lieutenant, Richard Wellesley, the Marquess of Wellesley, was the elder brother of the Duke of Wellington and a strong supporter of Catholic emancipation. He too had taken up his position in December 1821 and he would remain on as lord lieutenant until 1828.

Nimmo's initial instructions from the lord lieutenant were to confine himself to works already sanctioned by the grand juries: 'to superintend the making and repairing the several roads and the carrying out of the other Public Works for which Presentments were made at the previous Spring Assizes'.[22] The government would initially finance these, so as to bring them speedily into operation, the funds to be retrieved from the county purse at a later date. To bring immediate employment to the poor was the first priority. In areas for which presentment works had not been sanctioned, he was to carry out repair and resurfacing works on existing roads, funding for which would also be retrieved from the Grand Jury. He was to act as a supervisor in conjunction with the overseers appointed under the Grand Jury presentments and the 'respectable gentlemen of the country', and to pay the laborers employed himself, or through some person responsible to him.[23] His presence would guard against corruption and ensure that government funds were used effectively.

Nimmo took out advertisements in the local newspapers announcing his arrival and calling on the overseers of presentments to signify what works were wanted immediately. He let it be known that he did not intend to interfere with the overseers other than 'to direct the mode in which the work should be executed so as to be effective' and to issue payments in proportion to the value of the work done.[24] However, he quickly discovered that the Galway Grand Jury had passed very few presentments at the last Assizes and that no works of any magnitude were under way in the county, but 'much [was] expected from the Aid of Government'. Some country parishes had difficulty agreeing on a project to propose for government funding, but he had received no reports of funds being directed to private projects.[25]

On 19 June, Nimmo wrote to William Gregory, giving his address as 'Cliffton in Conamara' (sic). He had spent the previous days travelling along the coast from Galway to Cleggan and Ballynakill and reported that 'death and distress in several places, particularly in Garomma [Gorumna] Island and in the parish of Ballynakill, far exceeds any thing I have yet witnessed. In An Trabain [Trawbaun] in Garomma on Sunday last four persons had just died of want, a great many others were confined by swollen limbs. In Ballynakill parish, last night and this morning, 6 persons died, making 22 in that parish who have perished and it is to be feared that a few days will make a fearful addition to the number'. Nimmo himself was confined for a day at Clifden, 'by indisposition arising, I believe, from the fatigue of travelling along the coast which you are aware is no easy matter'. He had spent the morning at Cleggan setting out the place for the fishing pier and was greatly affected by what he found there: 'The more dejected aspect of the poor creatures, struck me as more remarkable than any I have seen. They were in hourly expectation of a boatload of provisions from Galway – which, after all, would not be more than two days supply'. He left a few pounds in advance of wages for work on the pier. The construction of the pier was expected to give work for some time to about a hundred, but this would 'afford little relief in a case of so immediate and extensive distress'.[26]

He was anxious to begin works in Connemara as quickly as possible and viewed road construction as the best possible means of employing large numbers in remote areas. He put forward a proposal for laying out new roads, similar to that suggested in his bog report. He felt that conditions in Connemara justified a considerable expenditure on new works, and that the free government aid should apply to those parts neglected under the Grand Jury system.[27] The Grand Jury could not be relied on to provide the funds for the construction of these roads, he warned, as few presentment roads existed in Connemara and 'no aid of consequence can be obtained'. Where presentments had been made, Nimmo found that the usual confusion arose over the use of funds and the payment of labourers, some of whom were being paid in food. He advised that presentment works of this type be avoided, as interfering with existing arrangements could create complications for repayments in the future.[28]

Instead, Nimmo had his own plans for Connemara. He was eager to put in place all the roads he had outlined in his bog report of 1813. He had no plans to interfere with the improvement of the old centre road from Galway to Clifden, going on under the management of the landed proprietors. In 1819, funds had been issued to the Galway Grand Jury from the Consolidation Fund for work on the old centre road. This was for a new stretch of road covering twelve miles (from the Halfway House at the west tip of Lough Shindilla to Emlaghdauroe, lying west of Ballynahinch Lake), which would replace some of the most difficult and mountainous stretch running through the Martin estate. The grant for this work, however, fell short of the estimated costs and the Grand Jury refused to provide the remainder. The overseers, Thomas Martin and James Val Brown, decided to go ahead with the works regardless and a considerable amount of work was carried out before the commissioners for the Consolidation Fund realized that the road would not be finished for the sum borrowed and refused to make any further advances. This brought the work to a halt and the road lay unfinished and unusable for several years. At the same time the Grand Jury was refusing to grant money for repairs to the old road, leaving Connemara without any central communication with Galway town for a carriage or cart.[29]

Nimmo preferred instead to concentrate on the more neglected coastal routes where only poor quality bridleroads, or in some cases no roads whatsoever, existed.[30] However, his proposal would commit the government to spending large sums of money on public works, money they had no hope of retrieving, and he seemed unsure as to how the government would respond. He was fully aware of the magnitude of the works he was proposing, even if Dublin Castle was not cognisant of their full extent.

From the start Nimmo sought clarification as to exactly what was expected of him in his new role. Alexander Mangin, first clerk of the Civil Department in the chief secretary's office, wrote informing him that the lord lieutenant 'entirely approves' of his suggestions for the works in Galway. The lord lieutenant, it would seem, had every confidence in his

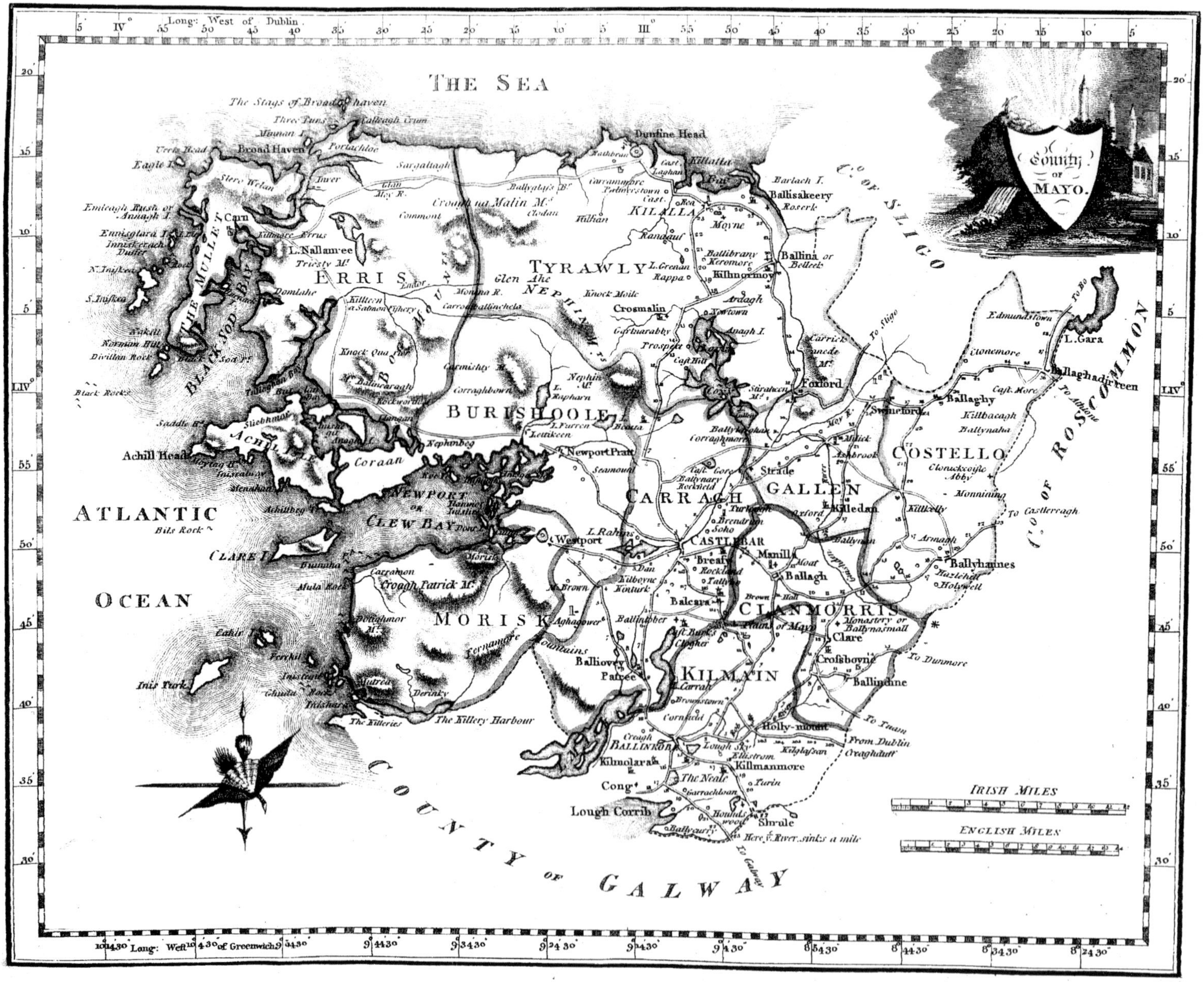

Map of County Mayo circa pre-1825
(New and Correct Irish Atlas)

ability, honesty and integrity, in choosing suitable works and in handling large sums of government money:

> It is His Excellency's particular desire that you will use your own discretion in conjunction with the overseers and respectable gentlemen of the county, in carrying on directly such works as will most immediately check the distress which is represented so inimically to prevail in that quarter.
>
> Continue to let me know as speedily as possible what effect these exertions have in supporting the peasantry and their families.[31]

Within a few days Nimmo wrote again to Gregory continuing his report on conditions and urging the government that, 'as the distress is so general and severe, I would earnestly beg of the government to consider the propriety of undertaking the works suggested in my last letter'.[32]Recognising the extent of his district and well aware of the dire need for such works as the government seemed prepared to fund, he was anxious to begin work on as many projects as possible, but not without government sanction. The roads, he informed them, would cover about fifty miles and would cost an average of £250 per mile, 'excluding bridges'. However, they would take a 'considerable time' to complete and he was only asking for £5,000 at this point.[33] Time would show this to be a very rough estimate and Nimmo would, in the coming years, eventually spend over £20,711 on the Connemara coast road alone.[34]

Identifying Public Works for Mayo

Two days later, Nimmo was in Castlebar assessing nineteen applications for public works in the Western District, forwarded to him by Dublin Castle. In Mayo, Grand Jury presentments had been passed for repairs and for some new roads in the north and east of the county. Nimmo advanced funds for payments to the workmen, but informed the lord lieutenant that for 'the vast tract between Killala Bay and Clew Bay there [was] no presentment works to be aided'. He had started work on the piers at Achill and Saleen, but the relief offered there was 'trifling in comparison of the extent and urgency of the distress'. He begged the lord lieutenant to sanction new works in this area: 'I should therefore beg His Excellency to grant permission to lay out and form the several lines of road recommended for opening up this highly improvable county'.[35] His plan for Mayo was similar to that for Galway; new coastal roads with cross roads linking them with the interior. He recommended that a new road be run along the north coast from Killala to Blacksod Bay and improvements be made to an existing road from Newport to Achill. He also recommended that they complete the central road from Castlebar to Erris, already started by Mayo Grand Jury. The Mayo Grand Jury had previously received funds from the Consolidated Fund and had started work on two roads into Erris and one from Westport to Killary Harbour.[36]

Nimmo's suggestion that he take over the Castlebar to Erris road was in stark contrast to his decision not to interfere with similar works under way on the old centre road in Connemara. The reasoning behind this may well have been that he favoured the Mayo Grand Jury over that in Galway, certainly the Mayo Grand Jury appeared to be more active in passing presentments. However, it could also be that in Galway there was some confusion over the method of paying the labourers, while Nimmo found the accounts for the Mayo road to be in perfect order. Major Denis Bingham, of Binghamstown, complained to Dublin Castle of the 'misapplication' of funds on this road, but Nimmo refuted this after examining the relevant documents and stating them to be 'remarkably correct and distinct'.[37]

Here again Nimmo was requesting free government funds for public works. Prior to this the government's response to periods of distress was usually temporary and lacked the commitment to long-term public works. Nimmo was mindful that there should be no ambiguity regarding any government sanction for his proposed roads. The roads he was proposing for Galway and Mayo would not be completed in the short term, they were long-term projects. He pointed out the difficulties he was already experiencing because of the extent of his district and the number of works requiring assistance. To comply with the government's desire to pay the labourers each week in cash would alone require several pay clerks, 'whose fidelity', he informed Gregory, 'in a temporary operation of this sort could not be easily secured'. He had already advanced money to overseers for works scattered throughout his district, 'as a visit to each of [the works] was more than one person can possibly make'.[38] Conscious of what he had witnessed while working for the Loan Commissioners, regarding the swearing of oaths on the validity of 'dubious' figures, Nimmo abandoned the procedure in the west and, on handing over the cash, he insisted 'on written vouchers [receipts] and regular witnessing'.[39]

The lord lieutenant grew impatient at Nimmo's reluctance to take action and ordered that he be issued immediately with a new letter of instruction. The letter, signed by the under-secretary, William Gregory, outlined Nimmo's powers and guidelines:

> I am directed to signify to you His Excellency's particular desire that you will forthwith make arrangements for employing such assistants as you shall conceive necessary and competent and in whom you can place full confidence for the immediate relief of the poor. The anxious desire of His Excellency that you should instantly employ the people in every possible way without any reserve as to the application of any particular funds it being no longer a legal question, as all funds are to be applied <u>generally</u> to the <u>great</u> objects of the immediate employment and support of the distressed peasantry. You will, of course, keep distinct accounts of the expenditure of the funds entrusted to you.[40]

With this letter the lord lieutenant had sanctioned Nimmo's free government roads for Connemara. Within a week he had approved all the works recommended by Nimmo for Mayo and ordered that they commence immediately.[41] This would appear to leave Nimmo completely in charge. It gave him the freedom to operate using his own judgement as to which works to undertake, who to employ in positions of authority and when and how they were to be paid. Detailed accounts were to be kept, but as the number of works increased and these were spread over several counties, this was not always adhered to and some lapses in procedure, particularly in the early years, would later bring him into dispute with Dublin Castle.

Nimmo held the opinion that Ireland had almost double the roads it required, if they were laid out correctly. In the Western District he set out to rectify this by planning to run main lines through the counties, linking the principal market towns with the canals and sea ports, 'and thereby supersede a multitude of crooked cross and branch roads, which form the principal communication at present'.[42] Nimmo's plan for the interior was to link the Royal Canal with the ports of Sligo, Ballina and Killala, and the Grand Canal with the town of Castlebar, Westport, Galway and Gort.[43] In the coastal regions, new roads would link the new fishery piers, and improved harbours, with the interior. The central Erris road in Mayo, and eventually the old centre road in Galway, would be upgraded and completed. However, the coastal ring roads of counties Galway and Mayo, identified by Nimmo from the start as being the most needed for the development of the west, would be the most extensive roads undertaken by him in the Western District. Limestone was the preferred stone for building roads in the interior, as these were expected to attract a good deal of traffic; for roads on the west coast Nimmo used the local granite and gneiss. The mountain roads he made with gravel only, but those on the plains were paved with broken stone.[44]

In the case of presentment or Grand Jury roads, the landed proprietors had already sanctioned the route, but Nimmo frequently improved on this. For the government roads Nimmo chose the route. A study of early maps would indicate that some form of paths, bridleroads and boreens already existed on the routes eventually chosen by Nimmo; there appears to be very little cutting through virgin land, except in the case of the Connemara south coast road. It would seem to be more a case of widening, realigning, and re-routing old routes, presumably with the cooperation of the local landowner. Compensation to the landowner was determined by the quality of the land through which the road ran.[45] However, as the construction of new roads and the realigning of old roads would take place in mostly underdeveloped districts and would, in time, greatly benefit the neighbourhood through which they passed, there is no evidence of any objections from the proprietors. Compensation was, however, sought for damage caused to fencing, trees and crops by the movement of men and materials through the countryside during their construction.[46]

Employees of the Western District

Armed with the clarification and sanction of the lord lieutenant, Nimmo laid out the plans, positioned his staff and set the wheels in motion. An office was rented at 56 Marlborough Street, Dublin, specifically for Western District purposes, but this was soon found to be inadequate and the lease was terminated in June 1823. A house was erected to fit the requirements a little further along the street, No 78, and this office remained the headquarters for the Western District until 1832.[47] Nimmo also lived at this address and declared in 1824, before a House of Lords inquiry, that he had just one property, 'a house in Dublin, leasehold of nine hundred years'.[48] This would indicate that Nimmo built the house for himself and was leasing office space to the government. However, this was later disputed by the government auditors and the house had to be handed back after Nimmo's death in 1832.[49]

Nimmo never married, but he kept close contact with his family in Scotland. At about this time, two of Nimmo's brothers, John and George, arrived in Ireland to assist him in his work. They brought with them their wives and children, and both brothers would remain in Ireland for the duration of their lives. It is not known how many children the brothers had, or if any of them went on to have children of their own; all that is recorded is the early death of two girls, named Margaret and Charlotte, believed to be the daughters of George and his wife, Charlotte, and Alexander, the son of John and Marion (Mysie).[50] George, a contractor, is credited with having constructed Nimmo's house in Marlborough St.[51] John moved west to assist Nimmo in Connemara.

Over the next nine years, Nimmo employed a number of men in varying capacities, from engineers to general dogsbodies. He kept a permanent staff in Dublin and in the west. In Dublin he employed an accountant and other administrative staff, along with a number of student surveyors and engineers who were frequently sent into the west to map various road routes. In the west, as well as the large numbers of poor labourers, he employed engineers, surveyors, pay clerks, overseers, foremen, storekeepers and a variety of artisans, such as stonemasons, carpenters, smithies, haulers, rock blasters, chainmen, tool makers, tool minders and even tool collectors, for tools were a precious commodity in the remote west at this period. As well as the large contractors who undertook the construction of bridges and some Grand Jury roads, there were many other individuals who benefited by providing boat hire, cart hire, stores, grazing for animals and lodgings for men. Added to this was the large amount of materials purchased locally and in the nearby towns. All of this gives some idea of the large sums of money being expended in the Western District at this time.

All funds allocated for public works in the Western District were transferred periodically to Nimmo's bank account in the Bank of Ireland, and he alone was accountable for expenditure. Nimmo instructed his pay clerks to pay the labourers and small contractors in cash; larger contractors and senior staff wages were paid by cheque. This necessitated the transport of large quantities of cash throughout the country, which, despite the wretched conditions experienced by the people at the time, appears to have taken place without interference of any kind. The total sum expended by Nimmo on public works in the Western District, between the years 1822 and 1830, is recorded as coming to over £167,430.[52] The benefits arising from such large-scale expenditure of public money in the remote west will be dealt with later.

Salaries and all administrative costs, for the west and Dublin, were charged separately for each county, and were included in the payments. For the ten years, 1822 to 1832, £10,910.16.10d was received for 'management, contingencies, Dublin office etc.,' for Mayo and £16,523.2.3d for Galway.[53] There is no record of the sums received for the three other counties in the district. As Engineer for the Western District, Nimmo was paid three guineas per day and this was to include all travel expenses.[54] There is, however, no record of the total he received while employed in this capacity.
Nimmo divided his district under the following headings:

North West Division comprising Mayo
South West Division comprising Galway
North East Division comprising Sligo, Leitrim and Roscommon.

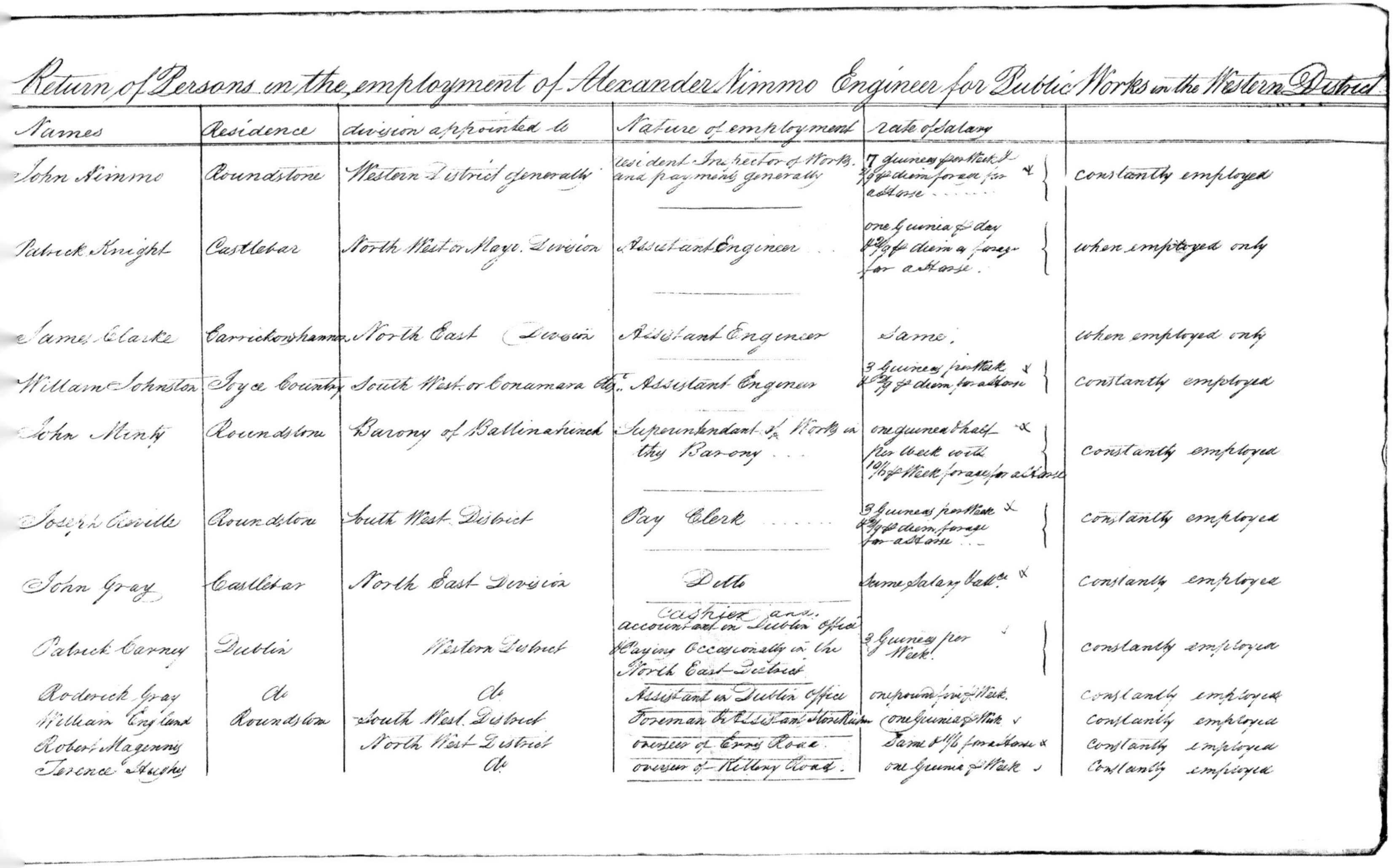

Return of Persons in the employment of Alexander Nimmo Engineer for Public Works in the Western District

Names	Residence	division appointed to	Nature of employment	rate of Salary	
John Nimmo	Roundstone	Western District generally	resident Inspector of Works and payments generally	7 guineas per Week & [illegible] per diem forage for a Horse	constantly employed
Patrick Knight	Castlebar	North West or Mayo Division	Assistant Engineer	one Guinea per day & [illegible] per diem a forage for a Horse	when employed only
James Clarke	Carrickonshannon	North East Division	Assistant Engineer	Same	when employed only
William Johnston	Joyce Country	South West or Conamara &c.	Assistant Engineer	3 Guineas per Week & [illegible] per diem for a Horse	constantly employed
John Minty	Roundstone	Barony of Ballinahinch	Superintendant of Works in the Barony	one guinea & half per Week with 10/6 per Week forage for a Horse	constantly employed
Joseph Coville	Roundstone	South West District	Pay Clerk	3 Guineas per Week & [illegible] per diem forage for a Horse	constantly employed
John Gray	Castlebar	North East Division	Ditto	Same Salary &c.	constantly employed
Patrick Carney	Dublin	Western District	Cashier and accountant in Dublin Office Paying Occasionally in the North East District	3 Guineas per Week	constantly employed
Roderick Gray	do	do	Assistant in Dublin Office	one pound five per Week	constantly employed
William England	Roundstone	South West District	Foreman & Assistant Storekeeper	one Guinea per Week	constantly employed
Robert Magennis		North West District	overseer of Erris Road	Same & 10/6 for a Horse	constantly employed
Terence Hughes		do	overseer of Killeny Road	one Guinea per Week	constantly employed

Nimmo's principle staff circa 1831 (NAI OP 1831 973/80)

He appointed his brother, John, as resident inspector of works and payments generally in the Western District. Assistant engineers were employed for each division and these resided locally. Nimmo had a high regard for his men, many of whom had received their training under Taylor and Skinner, surveying roads for the post office. The younger men he trained himself.[55]

The principal men employed in the west were John Nimmo, Patrick Knight, William Johnston, Daniel Mackin, Jamie Clarke and John Minty. John Nimmo received seven guineas per week and two and nine pence per day for forage for a horse. John resided at Roundstone. Patrick Knight was appointed assistant engineer for the North West Division. He resided in Castlebar and was paid one guinea per day and allowed two and nine pence per day for a horse. Knight had worked as a surveyor with the Bog Commission and had surveyed Erris for inclusion in the William Bald map of Mayo. William Johnston, assistant engineer for the South West Division, resided in Joyce Country and was paid three guineas per week with the usual allowances for a horse. He too had worked as a surveyor with the Bog Commission and had trained under Nimmo.[56] Daniel Mackin was employed as assistant engineer in the North East Division and was described as Nimmo's 'principal clerk or secretary' from 1822 to 1829.[57]Jamie Clarke replaced Mackin and he resided in Carrick on Shannon. Clarke received one guinea per day plus the usual allowances for a horse. Clarke had previously worked as an assistant to William Larkin. Knight and Clarke were employed as needed; Johnston was in constant employment.[58]

The pay clerks were; Patrick Carney (Western District generally), resident in Dublin, Joseph Reville (South West District), resident in Roundstone and John Gray (North East Division), resident in Castlebar. All received three guineas per week and both Reville and Gray received two and nine pence per day for a horse. Roderick Gray, Carney's assistant in Dublin, was paid one pound five shillings per week.[59]

The following are the names of surveyors who worked in the Western District, some of whom later went on to become engineers and surveyors of note. William Grant, as previously mentioned, had worked with Nimmo from the start. Grant had also been a surveyor with the Bog Commission and had worked with Nimmo on the coastal survey for the Fishery Board. He contracted fever while working in Sligo and died in 1824.[60] Alexander Hay, architect and engineer, was employed on Rossaveel, Cleggan and Roundstone piers. Hay also superintended the construction of the Market House in Galway city.[61] Michael Logan drew up plans and surveys for the Spiddle to Moycullen and Menlough roads. William Bermingham was assistant surveyor on the Clifden to Killary road (1822–1824). Robert St Ledger was surveyor and superintendent on Killary to Westport road and roads in Achill and Erris (1826–1829). Samson Carter was surveyor and superintendent on central Erris road (1828–1830). George Halpin Junior surveyed and mapped the central Connemara road and the road from Kilkieran Bay to Killary.[62] Halpin was later appointed engineer to the commissioners for improving the port of Dublin and was responsible for seeing Nimmo's work on sailing directions for St George's

Channel and the Coast of Ireland through to publication after his death in 1832.[63] Hugh Grant was surveyor and superintendent on the Galway coast road and on the Westport to Killary Harbour and Joyce Country roads (1825–1826). John Edward Jones was surveyor and superintendent on Westport to Killary road (1825).[64] Jones (1806–62) also worked with Nimmo on Waterford bridge (1829–32). He later retired from engineering to become a sculptor and produced a marble bust of Nimmo, now on display in the head quarters of the Royal Dublin Society, Dublin.[65]

John Minty was employed as superintendent of works in the Barony of Ballynahinch. He resided at Roundstone and was paid one and a half guineas per week, and ten shillings and six pence per week for a horse. William England was also resident in Roundstone and employed as foreman and assistant storekeeper at one guinea per week. England, a stonemason, had constructed the pier at Clonakilty and had worked with Nimmo at Dunmore. He was employed as site manager at Raghly Pier and harbour works before moving to Roundstone.[66]

The rate of pay for overseers was one guinea a week and eleven and six pence for a horse. The following names appear for overseers on the following roads: North West Division – Robert Magennis (sometimes written McGuiness) on the Erris road, Malcolm Walsh on the Mayo north coast road. Terence Hughes on the Westport to Killary Harbour road (October 1830 to June 1831) and Michael Hughes.[67] South West Division – Michael Plunket on Galway roads, Peter Mullen on Keelogues to Newforest road, Anthony Dwyer and Robert Sullivan on Oughterard to Clifden road (1830).[68]

Other men said to have worked with Nimmo are, D. McCallum, John Hampton, surveyor with the Bog Commission, and Sir John Hawkshaw. Hawkshaw worked with Nimmo early in his career before going on to build railways in England, Russia, Mauritius and India.[69] The following names also appear in Nimmo's employees' list, but about whom little is known; William O'Hara, Brinsley Fitzgerald, John Kilroy and Robert H. Minty.[70]

Nimmo's brother John was not an engineer, but John's son Alexander was.[71] It is difficult to say for how long Alexander junior worked with his uncle in Connemara, but it is known that he worked for a time as overseer at Maum and was resident engineer on the Dromahair Colliery road in 1830 and 1831.[72] Alexander had attended school with Maria Edgeworth's brother, Michael, and was, according to her, a bit of a wild 'lod', who spoke with a strong Scottish accent and was 'very clever'. According to Maria he 'learned engineering, dyke-making and common sense by hard work in Holland and experience in America, where he was employed on the Welland Canal', a canal linking Lake Erie to Lake Ontario in Canada.[73] He was also employed erecting harbours and quays on the east coast of Ireland and the south coast of Wales. In the past some people have confused the nephew with his more famous uncle and attributed evidence given by him before select committees in 1835 to Alexander senior, who was in fact already dead at that time. Alexander Nimmo jnr never married. He died just seven years after his uncle in Dublin on 29 December 1839.[74]

It is impossible to gauge the impact Nimmo's assistant engineers had on the overall construction of the government road and bridges in the west; how much influence they had on the choice of route taken for the roads or the design and construction of each and every bridge. Certainly Nimmo projected the direction the routes were to take and young student surveyors were sent into the field to map the intended route. However, his assistant engineers may well have determined the exact route on the ground, following a close study of the terrain etc. The assistant engineers may well have designed the smaller bridges, but it seems safe to assume that the larger bridges encountered on the western roads were originally designed by Nimmo. However, as will be seen later, many of these sustained damage from flooding soon after erection and had to be repaired or replaced in later years.

Local labour force

Nimmo and his staff appear to have had little difficulty in dealing with the people within his district. Nimmo himself kept in close contact with the workers. He seems to have enjoyed their company and admitted to having sometimes slept in their cabins. Just two years after starting public works in the west, he can be found before a House of Lords committee in 1824 defending the name and reputation of the Irish labouring poor. The Irish were not an indolent people, he told the committee. Being of independent spirit they preferred employment to charity, but he admitted to having 'a far higher opinion of the spirit of independence of the people than perhaps many persons who are immediately connected with the country'.[75] 'Nothing', he declared, 'gives an Irish peasant more happiness than to have some prospect of employment', and, once employed, his staff found they had 'no reason to complain of them in any way'. It was Nimmo's experience that employment made the people 'industrious, sober and honest'.[76]

However, they were greatly abused by their landlords. From his observations, while travelling the country during the course of his work, Nimmo concluded that the poor of Ireland were 'general in almost the lowest possible state of existence, their cabins are in the most miserable condition, and their food [was] potatoes with water, very often without any thing else, frequently without even salt'.[77] Irish landlords held great power over their tenants, he explained to the committee, without accepting any of the responsibility: 'in Ireland the landlord has greater power than in any other state that I know; he is not bound to protect the tenant in case of distress or starvation, as he is in England, or in the countries, such as Livonia and Germany, where they cultivate the land by predial slaves, or as the negro slaves in the West Indies'.[78]

Nimmo gave 'the management of the land', or perhaps more correctly the mismanagement of the land, as the principal cause behind the miserable conditions of the poor. With so little industry in the country, the land was the only resource available to the people:

> There is no other means of employment for the Irish peasant, nor no certainty that he has an existence for another year, nor even for another day, but by getting possession of a portion of land on which he can plant potatoes; and in consequence of the increase of population, which does not seem to be at all checked by the misery which they undergo, the competition for land has attained an appearance something like the competition for provisions in a besieged town, or in a ship that is out at sea; and as there is no check to the demand which may be made by those who may possess the land, the land appears to have risen to prices far beyond what it is possible for the poor peasant to extract from it.[79]

This competition for land left the tenant at the mercy of his landlord. However, the subletting and subdivision of holdings frequently meant that more than one landlord held an interest in the land. In the south of the country, he discovered that, between the principal landlord, who was the proprietor of the land, and the worker of the soil, there could be as many as four middlemen, each extracting a profit from the rent before the principal landlord got paid. In many cases the middleman did not pass on the rent collected and this left the tenant liable for a second collection from the principal landlord.[80] There was no check on the power of the landlord, Nimmo told the inquiry, because he had, in the eyes of the tenant, the right to take everything he possessed when rent was outstanding. Under the cover of law he exerted a reign of terror over the tenant, keeping him poor and in fear for his very survival:

> when he wishes [the landlord] can extract from the peasant every shilling beyond bare existence, which can be produced by him from the land. The lower orders of peasantry in Ireland can never acquire any thing like property; they are always in a state of beggary; and the landlord, or the middle-man, [regardless of the cause for arrears] ... has it in his power, and does come and seize his cow, his bed, and his potatoes in the ground, and every thing he has; and without referring it to any tribunal, which might perhaps justify resistance, or the impossibility of paying all instantly, can dispose of his property at any price.[81]

The power of 'distress', the seizure of the tenant's property in lieu of rent arrears, was, Nimmo explained, 'never resisted by the peasant, unless in cases of actual insurrection'. While declaring himself no lawyer, Nimmo recommended to the committee that a check be put on the behaviour of the landlord and that legislation be introduced to establish an arbitration court or tribunal to offer the tenant an opportunity to state his case. This he felt would go a long way towards 'ameliorating the condition of the people of Ireland'. He also suggested that each middleman be obliged to clear off their own rent before they had the power of distraining.[82]

The majority of the labourers employed on the public works were paid as day labourers, at the usual rate of 8d per day. Typically two out of a family of six persons were employed, the father and one son. In cases where the father was too old, two sons were taken. Nimmo was more successful in introducing work by measure in the west than he had been in Kerry. In this way he endeavoured, whenever possible, to employ the entire family 'breaking stones and covering roads with clay'.[83]

Nimmo was conscious of the large numbers dependent on the income derived from just one person employed on the public works. He observed that 'on almost all works the advantage to the people extends much farther than the immediate people employed'. The employed person usually supported an extended family. This was particularly so with regard to artisans; they frequently demanded high wages so as to support the many relatives dependent on them. While working on Dunmore Harbour, Nimmo employed about three hundred people, but was aware that a far greater number was supported by the wages received by the workforce: 'I have heard calculations frequently made that there were at least sixteen hundred people actually subsisting upon it, as the relations of the person employed, in every sense of the work, came and "coshered" upon the persons employed'.[84] Even the promise of employment was often sufficient to sustain a family through hard times, as Nimmo explained: 'I have frequently had occasion to meet persons that begged of me on their knees for the love of God to give them some promise of employment, that from the credit of that they might [be given] the means of supporting themselves for a few months until I could employ them'.[85]

However, even on the public works the labourers were frequently subject to intimidation from the landlords' bailiff, and those in arrears of rent found it difficult to hold on to their wages. Despite Nimmo's efforts to ensure that the labourers received their wages directly into their own hands, to spend as they saw fit, the bailiff was sometimes not far behind: 'and with all the exertions I could make, I cannot but say, that frequently we have had, upon paying the persons employed in this way, the drivers alongside of them to receive the money from the one hand as fast as we put it into the other. The driver is the bailiff on the estate'.[86]

Galway works started

Works in Galway and Mayo were opened almost simultaneously and the initial spadework, surface work, draining and fencing, etc., was started all along the proposed routes. This early spadework facilitated the employment of large numbers and brought instant relief to the remote regions. From the start there were two types of works in operation in the district. As previously stated, presentment works (Grand Jury roads and other construction works) already begun under local management were to continue, but these would hereafter receive advice and financial support from Nimmo. All other works came directly under the control of his team. Some of these works involved resurfacing and repairing existing Grand Jury roads, others required realigning old county roads and in others it was laying down new roads where before there was perhaps paths or bridleroads. Funds expended on presentment works, and on the resurfacing and repair work, would, in time, be paid back by the Grand Juries. However, realigning and new roads would be paid for with free government aid. The construction of these roads would, Nimmo told Dublin Castle, 'not only relieve [the people's] present necessities, but ultimately be a means of providing against future returns of death, by laying open there great wastes to the purposes of agriculture'. As the works in Connemara were to be carried out in remote, unpopulated districts, Nimmo requested fifty tents and cooking utensils for the labourers use.[87]

Road communication in Connemara had improved little in the ten years since Nimmo's first visit in 1812 and goods and people were still transported by sea. The road from Cashla Bay to Galway was still the only good road to be found. The work started under Thomas Martin on the old centre road from Galway to Clifden remained at a standstill, with both the Grand Jury and the commissioners for the Consolidation Fund refusing to provide funds for its completion. And almost every bridleroad in the district was impassable in the winter. Some of the recommendations made in his bog report had been acted upon. New stretches of road were built with county funding under the supervision of local landlords, but these usually ran through the landlords' own lands and were frequently unconnected. They were, for the most part, poorly constructed and in bad repair, and in Nimmo's opinion they were nothing more that bridleroads.

Lord Sligo had constructed a road from his lodge at Delphi to the middle of the north shore at Bundorragha on Killary Harbour. This was the only road on the northern shore of Killary. On the southern shore there was 'a dangerous bridleroad', coming from the centre of Connemara, across the Maumturk Mountains, to the village of Bunowen, situated at the mouth of a river of the same name. From here a rocky bridleroad led along the south shore and across a mountain pass to the inlet of Salruck, and the Little Killary. From Salruck there was a ferry over the mouth of the great Killary to Mayo. The bridleroad continued along the shore, as far as Renvyle House, where Henry Blake had recently taken up residence.[88] Henry was the first of the Blake family to reside at Renvyle. He had taken back the lease of the Renvyle Estate when it fell due for renewal, despite the opposition of the son-in-law of William O'Flaherty, his former tenant. Blake's descriptions of his early journeys through Connemara in his *Letters from the Irish Highlands*, give a good indication of the direction and state of the roads in the district at the time.[89]

John D'Arcy had laid out a road on his estate, from Clifden to Streamstown, with Grand Jury funding. South of the town, Bunowen, on the O'Neill estate, was connected to Clifden by a good bridleroad. But the heavily populated bays of the south coast were still without road communications. These bays were considered by Nimmo to be 'the very finest harbours in Ireland for great ships, which [were] as completely inaccessible by land as if they were situated in America, and unknown even to the gentlemen of the country'.[90] But the district was not as isolated as one might assume, a good trade continued in turf, kelp and fish with Galway town and the coast of Clare. The larger islands between Kilkieran bay and Greatman's bay were accessible on foot in low tide and by boat in high tide: 'A series of shallow fords, passable at low water, connect the isles ... so that one may walk from Bealadangan into Annaghvaan Island, Lettermore, Garumna, Lettermullan, Knappough

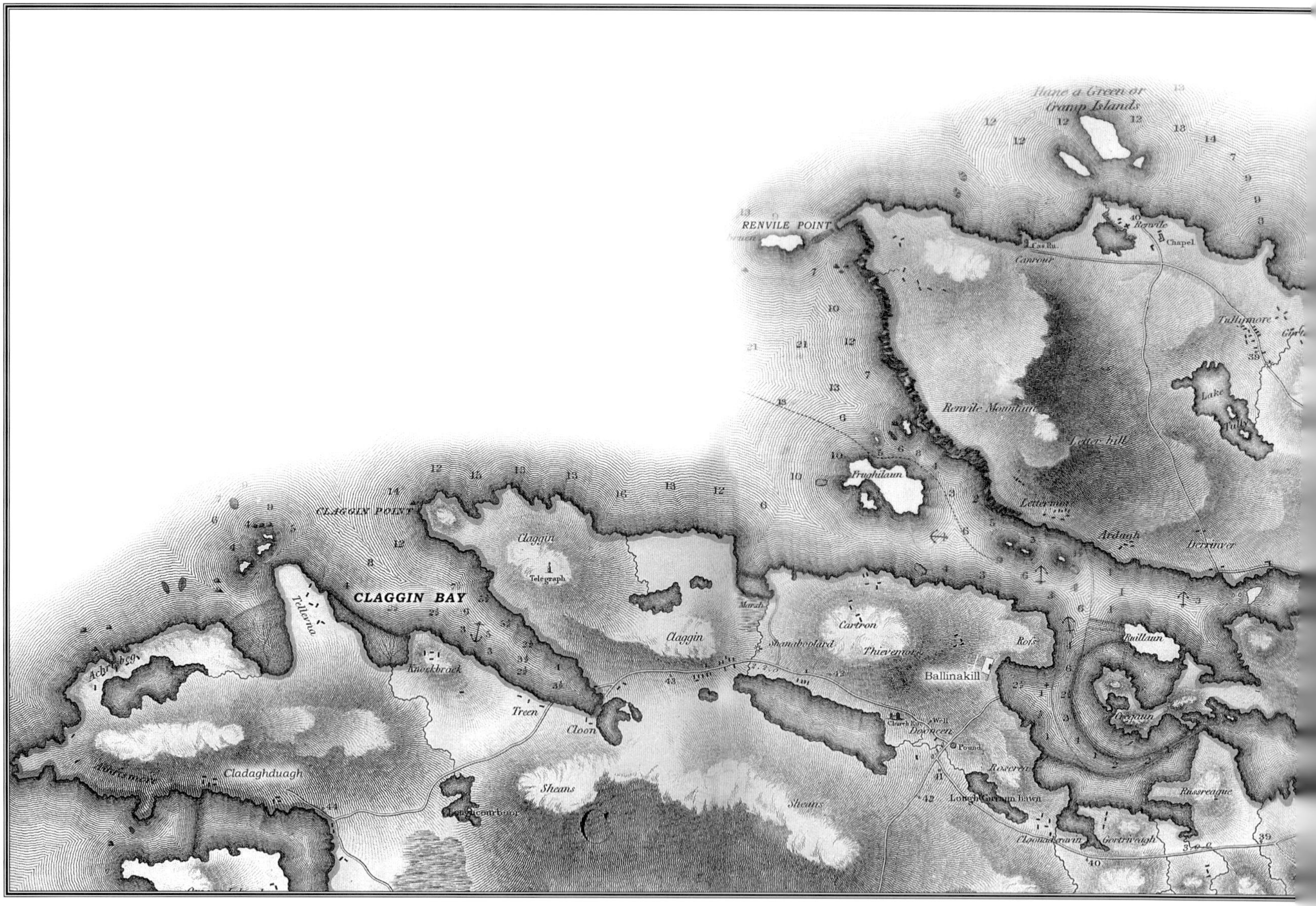

Roads on the north coast of Connemara, from Claddaghduff to Killary Harbour
Also showing is the bridleroad (the Western Way) coming across Maumturk Mountains from Lough Oorid, this divided in two at Tooreenacoona with one branch going west to Ballynakill Bay and the other going north to Killary Harbour.
(William Larkin's 1819 Map of Galway)

(sic), Furnish etc.; these fords are passable by boats at high water, so as to communicate with Kiggaul and Greatman's Bay; and this is the common route pursued to Galway by the small craft with turf, etc. from Kilkerran Bay'.[91]

It was still Nimmo's intention to create one continuous road along the entire coast of Connemara, from Galway westwards to Clifden and then northwards to Killary Harbour, with crossroads running through the centre of the district, connecting the deep bays on the south coast with Killary Harbour on the north. However, on reflection, he chose slightly altered routes from those first outlined in the bog report; branch roads would now connect with the many new piers under construction. The first roads to get under

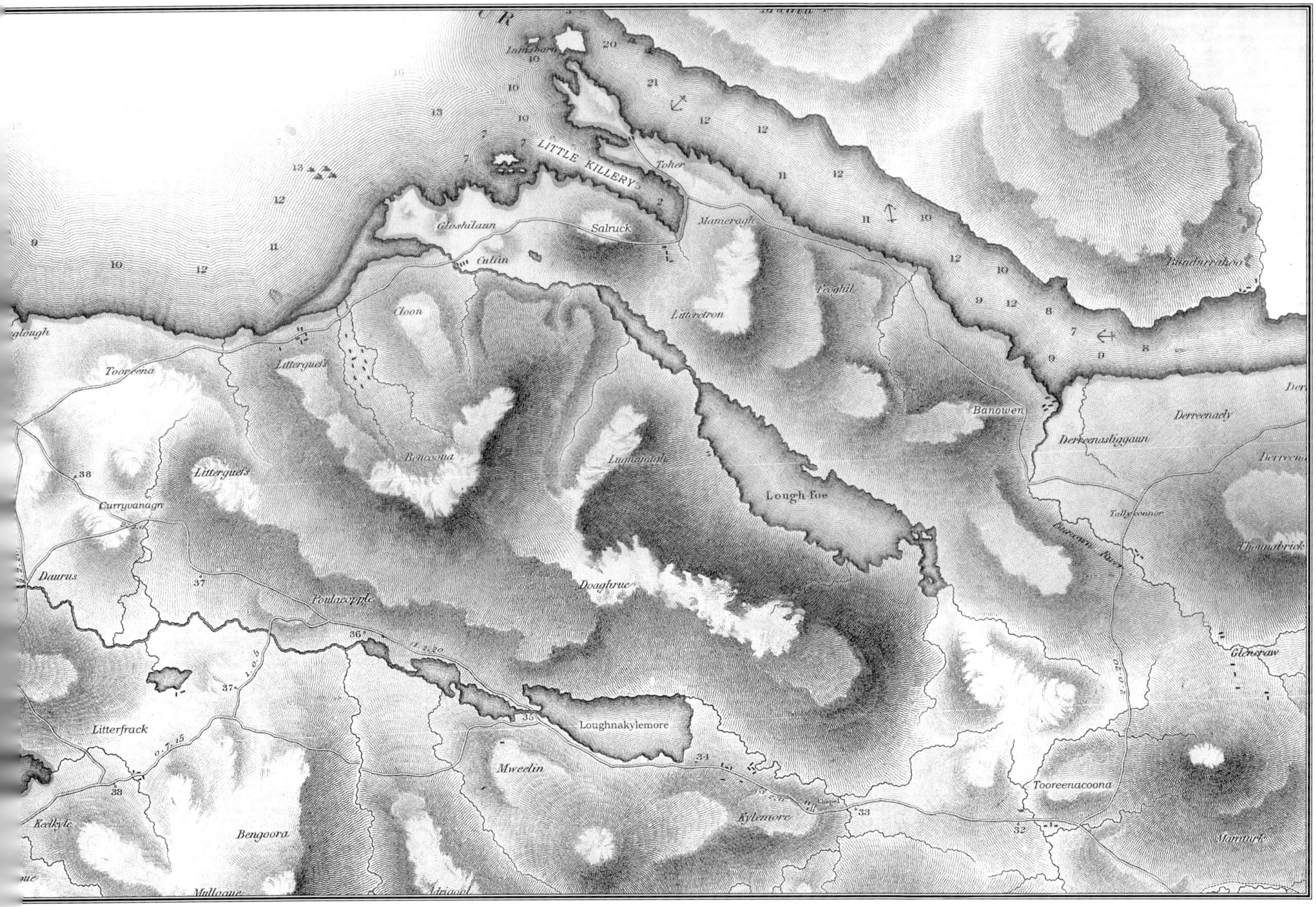

way were a road from the proposed pier at Cleggan to Ballynakill Bay, the coastal road from Galway to Clifden and a crossroad linking Kilkieran Bay with Killary Harbour.[92] The road from Cleggan to the head of Ballynakill Bay, estimated to cover about seven miles in total, would replace an existing road, which was scarcely passable on horseback and 'too hilly to be made a carriageways by any kind of repair'.[93] Improvements were under way on the coast road from Galway to Cashla. A new line would extend this road north, to the head of Camus Bay, and then west, along the heads of all the bays and through the bogs, to connect with the Bunowen to Clifden road at Ardbear Bay. A new approach to Clifden would be laid out and Weir Bridge built, creating Salt Lake where before there

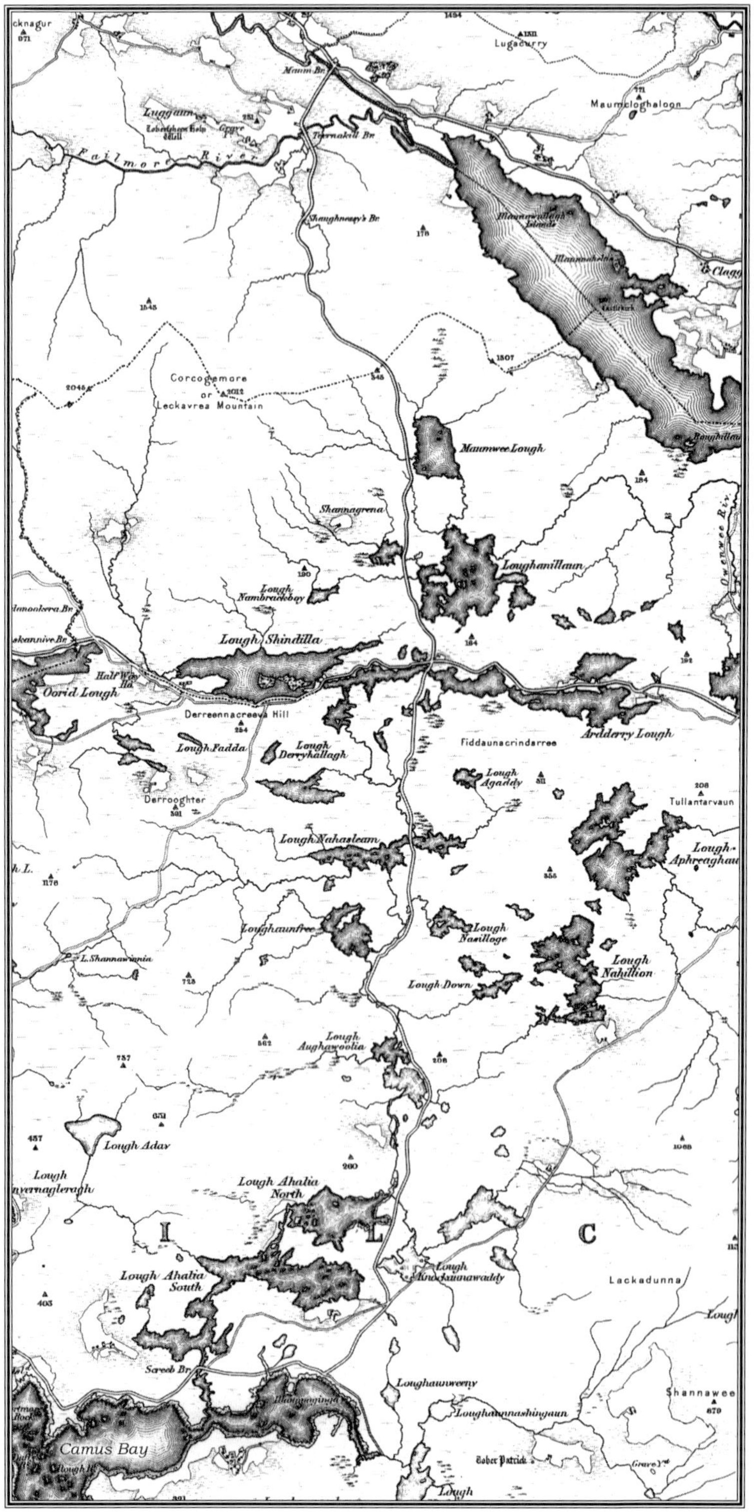

was an inlet. At Camus Bay, a branch would be taken north to Killary Harbour, connecting with the old centre road at Ardderry Lough. A new Grand Jury road was planned to pass through the townlands of Shannaunnafeola and Teernakill South [Tir na Cille Theas] and this would link the old centre road at Ardderry Lough with Lough Corrib at Maum [An Mám].

This latter presentment had been granted to Henry Blake and William Bermingham and Nimmo would incorporate it into his Kilkieran Bay to Killary road. The specifications for this road give an indication of the standard demanded by the Galway Grand Jury for road making at the time: 'To Henry Blake and William Bermingham, to repair 2,224 perches of road from Galway to Ballynakill bay between Shankeelagh [Shannaunnafeola] and Toorenacoony [Teernakill South]. This was to be 21 feet wide in clay and 14 feet wide with gravel or small stones at 5/2d per perch = £376.15'. The road from Streamstown to Cleggan was estimated to cost £488.5s.[94]

A new presentment road from Streamstown to Cleggan pier, already under way under the management of Thomas Martin and John D'Arcy, received funds from Nimmo. Other presentment works, sanctioned but not started, were given funds to get them quickly under way. Advances were made towards improving a section of the Galway to Moycullen road, at Woodstock. This was to connect with improvements about to start at Moycullen, where a local charity was providing £200 for repair works in the district. Similarly, he contributed to the laying down of a branch road from Moycullen to Knockferry, on the shore of Lough Corrib. A new pier was planned for Knockferry and one for the opposite shore. The road on the north shore was continued to Headford. Work was also started on a cross road from Moycullen to Spiddle.[95]

For the employment of the people in the mountain district of Connemara, Nimmo began a road from Killary, through Joyce Country, and along the head of Lough Corrib and on to Mayo.[96] This passed through 'Ecclesiastical or College Property', owned by Trinity College Dublin, linking Lough Corrib with the sea at

New road from Maum Bridge to Camus Bay circa 1862
This new road started by Nimmo in 1822 connected Camus Bay with Lough Corrib and Killary Harbour. It intersected with the central Connemara road at Ardderry Lough forming the junction presently known as Maam Cross. (First Edition 1 inch Ordnance Survey map 1862)

Killary Harbour and Westport in the west, and with 'the great plains of Connaught' in the east.[97] The new Maum road, under the management of Blake and Birmingham, would link this with the old centre road from Galway to Clifden and consequently with Kilkieran Bay as previously indicated. On the south coast of Killary (County Galway coast), Nimmo brought the old county bridleroad down to the shore and built 'a rough jetty pier at the head', which proved to be 'an excellent boat quay', and of considerable value to the fishery. Nimmo considered the site 'well adapted to become a thriving village', but felt that being in the ownership of the provost of Trinity College might restrict its development.[98] In time the village of Leenaun developed to the east of Nimmo's pier, at the junction of the Westport, Clifden and Maum roads.

In districts removed from the proposed government works and where no presentment work had been sanctioned, Nimmo provided government funds for work to begin on the following piers where the local contribution was not forthcoming: Clifden, Inishmore, Dooras and the Claddagh. He also constructed two piers on Lough Corrib, Annaghdown and Oldbury.

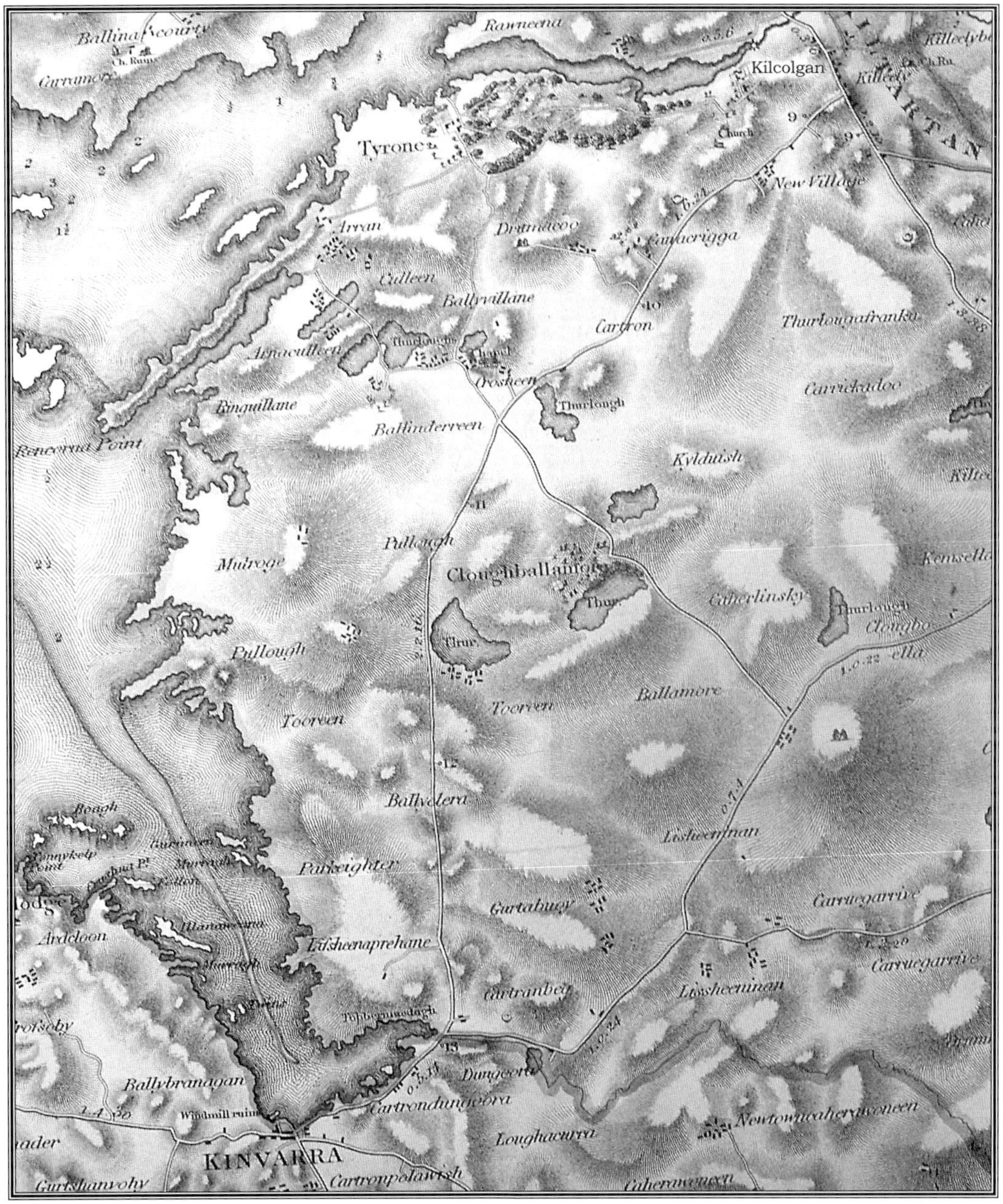

Kinvara to Kilcolgan Road
(William Larkin's 1819 Map of Galway)

Work was also started on roads in other parts of Galway; some of these were presentment work others were free government roads. The principal works are listed below. However, there were several other short roads, along with localized resurfacing works, which are not listed. In south Galway, works were opened up on the road from Kinvara to Kilcolgan. Arthur French St George, had requested repairs for this road but, on investigation, Nimmo decided that a new stretch of road was needed, covering two miles to the south, as the old road ran 'over hills and flooded

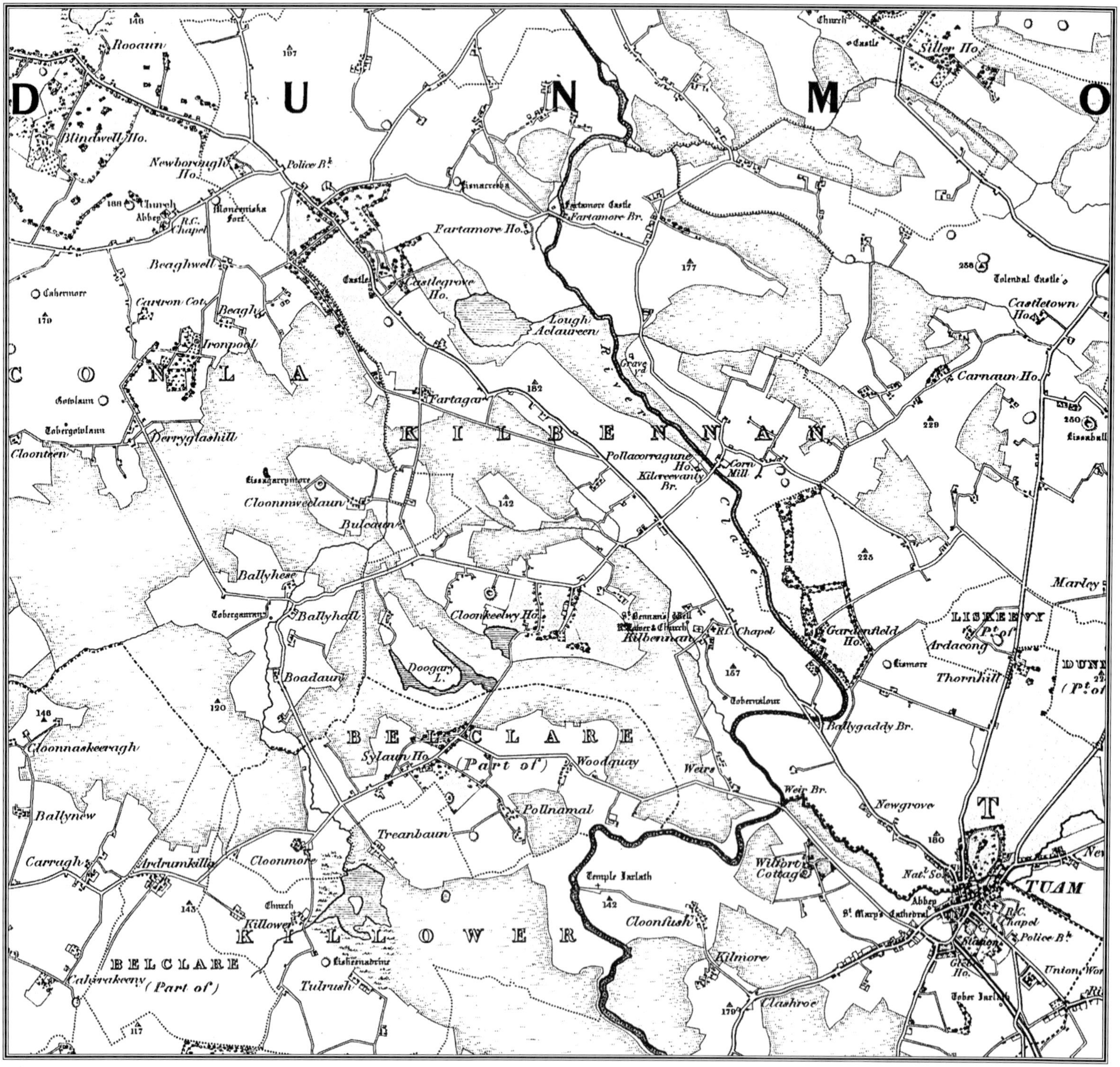

Tuam District circa 1862

(First Edition 1 inch Ordnance Survey map 1862)

marshes in an absurd way'. He gave a rough estimate of costs at between £600 and £700.[99] North of the county works were opened on the roads from Headford to Galway and from Tuam to Galway. Other roads in Tuam district were from Tullinadaly Hill to Beagh, works in the Blindwell Hill, Ironpool district and the Tuam to Cloontooa road. Work was also started on Ballinderry Bridge crossing the Grange river and Corofin quay and cut on the Clare river. East of the county, works were opened on the Crushna road, on the road from Creggs to Galway, from Killagh to Kilconnell, the Menlo road, the road from Keelogues to Newforest and the Castlekelly road. Money was also advanced for the construction of a sewer at Tuam and the Market House in Galway.[100]

Mayo road works started

Nimmo arrived in Mayo fully acquainted with Bald's survey and map. He was also familiar with the maps and the many suggestions and recommendations made by both Griffith and Bald in their respective reports to the Bog Commission. All of these would prove useful in prioritising works and deciding routes for new roads. Work soon got under way in almost every district in the county.

The publication of Bald's survey and map had brought a good deal of attention to the more remote regions of the county and given the Mayo Grand Jury and landlords a sound basis from which to consider projected improvements. The Grand Jury had constructed a road from Castlebar to Ballina, 'across the narrow part of Lough Con', and another along the east of the Moy river from Ballina to Swinford. Both of these appear to have been completed by 1822. However, a carriage road from Castlebar, intended to run northwest to Bangor and on into Erris, begun in 1819 with an £8,000 loan from the consolidated fund, was nowhere near completion.[101] Nimmo took over this road and had it extended to the Mullet peninsula. He gave a rough estimate of £6,000 for completing the work, pointing out to the lord lieutenant that the construction of the road would offer 'useful employment for all the spare and distressed population of this distressed district', but warned that 'the future public advances in the improvement of so extensive a tract of country [could] not be easily calculated'.[102] This road was referred to in Nimmo's reports as the Central Erris Road.[ii] Branches were later added, running down the Mullet to Elly Harbour and Blacksod Point and north along the east coast of Broad Haven. Two branch roads running from Bangor to Ballycroy and along the north bank of the Owenmore River to Tullaghan ferry were also started in 1822.[103] Nimmo would, in time, spend £14,714.18s on this road and its branches.[104]

Nimmo's assessment of road conditions in northwest Mayo was even more unfavourable than that provided by Griffith ten years earlier. Two roads, leading from Ballycastle to Killala and Crossmolina, were all that existed in the region. West of Ballycastle, Nimmo pronounced, 'as far as Broadhaven (sic) had not even a footpath before

[i] R312 in 2006

the year 1822', although McParlan's map would indicate a bridleroad running through this district in 1802.[105] Nimmo set the people to work extending the Ballycastle road westwards to Belderg [Béal Deirg], with the intention of continuing it to the entrance of Broad Haven at Doonkeeghan, as suggested in Griffith's bog report. He estimated a little over £6,919 for completing the work, but, as will be shown later, he was never given sufficient funds to do so; he eventually spent £3,206.9s on this road.[106] Nimmo referred to this road in his reports as the North Coast Road.[iv] This, he told Dublin Castle, would greatly assist the 'distressed peasantry' in the vale of Glenamoy, where the poitín makers mentioned by Griffith were still in operation and the people were still 'in a state of insubordination to law, and exposed to the machinations of the evil minded'.[107] He began improvement works on Killala pier and on an embankment wall in the harbour. Road works were also started on the east coast of the bay near Inishcrone and Castleconor, and on a new line from Ballina to Foxford and on to Swinford. The road from Ballina to the quay, about one mile, was levelled and similar work was done on the quay road at Killala.[108] While in the district, Nimmo was employed privately by the merchants of Ballina to survey the Moy River and to suggest ways of improving the navigation to the town. He began constructing a ship canal on the river as public works, but this was later abandoned when funds ran out.[109]

In the south of the county, works were started on roads from Newport to Achill Head, Westport to Claremorris, Ballinrobe to Cong and from Westport to Killary Harbour. The latter two would link up with the Killary to Cong road, creating an extensive network through the interior and providing access to the port of Westport.[110] Bridges at Bundorragha on Killary Harbour, Annios at Lough Carra and Kilnacarra were started, and works were also under way at Ballyglass, Ballintober, the road from Ballina to Moyne and the road from Ballinrobe to Ballyhaunis, by way of Hollymount and Claremorris.[111] Patrick Knight, the assistant engineer for this region, as will be seen later, was also responsible for the design of Belmullet town and pier, and his knowledge of the county would have been a tremendous asset to Nimmo.

By August the counties of Galway and Mayo were a hive of activity. Almost every district had some form of public works underway and thousands of people were receiving regular wages or food in exchange for labour. The Grand Juries of both counties passed very few presentments at the August Assizes. Many of the magistrates did not even bother to turn up, deciding instead to leave the matter of public roads to the government. This was precisely the sort of attitude the government was anxious to discourage. Just the month before, on 16 July, they had refused Nimmo new funding, advising him that the funds already issued to him were 'sufficient for the execution' of works in his district. Nimmo, however, was continuing to receive applications for more roads and bridges and was anxious to proceed. He was aware that the lack of employment in regions of Galway and Mayo not affected by the public works was causing great hardship. He wrote to the chief secretary acquainting him with the situation and pointing out that other counties in his

[ii] R314 in 2006

district, Roscommon and Leitrim, had been totally neglected. He had given no employment whatsoever in both counties and the latest reports from his man in the region stated that the labourers there were 'quite unable to work being absolutely starving'.[112]

Road works in Sligo, Leitrim and Roscommon

Nimmo later took over county presentment works in Sligo, Leitrim and Roscommon. Here again it was resurfacing and realignment existing county roads and laying out new roads with free government aid. In Roscommon a new road was laid out from Castlerea to Ballymoe with a branch leading to Ballintober. A road from Boyle to Termonbarry, from Carrick-on-Shannon to French Park, from the Arigna iron works to Mountallen and from there to Keadew.[113] In Leitrim, works were opened on the road from Cloone to Mohill and from Ballinamore to Belturbet.[114] Later (1824) Nimmo was employed privately by the Leitrim Grand Jury to oversee improvements on the principal road running through the county from Bundroes to Manor Hamilton and on by Lough Allen to Carrick-on-

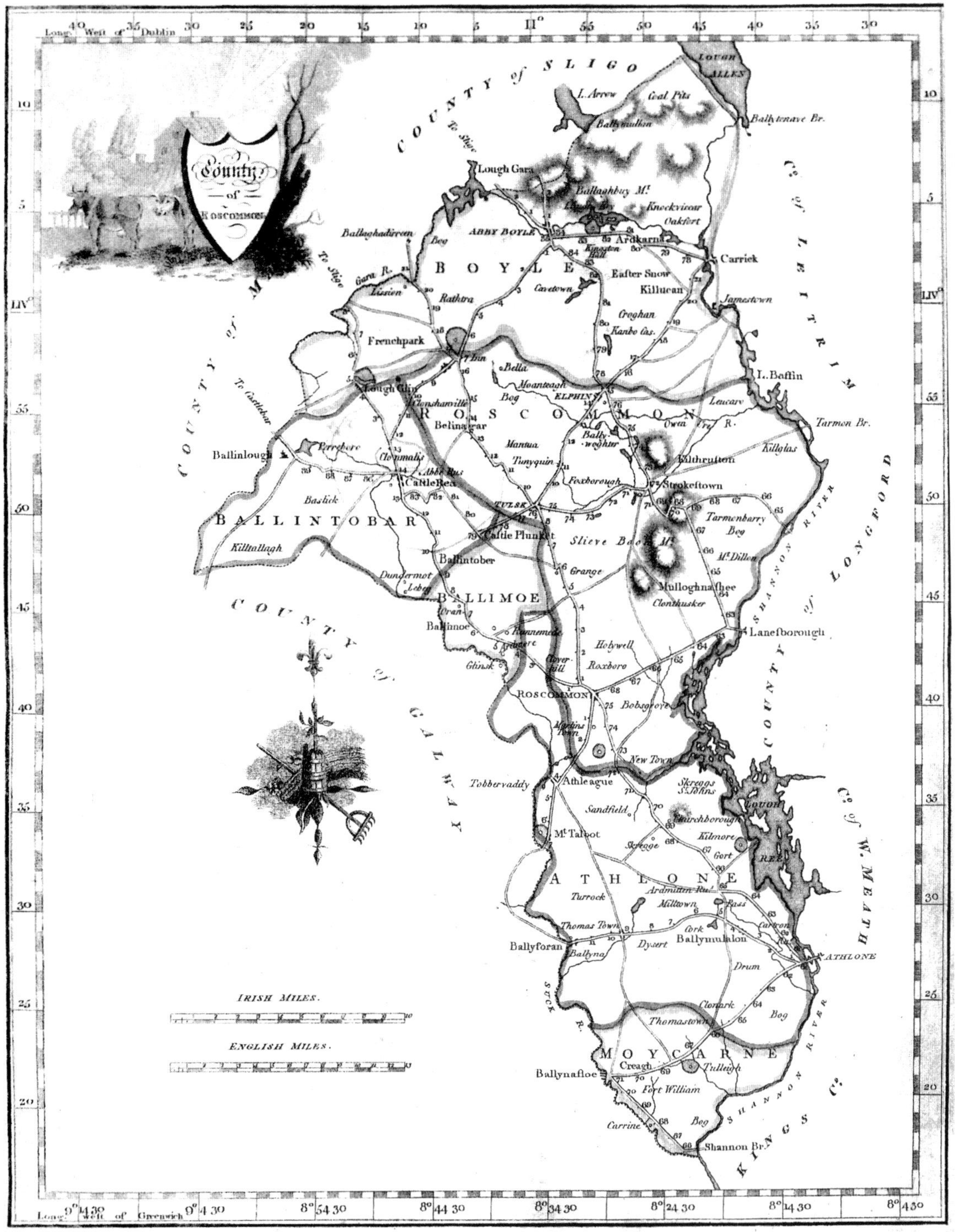

Map of County Roscommon, circa pre-1825
(New and Correct Irish Atlas)

Shannon and Drumsna. The work was carried out by his men at the expense of the county and was completely independent of government funds, although public funds would be used in 1830 to construct a bridge over the Shannon, linking this road with a new road on the Roscommon side of the river.[115] Works were also opened on the Geevagh to Ballyfarnan section of the county road from Sligo to Leitrim, giving improved access to the valley of Glennead, 'abounding in freestone and limestone of an excellent description', and to the Connaught coalfields at Arigna.[116] Nimmo provided funds for road works and harbour works already begun under the direction of charitable committees in Sligo. Work on the construction of Mullaghmore, Raghly and Pullagheeny piers was providing employment along the coast. He began work on a road and wall at Cartron, [believed to be the sea barrier wall on the road from Sligo to Rosses Point[iii]], and also roads at Drumcliff and Gortlownan, the Circuit road at Templehouse and the road from Bunnyconnellan to Pullagheeny pier.[117]

Charitable funding cut

Later in August 1822 the London Committee, on the advice of their correspondents in Ireland, brought their activities to a close. Public works had been in operation for over two months and conditions were believed to have improved considerably. This decision by the London Committee cut off the supply of aid to the local committees. During its brief existence the London Committee had granted funds to a wide variety of state and privately run organizations and had proved an invaluable source of funding for the local committees. Dublin Castle was also restricting the issue of funds and Nimmo was finding that government grants were proving insufficient to pay off all the claims on the works still in progress in his district. In his reports to Dublin Castle he was making every effort to distinguish between advances made on presentment works and funds issued to government works – distinguishing between funds that were redeemable and those that were not.

Nimmo issued instructions to his assistants to reduce expenditure on those works that were 'the sole expense of the government', but presentment works were to continue.[118] At the same time he continued to insist that the labourers be paid regularly and his officials and suppliers would have to wait; this left outstanding bills throughout the district.[119] In the two months since his appointment, he had advanced £13,300 to get works quickly under way in the district, but he was only issued with £10,300 from the treasury, leaving him with an overdraft of £3,000. He wrote requesting that funds be issued to his bank in Dublin or he would be forced to bring all works to an immediate stop.[120]

Reports of a bountiful harvest in the west were reaching Dublin Castle and Nimmo confirmed these with his own report later in the month. On a tour of inspection in early September, Nimmo passed through Cavan, Leitrim and Fermanagh on his way from Dublin. Along the way he observed that, 'the harvest in that quarter is considerably

[iii] R291 in 2006

advanced, the crops both of corn and potatoes abundant and of good quality'. Two cargoes of meal from Scotland were at Sligo quay and potatoes were selling at 3d per stone. The poor are 'now comparatively comfortable' and he gave orders to suspend operations until after the harvest.[121] The chief secretary, Henry Goulburn, instructed that all works be scaled down, and several were brought to a close. Presentment works, which had been postponed while government works were under way, were again activated and demands for payments for wages and contractors were coming in. By 26 September Nimmo reckoned that he had expended, in total, £17,620 on public works in his district and had received £14,300, leaving a balance outstanding of £3,320.[122]

Although conditions had improved for many, for others the agony continued. September also brought reports that fever was rampant throughout Connemara and the fever hospital in Galway was already full. Dr T.L. Whistler, working for the Board of Health in Galway, sent Dr J. Mahoney into Connemara to ascertain the extent of disease and famine in the area. Mahoney reported that having spent three days inspecting the district of Errismore, south of Clifden, he found the people to be in appalling conditions. Many deaths had occurred and large numbers of the population were suffering from sickness, starvation and nakedness. He accused their landlord, John Geoghegan O'Neill of Bunowen Castle, of being continuously absent and exacting rack rents on his estate. Out of a population of five to six thousand, he lists one hundred and twenty-two suffering from fever, thirty with dysentery, forty with colic and dropsy and about eighty with convulsions from fever. Large numbers were suffering from malnutrition, some had lost toes during their illness, and many were without clothes of any sort: 'I have seen vast numbers of sick, as well as those in health without bed or body clothing; in a state of absolute nudity'. He requested medicine and clothes.[123]

Whistler passed Mahoney's letter to Goulburn, informing him that famine had driven the people out of the mountains and into the town where they were now spreading misery and disease among the people of Galway. He purchased blankets and some coarse flannel and issued them, along with food and medicine. However, he confessed that nakedness was also a problem in Galway and that they were unable to dismiss from hospital 'the numbers who are literally destitute of any species of bodily covering'.[124] Goulburn immediately issued orders to the barrack master at Galway to supply Dr Whistler with sheets, rugs and blankets. He informed Whistler that the lord lieutenant was prepared to sanction any sum of money that he might consider necessary, or take any other steps that might alleviate the suffering of the people. Twenty soiled blankets in Galway were sent immediately to Dr Whistler and four hundred blankets in stores in Dublin were to be sent the following day.[125] In response to requests for assistance from Dr Whistler, the Board of Health dispatched five doctors to help bring the fever under control.[126]

Nimmo was back in Dublin in early October working on a map of the Western District, at four inches to the Irish mile, to show the 'several works undertaken there for the employment of the poor'. Due to the remoteness of the region the map was taking longer

Map of County Leitrim circa pre-1825
(New and Correct Irish Atlas)

and proving more expensive than anticipated.[127] Nimmo eventually charged the government £300 for making this map, although there is no record of its having been handed over; nor has it proved possible to locate to date.[128] As the year drew to a close Nimmo was showing obvious frustration with Dublin Castle's slowness in crediting his account. Labourers were owed money for work already carried out and he was unable to meet the demand. In all, he told Goulburn, £7,000 was required to clear off all outstanding commitments, 'even if the whole of the works should be immediately stopped'.[129] At the end of October he embarked on a general tour of inspection. The good harvest had brought a falling off in labourers and, although the roads were nowhere near finished, he suspended many of the works for the winter.[130]

Progress reported on Nimmo's piers

The Fishery Board was anxious to complete as many of the piers as possible before the onset of the winter storms. On a tour of the Western District in November 1822, the Board's inspector, William King, reported that twenty-nine piers alone were under construction on the coast of Mayo, Galway and Clare. On the coast of Mayo and Galway, work was under way at Bunatraher, Saleen, Blacksod Point, Achill, Inishturk Island, Old Head, Bundorragha, Leenaun, Derryinver, Cleggan and Clifden. Work on the piers at Blacksod Point, Old Head and Cleggan had begun late in the season and these were expected to come under attack from winter storms, which would add to the costs and delay their completion. Thirteen piers were under construction on sites 'so judiciously chosen by Mr Nimmo' in Galway Bay: Roundstone, Máimín pier on Gorumna Island, Rossaveel, Spiddle, Furbo, Barna, Slate Rock in the Claddagh [Nimmo's Pier], New Harbour [Rinville], Ballynacourty, St. Kitts [Killeenaran], Dooras, Burrin New Quay, and Killeany on Inishmore. Of these, King considered Roundstone, Gorumna, Rossaveel and Killeany to be the most essential to the fishery and the adjacent country, and from which kelp could be exported to Liverpool and other markets, bypassing Galway port.

In the past there was not a single safe harbour on the coast of Clare from Black Head to Loop Head, but this was rectified with the erection of piers at Liscannor, Seafield and Doonbeg. These new piers were expected to be of great benefit to the agricultural interests as much as to the fishery. At the mouth of the river Shannon, the new piers at Kilbaha and Carrigaholt would be of little benefit to the fishery but would facilitate the turf trade with Limerick.[131]

The Fishery Board had received a grant of £5,000 from the London Committee in 1822; this was used to provide loans for the fishermen of the distressed districts for boat repair. The aid was given in material rather than money, with the landlords going guarantor. Fishermen in the Westport and Galway districts availed of the loans and soon it was reported that there was an increase in the number of fishermen and boats of all classes

Map showing the west coast fishery stations
(First report from Irish Fisheries inquiry 1837)

fishing off the coast in areas where the new piers were erected.[132]

For the functions of the Fishery Board the coast was divided up into districts. Westport District extended from the western point of Killala Bay, 'Lacken Point', to Slyne Head and the Galway District extended from Slyne Head to Black Head, Co. Clare. In the early years Connemara was divided in two; north of Slyne Head was included in Westport District and south of that was included in Galway District. Clifden was eventually established as a distinct fishery station in 1826. The increase in boats resulted in large quantities of herring being landed on the new piers, and a number of curing houses were set up. The success of Mr McLaughlin's curing house in Galway, operating since 1820, was encouraging others to do likewise, and King was happy to announce that locally cured herring were replacing Scottish herring in the adjacent towns and counties. McLaughlin was employing a number of coopers to build new barrels to contain the fish and this had increased the demand for local timber and sally rods. Prior to this the sally had been imported from England and Scotland, but was now being raised on marshy waste-ground locally.[133]

However, there were still very few salt stores on the west coast; this created a monopoly among suppliers and caused the price of salt to rise with the demand. The duty on salt was removed in 1822 and this was expected to greatly benefit the poor fishermen. Prior to this it was not uncommon for shoals of herring to come into the bays during the season, where they were fished with great enthusiasm, only to be left rotting on the shore for want of salt to preserve them. The price of

AN ACCOUNT of the total number of Decked and Half-decked VESSELS, together with Undecked or Open Boats of every description, registered and employed in the Fisheries; also their Tonnage, and number of Men employed therein.

	STATIONS.	Aggregate Number of Decked Vessels.	Aggregate Quantity of Tonnage on D°.	Aggregate Number of Men employed in D°.	Aggregate Number of Half-decked Vessels of 15 Tons & upwards.	Aggregate Quantity of Tonnage on D°.	Aggregate Number of Men employed in D°.	Aggregate Number of open Sail Boats, under 15 Tons.	Aggregate Quantity of Tonnage on D°.	Aggregate Number of Men employed in D°.	Aggregate Number of Row Boats.	Aggregate Number of Men employed in D°.	Total Number of Men employed.
Irish Fishery Office, Dublin, 9th May 1822.	Westport - -	4	96	22	2	31	10	135	825	675	725	2,900	
	Galway - -	1	20	5	3	45	15	460	3,680	2,300	826	3,304	
Irish Fishery Office, Dublin, 1st May 1823.	Westport - -	5	111	25	1	17	5	110		440	885	3,540	4,016
	Galway - -	6	160	38	2	31	10	524		2,620	890	3,560	6,228
Irish Fishery Office, Dublin, April 19th, 1824.	Westport -	5	114	23	1	17	4	119		476	1,117	4.468	4,971
	Galway -	6	-	-	8	-	-	519		-	1,169	-	6,975

Table giving the number, type and tonnage of boats working in the Westport and Galway fishery stations 1822-1824
(Third, Fourth and Fifth Report of the Commissioners of Irish Fisheries)

salt was immediately raised on such occasions and, when added to the cost and difficulty of transporting the fish by the old bridleroads across the country to Galway or Westport, 'or round the headlands, in winter weather', the whole business proved unprofitable. This resulted, Nimmo told the Fishery Board, in the herring being 'left to rot on the shore in heaps of many thousands, and the unfortunate fishermen, whose little stock has been expended during summer in fitting out his boats and nets, sees himself ruined, with abundance apparently in his grasp'. Such scenes were not expected to reoccur, now that the salt tax had been removed and salt stores were set up in Clifden and Belmullet, with other stores under consideration.[134] King recommended that the board set up its own salt depot at Roundstone or Rossaveel, as this would greatly assist the establishment of fishing villages and encourage the export of fish from that stretch of coast. The Fishery Board on consideration felt their involvement was unnecessary, as the abolition of the salt tax would encourage local businessmen to take up the opening.[135]

Government funding extended

The records show that, between 12 June 1822 and 5 January 1823, Nimmo was issued with a sum of £25,500 for works in the Western District.[136] The distribution of the aid had brought immediate relief to the poor throughout the district, but most particularly in Mayo and Galway. Following a further tour of Connemara in December Nimmo reported, with great satisfaction, that 'the change in the appearance of the peasantry in Conamara (sic) since my visit in September is quite wonderful – as provisions are now plenty – great parts of the wages every pay day are immediately laid out in the purchase of clothing. There is a considerable herring fishery now in Conamara (sic) – the people derive great benefit from the piers that have been lately constructed, as the boats are available at all times, and can be moored in safety'. The fishermen were, however, hampered by poor tackle, most of the nets were worn out and they had no means to repair them. Some of the piers were likely to be left unfinished, due to the Fishery Board's lack of funds, and Nimmo recommended that aid be granted to them. But there were still 'a great many poor objects who must depend in a great measure on the charity of their neighbours, as they had no means of sowing their potato plot last year – as much as possible of our employment is given to this class'.[137]

Nimmo closed his letter, outlining his travel plans for the next few days, which were as grueling as ever: 'I am just about to depart from Tuam and the county of Mayo then to Sligo and Roscommon'.[138] Then it was back to Dublin, where, on 28 December 1822, he was seeking an interview with the chief secretary, awaiting instructions for the ensuing season. Although the distress had abated, nevertheless, because of the magnitude and extent of the works opened up by Nimmo in 1822, it would be necessary for the government to continue its commitment to the west a little longer than planned. The meeting with Goulburn took place on 4 January 1823, but nothing could be decided until the government grant for public works in Ireland became known.[139]

Nimmo supplied Gregory with his list and estimates for continuing various works for the coming year and was back again in April seeking instructions and requesting a meeting, this time at Goulburn's office in London.[140] Goulburn, on the other hand, was having difficulty prioritising the many works put forward by Nimmo. The realisation of the extent of works undertaken by Nimmo in the west was beginning to dawn on the chief secretary. Nimmo, he told the lord lieutenant, had used his position in the west to advance some personal agenda, which could, potentially, commit the government to long-term spending on works that should be the responsibility of the county Grand Juries:

> Mr Nimmo certainly misunderstood the views of the Government in his original appointment. He embarked on works not so much calculated to afford employment to relieve immediate distress as to promote … more what appeared to him great national improvements and the misfortune is that he has thereby created an expectation that the Government considers itself bound in all time to come without reference to reasons of distress or of plenty to expend a large annual sum in works which ought in fact to be defrayed by the several counties.[141]

He did admit, however, that there was 'certainly great want of employment' and 'more than usual deficiency of food' on the west coast, but suggested that if the sums issued by government for relief fell short of expectations, and if, 'hopes of future aid [were] properly discouraged', in time, the country would 'gradually return to a reliance on its own means and resources'.[142] In the ensuing year, Nimmo would eventually receive £28,340 from the treasury, a small increase on the sum received in the previous seven months.[143]

The Galway newspapers had reported in March that the people of Connemara were again showing early signs of distress and were expected to flood into the town seeking relief: 'Hopeless and desperate, they will be driven from their homes by pestilence and famine, and bearing contagion along with them, will turn towards Galway, to beg relief at our doors'. Fearful of the spread of typhus fever, they had suggested that measures be taken to protect the people of the town by the removal of all strange beggars from the streets, who were viewed as 'objects of terror', and that the government be made aware of the distressed conditions of the people. Provisions were available but again the people were 'completely without money'. In April the papers announced that the government 'has sent down upwards of Four Thousand Pounds in Silver to Mr Nimmo to enable him to continue the prosecution of the Public Works in Connemara'.[144]

Over the next few years funding for the continuation of the roads was sought each year from parliament and the work undertaken depended on the sums received. Nimmo's permanent staff remained in place in Dublin and in the west, while works in the Western District were subject to stops and starts depending on the weather, the availability of labour, numbers dwindling during planting and harvesting times, and the allocation of government funds.

MONIES expended under the direction of Alexander Nimmo, Civil Engineer, for carrying on certain Public Works in the Western District of Ireland.

	£.	s.	d.
From 12 June 1822 to 5 January 1823 - - - - - -	25,500	–	–
5 January 1823 to 5 January 1824 - - - - -	28,340	–	–
5 January 1824 to 5 January 1825 - - - - -	23,000	–	–
5 January 1825 to 5 January 1826 - - - - -	29,467	13	4
	106,307	13	4
Deduct Treasury Fees - - - - -	3	13	4
Net sum from His Majesty's Treasury from June 1822 to Jan. 1826 - £.	106,304	–	–

Table showing monies expended under A. Nimmo on public works in the Western District from June 1822 to January 1826
(BPP 1830 [624], xxvii)

Each year Nimmo produced detailed costs and estimates for Dublin Castle to include in the annual budget to be taken before parliament. He also reported on the works in progress and proposed projects for the following year. All of Nimmo's letters and reports were written in plain English. They were written for civil servants, not fellow engineers, and included very little technical data regarding the construction of the roads and bridges, much to the regret of latter-day engineers. In all, Nimmo was positive in his assessment of the progress and extent of the works under way in his district. He never seemed overwhelmed by all that he had taken on or frustrated with the difficulties he encountered. He was, however, frustrated by the government slowness in providing funding and he continued to complain that his bank account was overdrawn. In the early years he rarely talked about completing the works; instead he used the funds issued to extend the works ever further, either to the coast or in the direction of a town, or to connect it with an existing road.

Works were continually evaluated on the immediate needs of the people, their long-term benefit to the district in which they were to be carried out and on the size of the grant allocated by Westminster for public works. However, the primary purpose of the grants was to bring immediate relief to the poor. This resulted in many works being brought to a stop due to lack of funds, only to be reactivated when isolated cases of distress arose in particular districts; some of the major works in Mayo were suspended in 1822 and would not reopen until 1831. Nimmo received £23,000 in 1824 and £29,467.13.4d in 1825, but would later complain that he was never allocated sufficient funds to complete the government roads begun in 1822. He was further hampered by the slowness with which the grand juries supplied funding for some bridges and small stretches of county road, which were necessary to open up completed sections. However, despite Nimmo's complaints, the benefits from continued government expenditure on public works were showing their effect all along the western seaboard.

The introduction of public works had advanced developments in the west far quicker and far beyond local capabilities. As Nimmo and Griffith had already pointed out, neither the landlords nor the Grand Juries were in a position to fund the extensive works needed to fully develop the west. Neither were they able to provide the expertise to oversee the works so as to attain the necessary standard to ensure their full utility and durability. But where the landowners and the Grand Jury had failed, the government, in the person of Nimmo, had succeeded. Under Nimmo, roads in Galway and Mayo were laid out in a more cohesive manner so as to provide interconnecting coastal routes for both counties, linked with crossroads that cut through the interior. Pier sites were chosen for their impact on the immediate coastline, so as to serve the interests of the fishery industry and local communities, rather than just those of individual landlords. The public works brought large-scale employment directly to the people in their own neighbourhood and the promise of future industry in agriculture and fisheries. Providing 'legitimate' occupations for the people would steer them away from poitín making and smuggling, which up until then had been the principal occupations along the west coast. And the money earned on the public works was expected to stimulate the local economy, assist local industry and eventually increase the revenue to the crown.

But still Nimmo argued that the government was not doing enough. He had in his district over eighty works in progress in the summer of 1824. These were scattered throughout the district and in his opinion did very little to improve the overall conditions of the poor.

Local measures needed to be taken to reduce the peoples' dependency on the land by providing alternative employment, he argued. Nimmo felt that providing employment for the poor ought to be the consideration of every individual in the country. Under the prevailing conditions, he complained, the poor were being left 'to obtain their subsistence by mendicancy'. He recommended that the government introduce some form of local parish poor law system, similar to that in force in England at the time. This would stimulate local taxpayers to provide employment in every parish in the country:

> It is my opinion, that nothing will ameliorate the condition of the people of Ireland, but some measure that will extend employment generally over the whole kingdom, by making it the interest of local associations, small communities, such as parishes or towns, or landed proprietors, to have the spare labour of the people generally applied to something that will be productive, if it were only by adding to their own comforts; and I conceive that the peasantry are in Ireland in want of almost every manufacture that tends to the comforts of life.[145]

The poor laws adopted in England since the time of Queen Elizabeth had made that country peaceful and prosperous, while the reverse was the case in Ireland: 'These measures having been neglected in Ireland have left her steeped in poverty, with an excessive population unemployed, and consequently unprofitable; destitute of property, and living on the very brink of want'.[146] Increased employment among the poor would bring overall prosperity and introduce habits of industry to the young. Without the example of employment from within their own families, the youth, Nimmo warned, 'generally grow up with the idle habits which prevent them settling to regular industry in their after life'. The adoption of a poor law system might also help to improve the peoples' perception of both the magistrates and the law, both of which were considered by the people never to be in their favour:

> if we had any thing similar to the parochial law in Ireland it would enable the magistrate to be known to the people by beneficent actions as well as by the terror of the law; that at present the tenantry in Ireland are uniformly raged against the law, because they know it only by its terrors; but if they had relief by a magisterial order, as in England, they would look upon the law as having the power of doing them good, - every magistrate and parish officer as the instrument and means by which relief could be communicated to them.[147]

The Poor Law was eventually introduced into Ireland in 1838 and was financed by a tax raised locally. In preparation for raising such a tax, a survey and valuation of Ireland was already under consideration by a government committee in 1824. Nimmo was one of those called to offer an opinion on how such a survey could best be conducted. From the outset Nimmo declared he was too busy to be even considered for a position with the survey, stating he was fully occupied, 'making harbours and roads for government and the board of fisheries'.[148] The valuation was later conducted under Richard Griffith and is commonly referred to as 'Griffith's Valuation'.

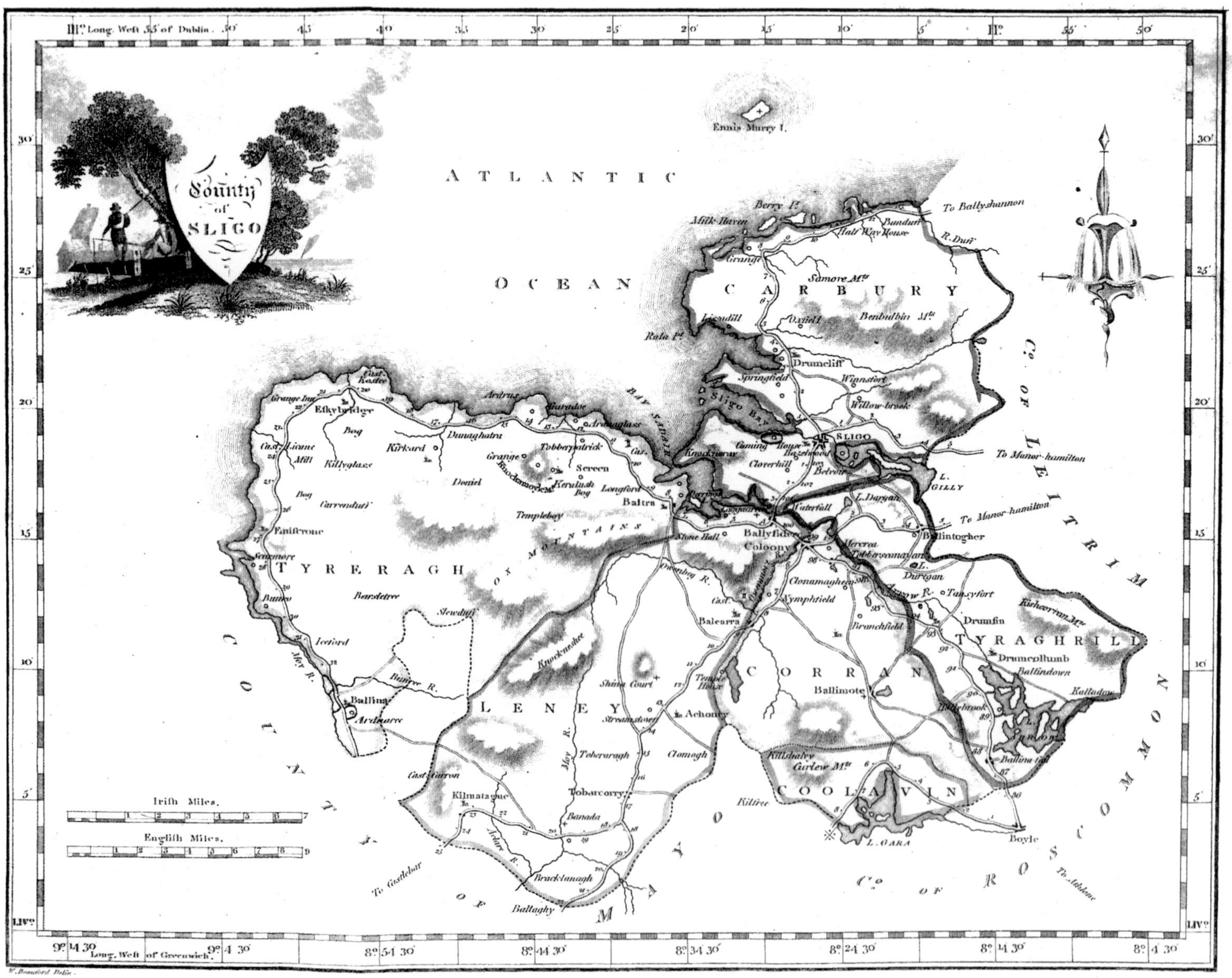

Map of County Sligo circa pre-1825
(New and Correct Irish Atlas)

Chapter 4

Nimmo's fishery reports and pier construction 1824–1826

Nimmo conducted his survey of the west coast while the public works were underway. His findings were outlined in two reports and these were turned in to the Fishery Board in 1824 and 1826. Plans for many of the piers dealt with by Nimmo were attached to the reports and these were later published.[i] In his reports Nimmo was able to give an overall view of the progress of the public works in the Western District and to outline their importance to the fishery. The first report began on the borders of Leitrim and Donegal and ended in Erris, the second covered the coast from Erris Head to Black Head, Co Clare. As with his survey of the east coast, Nimmo described the variety and type of fish caught and the boats used by the fishermen for each district. He emphasised the fact that, although the west coast abounded with fish, the majority of the fishermen fished for their own consumption only and took no part in the 'business' of fishing. The exception to this was the fishermen of Galway Bay. The principal reason for the underdeveloped state of the fisheries on the west coast was the size of the fishermen's boats and the poor condition of their tackle.

Nimmo's reports give a full account of the many new piers built with funds from the Fishery Board and charitable bodies; giving their measurements, costings, the type of stone used and in many cases the contractor or overseer responsible for their construction. Several of the piers were completed on time and within Nimmo's estimates, but a good many others ran into difficulty and on some the works were later abandoned due to lack of funds, bad weather or disagreement over rising costs. In many cases the Fishery Board eventually provided the necessary funds to complete the works; in others, as at Clifden and Gorumna, it took years to reach an agreement and the piers were long left exposed to the winter storms and a hazard to fishing and commercial vessels.

Nimmo outlined the difficulties experienced during the construction of the piers and explained why some were still unfinished and others were in need of further work. He is very clear in his criticism and where the workmanship had failed he usually gives the cause. This varied from inexperienced workmen or manager, bad weather, lack of judgement on the part of the contractor, the use of expensive stone and, as was so frequently the case, the contractor or overseer deviating from Nimmo's original plan and subsequently running out of funds. He provided descriptive details of the many harbours and bays, frequently making suggestions for further piers and improvements and including the estimates. His report on the making of Raghly Pier is a good example of the sort of agreements that were reached, and the difficulties encountered, in constructing piers and roads in the Western District.[1]

[i] Nimmo's Reports on the coast from Sligo to Clare are reproduced in the appendices to this book.

He described the roads that existed prior to his appointment to the Western District in 1822 and outlined how these had been improved, rerouted and extended to provide access to the coast. He pointed out where new lines were laid out in areas previously inaccessible and how these had opened up the mountain and bog districts and linked them with the major towns and the new piers. Large sections of his report on Connemara were 'lifted' from his bog report, written more than ten years before. Both reports carried the same breakdown figures for the kelp and turf trade, which must have surely improved with the erection of the piers in the neighbourhood. He also repeated the figures for the size and cost of boats. He says little about the type of fish available off this coast or the numbers involved in the fishery; these figures were by then appearing in the Fishery Board's annual reports. Instead he concentrated on the progress of the public works and the resulting advances made in the district.

Nimmo emphasised the benefits the piers and roads had brought to the coastal regions and explained how much more could be derived from more government spending in the future. He gave a very positive account of the state of the roads in both Galway and Mayo, giving the impression that these were complete and in full use, when in fact they were a long way from being so. This was to reassure the board that full use was being made of the piers and to explain the benefits they brought to the interior, as well as to the coast.

The improvements had, in many cases, exceeded all expectations and were attributed to the new piers and quays, along with the connecting government roads. The improved communications usually produced an increase in population and trade. As anticipated, the advantages extended beyond the fisheries to agriculture and industry, and resulted in the growth of new villages and towns close to the new quays: 'Houses and Villages of an improved class, with schools, public establishments, post-office communications, and other proofs of civilization, are progressively appearing in those uncivilized and sequestered districts, heretofore only known as safe places for illicit distillation, smuggling, and other unlawful practices'.[2]

Having a pier close to one's holding brought greater prosperity to both tenant and landlord. With a pier the tenant was able to combine fishing and farming, depending on the season. The landlord increased the value of his property by building stores and villages, thereby encouraging the development of trade and industry. In the immediate neighbourhood of the piers more land was brought under cultivation as sand and seaweed were easily landed for use in reclaiming poor quality soil. The construction of the roads had opened up previously inaccessible districts, bringing more land on to the market. The increase in cultivated land resulted in an increase in crop production and the new villages and towns provided a market for locally grown produce, with the surplus going to export.

Corn stores and corn merchants were becoming a prominent feature where, prior to 1822, the only market available to the corn grower was the local poitín maker: Clifden in 1824 boasted a brewery and distillery and both were thriving.

The increase in traders broke up monopolies and this, along with improved communications, encouraged the farmers to take their own goods to market and seek out better prices. The principal export from Westport and Galway was corn, both of which were exporting about 10,000 tons annually. And although the price of kelp had fallen dramatically in recent years, 7,000 tons was still exported annually from Galway. Previously the kelp had been brought to the ports in small open boats, but with the construction of the new piers, it was more frequently exported locally from where it was manufactured and this was producing a better quality crop.[3] Other locally produced materials, such as limestone and marble, were also exported from the new piers. However, many of the fishermen still relied on the 'carrying trade', usually the owners of decked boats, who carried their own and their neighbour's fish to the markets at Galway and Westport.

History of individual piers

Initially Nimmo designed functional piers to serve the needs of the local fishermen and to suit the sites chosen, but always with an eye to keeping costs low so as to win the Fishery Board's sanction. Frequently he suggested improvements that could be made later as trade on the piers increased. Most of this work was carried out after the coastal survey was suspended in 1824 and Nimmo's term with the Fishery Board had come to an end. For an independent assessment of the works carried out on the piers, the Fishery Board employed a second engineer, James Donnell, to make a general inspection of all the piers that had received board funding. Donnell was the Harbour Engineer and Inspector of Harbours to the board. He conducted his surveys in 1824, 1826 and 1828 and his reports to the board gave a history of the individual piers, their cost and progress. Although he was frequently critical of the standard of workmanship employed on the piers, almost all of which were in need of further funding for repairs or improvements, he strongly defended the board's decision to build several small piers over one or two big harbours. The funds allocated to the Fishery Board, he argued, prevented them from constructing anything greater than the small piers undertaken. He also defended Nimmo's choice of site for the piers. In Donnell's opinion, the site chosen for each pier ought to be judged against the 'geographical peculiarities of the coast'. The piers were built to facilitate the coastal population, to enable them to keep their vessels close to their dwellings, so as to be ready 'without loss of time or labour, to step on board and start for the fishery when weather and other inducements offer'.[4]

In the years that followed many of the piers were enlarged and extended. Parapets and capping were added and in some cases the basins excavated and the quays extended to accommodate the erection of stores. Many of the improvements were those suggested by Nimmo and included in his original plan, others were recommended by James Donnell.[5]

Saleen, Tarmon & Inishturk Piers

The three piers, Saleen, Tarmon (Blacksod) and Inishturk (north), had, according to Nimmo, contributed greatly to the development of the Mayo coast. All three were constructed under the management of William Bald. Bald deviated from Nimmo's plans on both Saleen and Blacksod, incurring extra expense, which irritated Nimmo and delayed their completion. Bald had faced the Saleen pier on each side with hewn granite from Tarmon Point, which Nimmo considered expensive. At Blacksod Point, he was again 'tempted' by its proximity to the 'excellent blocks of granite' and again faced the pier with hewn stone.

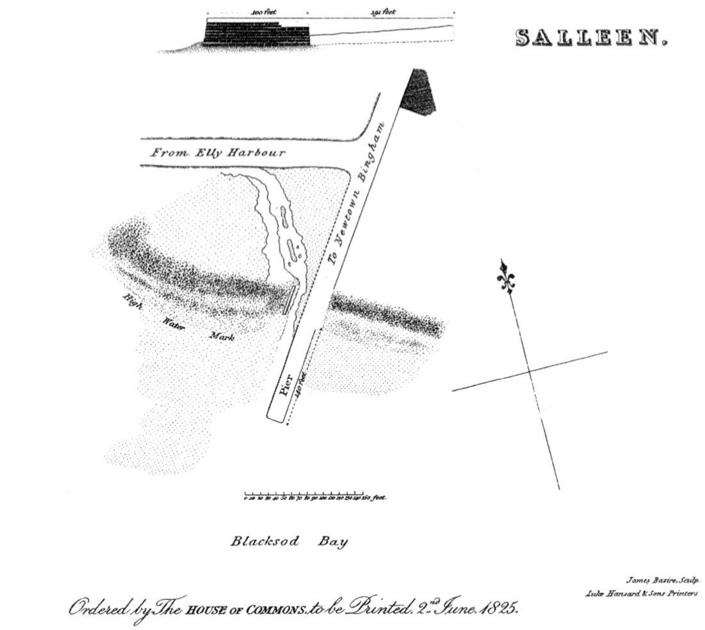

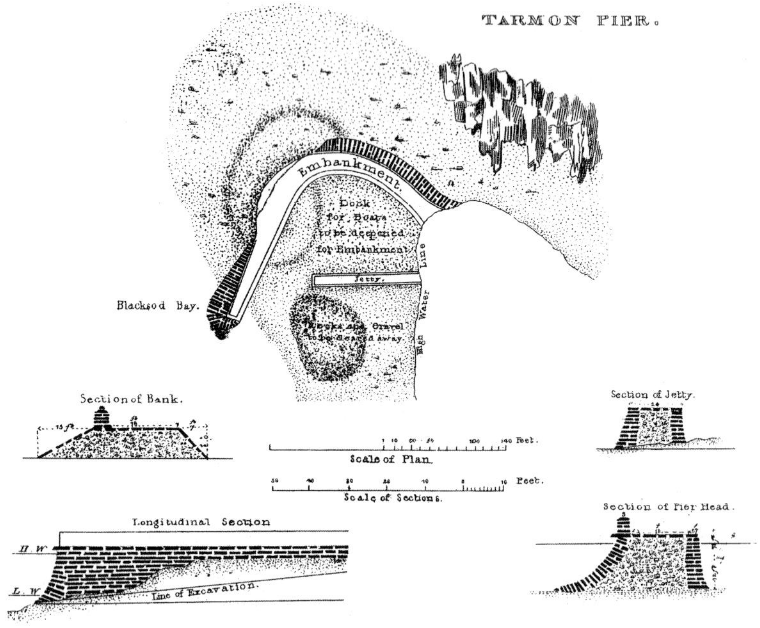

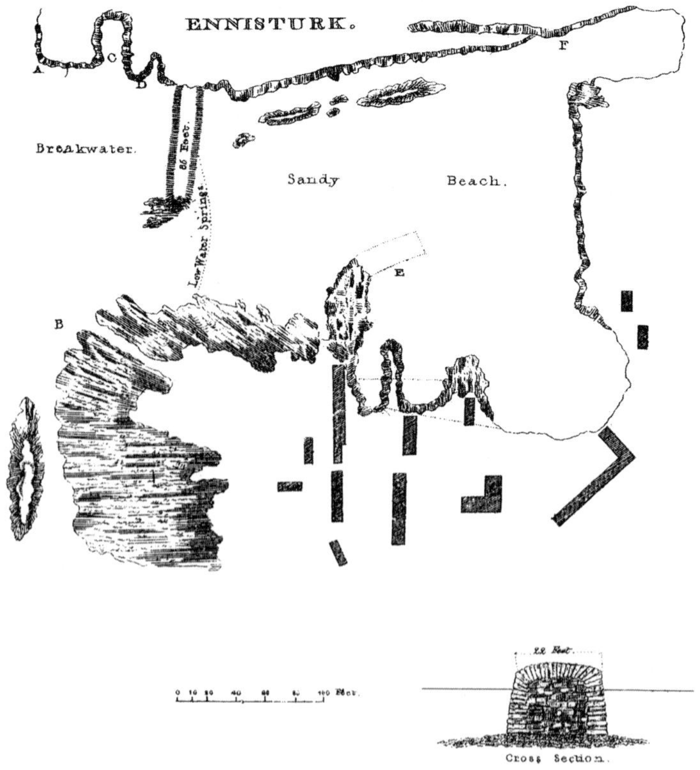

Nimmo's sketch of Saleen, Tarmon and Inishturk Piers
(Nimmo's Reports on the Fishery Stations of the Coast of Connaught and of the West Coast of Ireland)

Nimmo did not approve of what had been done: 'as might have been expected, the sum allotted for the purpose was exhausted ere the pier was made of much use'. Inishturk pier also proved unsuccessful and expensive, and Nimmo was forced to send in Alexander McGill, his trusted friend from

Waterford, to render the quay safe. Nimmo was, however, completely satisfied with Bald's pier on Achill Island, situated opposite Achillbeg.[6] His assistant, Patrick Knight, later wrote that Saleen, in his opinion, was badly positioned and Blacksod Point was seldom used. The Board of Public Works, however, struck a more positive note in their report in 1833.[ii] Knight's pier at Belmullet will be dealt with later.

Bundorragha & Old Head Piers

The pier at Old Head was expected to be of great benefit to the fishermen of Clew Bay, as it was much nearer to the fishing grounds than Westport. Here the works had run over budget and it was some years before the extra funds needed were allocated for its completion.[7] The contract for Bundorragha pier, Killary Harbour, although publicly advertised, was awarded to one of Nimmo's assistants, William O'Hara. Work on the pier began in 1824 and it was finished, 'in a very substantial manner', in 1825.[8] Nimmo's 'rough jetty-pier' at Leenaun was reported as being frequented by about thirty fishing boats in 1833. The site was later improved during the Great Famine with the addition of the west pier.[9]

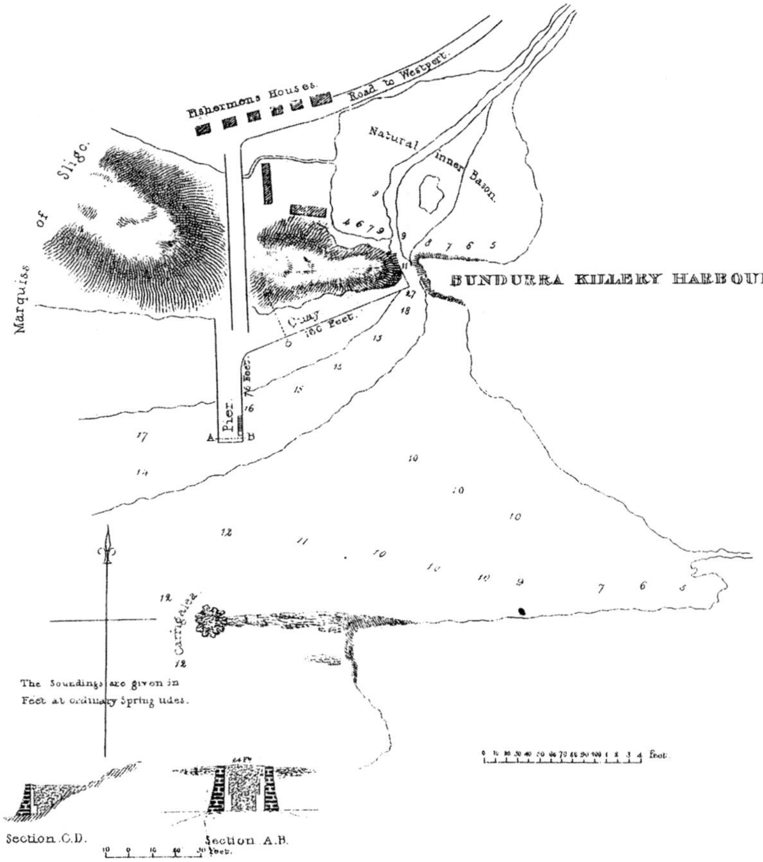

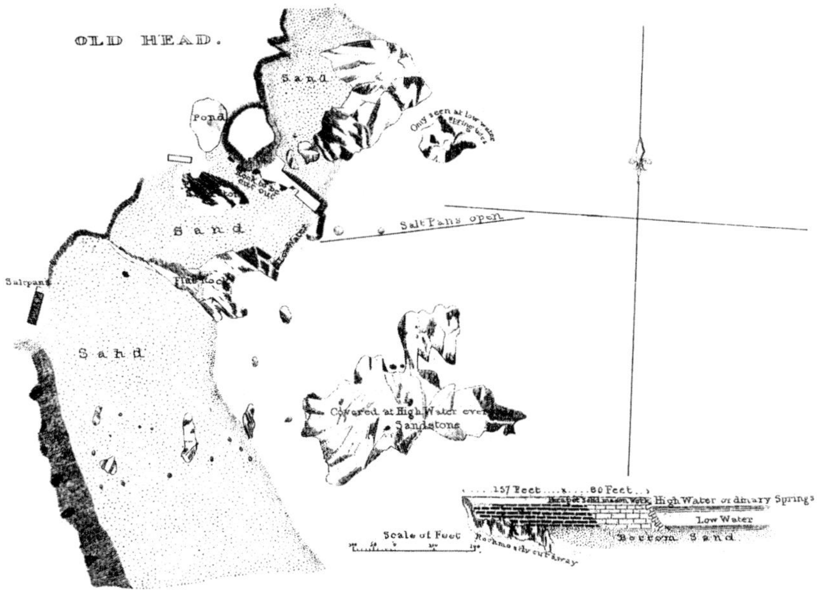

Nimmo's sketch of Bundorragha and Old Head Piers
(Nimmo's Report on the Fishing Stations of the West Coast of Ireland)

[ii] See page 151

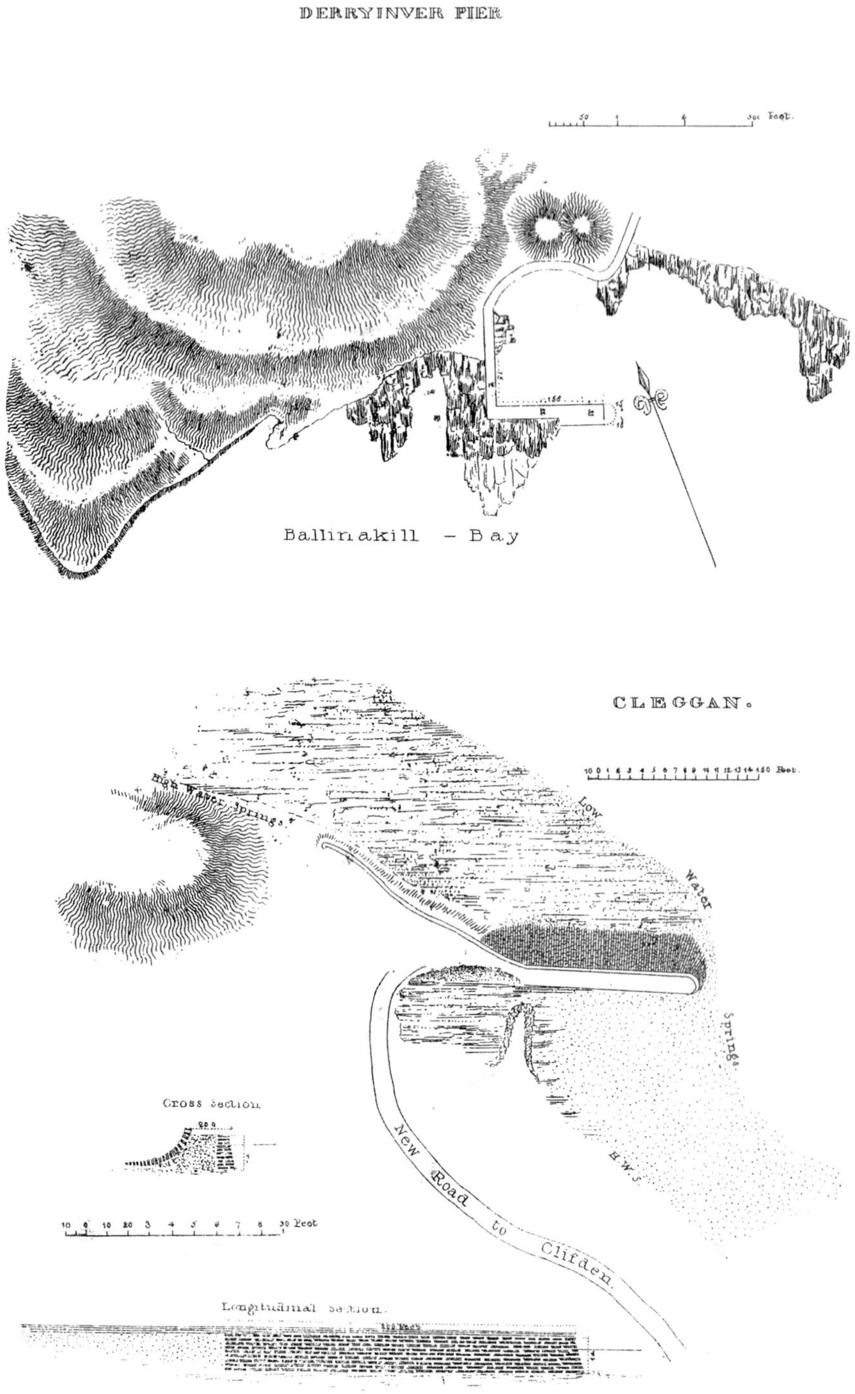

Nimmo's sketch of Derryinver and Cleggan Piers
(Nimmo's Report on the Fishing Stations of the West Coast of Ireland)

Derryinver, Cleggan & Roundstone Piers

The pier at Derryinver provided a 'convenient point of access' for Ballynakill Bay and was, in Nimmo's opinion, a good site for a fishing village. The piers at Cleggan and Roundstone were started in 1822 under the management of a young architect, Alexander Hay, whom Nimmo described as a man 'of talent and good character, though not of great experience in such works', which, he explained, was difficult to acquire at the time. Work on the piers was progressing satisfactorily until October when an unusually high tide damaged Cleggan pier and winter storms caused damage at Roundstone. Nimmo decided to transfer the works to a person of skill and experience and sent for Alexander McGill. McGill was instructed to put both piers, and the one at Inishturk (north), 'in a state of safety', and it was due to 'his skill and exertion' that the piers were preserved.[10] However, it would be some years before both piers were finished and only after some alterations were made to Nimmo's plans by James Donnell.

Even before Cleggan pier was fully completed it came into regular use and was seen as an important landing place for the boats of Inishbofin and the western fishermen. The adjoining farm was the property of Thomas Martin of Ballynahinch Castle and was leased out to a tenant, Thomas Evans.[11] There were no buildings erected on the quayside and no sign of a village developing, but Nimmo was convinced that, in time, the site would become 'a thriving village'.[12] However, by 1836 there was still no sign of a village and the land close to the shore was occupied for the most part by desperately poor fishermen: 'All the male inhabitants of Claggan (sic) are more or less fishermen. They are destitute of every convenience; they live in the most wretched hovels that can be described; they are yearly tenants, generally holding land under freeholders, who are themselves miserably poor ... The aged are supported by their neighbours, and the widows by begging'.[13]

Clifden Pier

Construction on Clifden quay began under Nimmo in 1822, with funds provided for the relief of the poor. Part of the upper quay was under construction at the same time as improvement works were underway at Dooghbeg, a little further down the harbour.[14] D'Arcy had previously built a road, at his own expense, from Clifden to the proposed site for the quay, and continued it along the coast to Dooghbeg and on to Clifden Castle.[15] He built a small storehouse on the shore at Dooghbeg, where the improvement works were expected to provide shelter for boats and a landing-place for vessels in the lower harbour. The Fishery Board took over the funding of the pier in 1823 and the work was handed over to D'Arcy. As the work progressed, the pier was soon brought into use. Nimmo's assistant wrote informing him that the people of Clifden were overjoyed when the first cargo was loaded for export at the quay: 'I am sure you will be gratified to learn, that this morning a brig of two hundred tons burthen came to the quay to take in a cargo of Marble, and so sensible were the inhabitants of the benefit they derived, and are likely to derive from this quay, that in the evening all the houses in the village were illuminated'. A shopkeeper from the town left on the same boat, bound for Liverpool with five hundred pounds in his pocket, 'to import goods direct from that place'.[16]

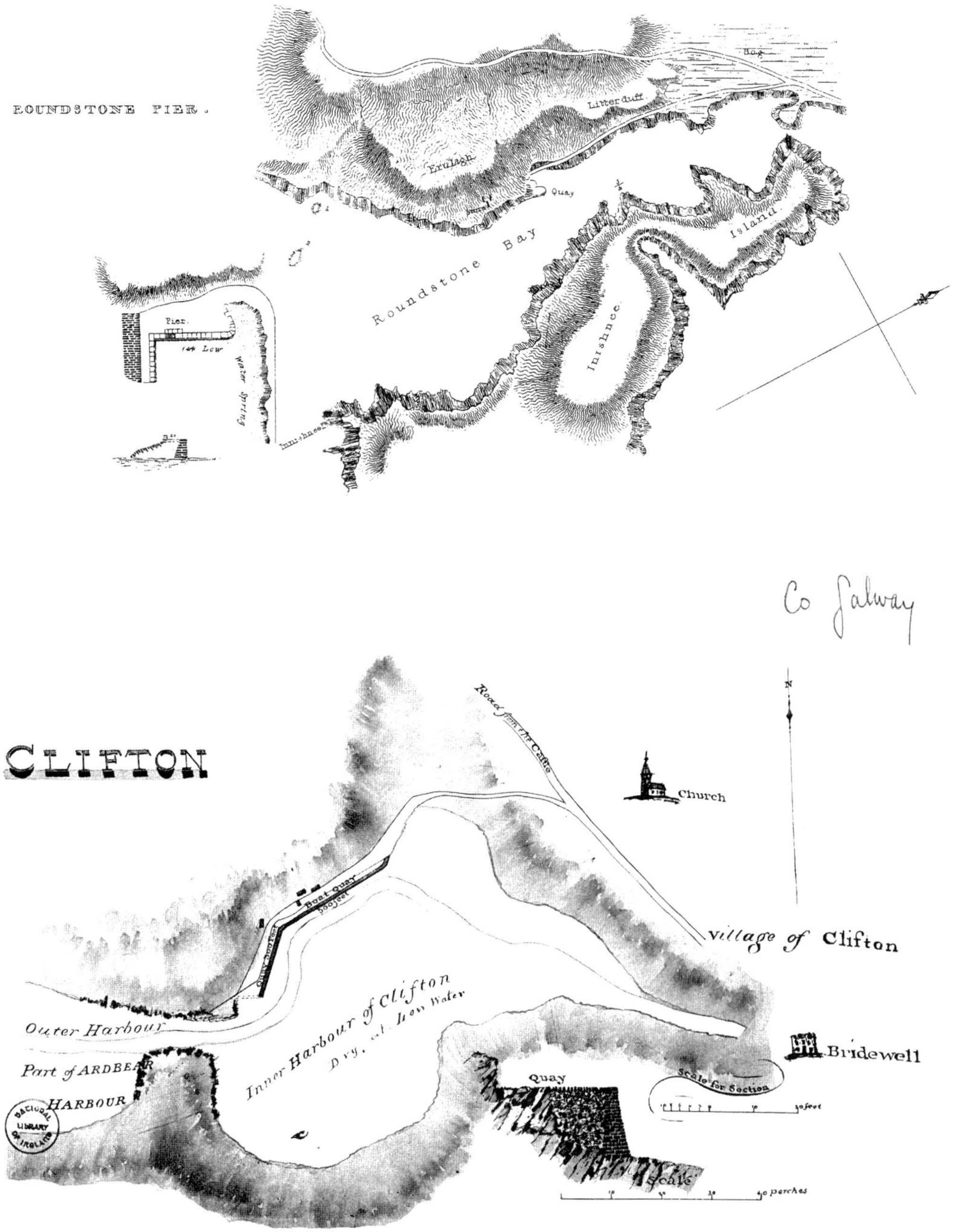

Nimmo's sketch of Roundstone and Clifden Piers
(Nimmo's Report on the Fishing Stations of the West Coast of Ireland)

The pier was still unfinished in 1826, but was said to be of enormous benefit to the rapidly developing town. D'Arcy, however, had deviated from the original plan and consequently the Fishery Board was withholding funds. Nimmo accepted D'Arcy's reason for doing as he did, pointing out to the board the necessity for extending the quay and the advantages it had brought, which were sufficient to convert the village into a town. The construction of the pier was not completed until the 1830s.[17]

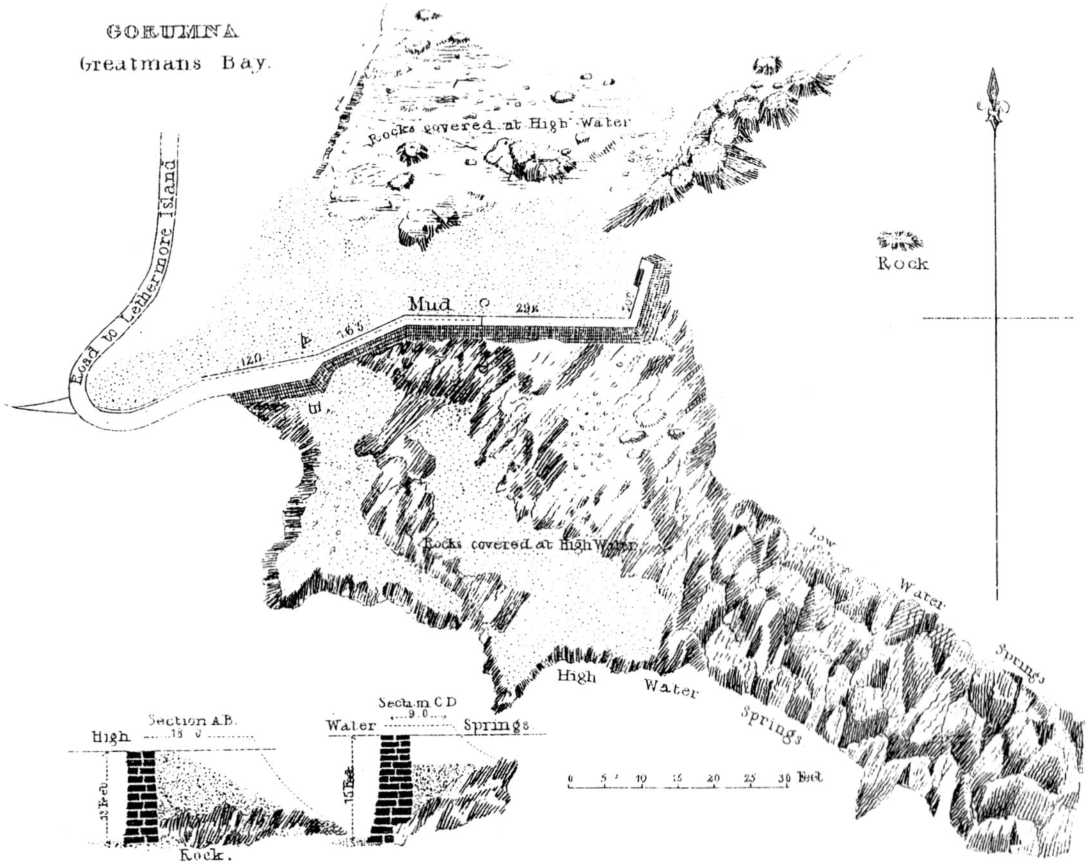

Nimmo's sketch of Máimín Pier
(Nimmo's Report on the Fishing Stations of the West Coast of Ireland)

Máimín Pier, Greatman's Bay

The piers on the Connemara end of Galway Bay were, by 1824, in various stages of completion. Nimmo had suspended works on Máimín pier on Gorumna Island, although it was not fully complete. The distress in the area had passed and he suspected that there was 'some attempt to enhance the expense of the labour performed there'. He decided to suspend the works until an extension of the new coast road could be brought to the neighbourhood to facilitate easier access for inspection.[18] However, in Nimmo's opinion, even in its unfinished state it formed 'a valuable place of shelter for the boats of the bay in their voyages to and from Galway, or to lay them up in'.[19] Several landowners, fishermen and some merchants of Galway disagreed, and in November 1825 they petitioned the chief secretary to have the pier removed. They complained that the unfinished pier was dangerous. Winter storms had washed many of the large stones into the sea, destroying a natural pool where 'craft could formerly lie secure'. King, the Fishery Board inspector, investigated the site and decided that it would cost less to finish the pier than to remove it. But his advice was ignored and the pier was abandoned by the Fishery Board and was almost destroyed before funds were eventually raised from charitable bodies to rebuild it.[20]

Rossaveel, Spiddle and Barna Piers

There was still no road connecting Rossaveel pier on Cashla bay with the new coastal road to Galway. It was, however, already proving convenient for vessels trading with the town and in 1824 was being used by almost one hundred fishing boats.[21] Donnell agreed that the pier was 'very useful to the fishery of this district', especially during the herring season, but he considered it to be 'insecure, from imperfect workmanship', and a contractor was hired to correct the faults.[22]

Other sites that received government funds in 1822 were Inveran, Spiddle, Furbo and Barna. Improvement works were carried out at Inveran, Furbo and Barna and a new pier was erected at Spiddle. But Spiddle village was slow to develop as the proprietor, named Cumming, was a minor and this made leases difficult to obtain.

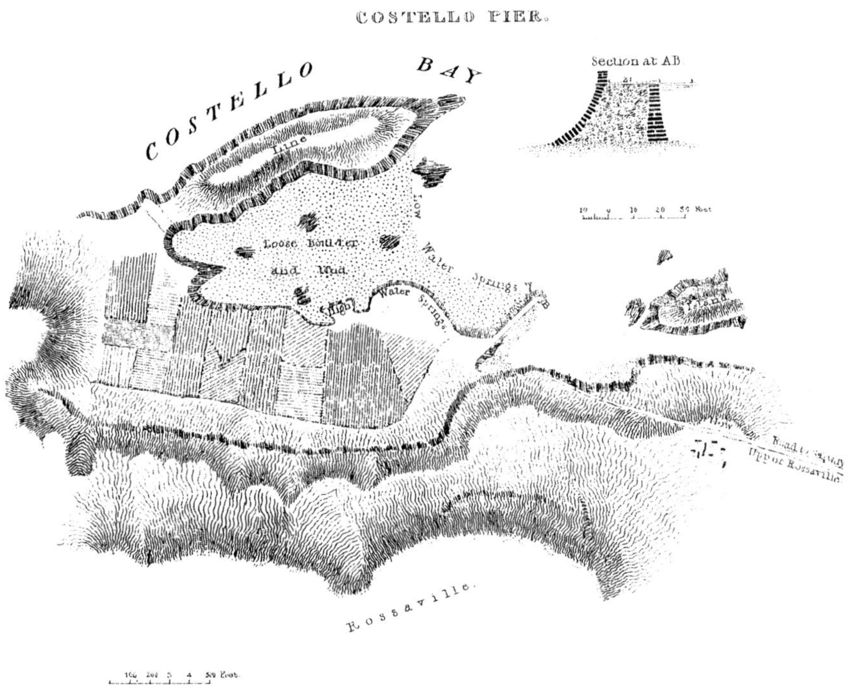

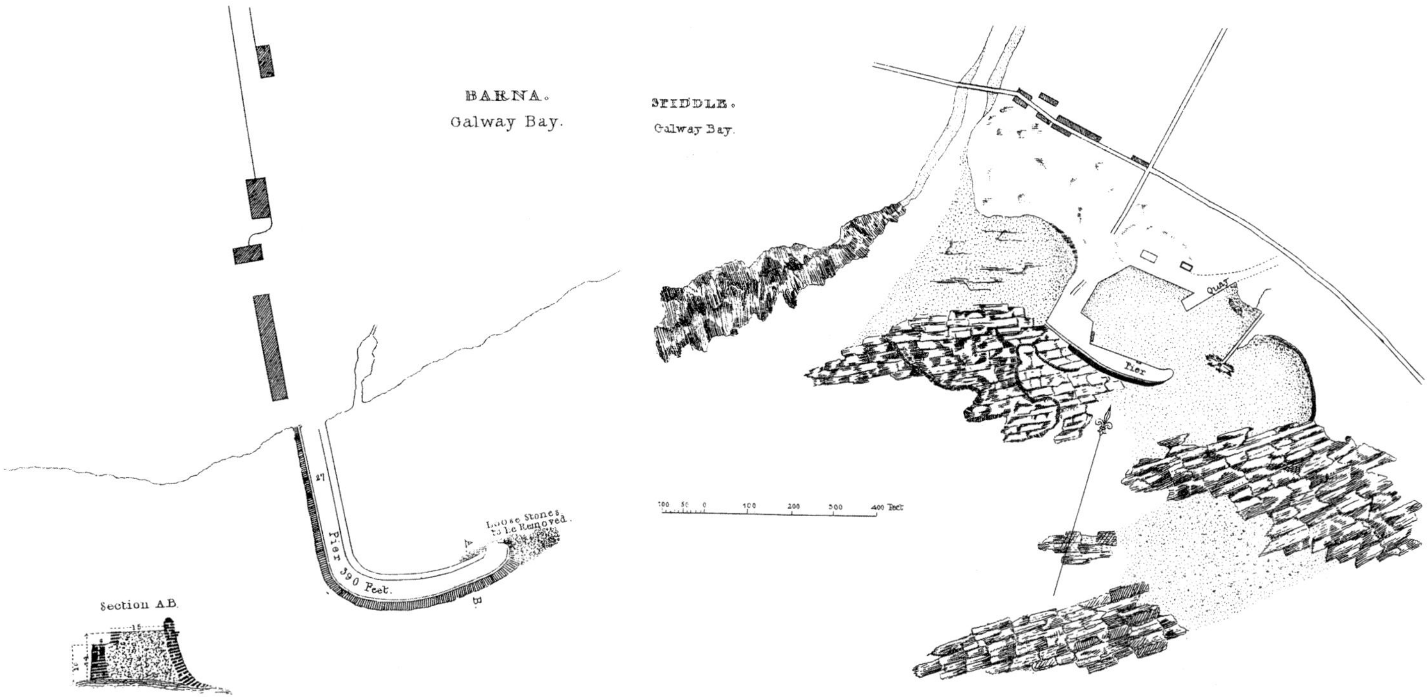

Nimmo's sketch of Rossaveel, Spiddle and Barna Piers
(Nimmo's Report on the Fishing Stations of the West Coast of Ireland)

Galway and the Claddagh

When writing about Galway harbour, Nimmo took the opportunity to outline his suggestions for improvements to the Slate quay, the town quays, the harbour, and the construction of a canal between Lough Corrib and the sea. The fishery at Galway employed a large section of the population, more than in any other town on the west coast and the fishermen almost all resided in the Claddagh, 'a beach on the west side of the river'. The Claddagh fishermen were an extremely close-knit community with strong traditions and practices and were frequently in conflict with the other fishermen of the bay. They had, what Nimmo described as, 'sundry peculiar customs', and were in the habit of forming laws and regulations for the conduct of the fishery, and enforcing these on the other boatmen of the bay.[23]

Nimmo's sketch of Claddagh Piers
(Nimmo's Report on the Fishing Stations of the West Coast of Ireland)

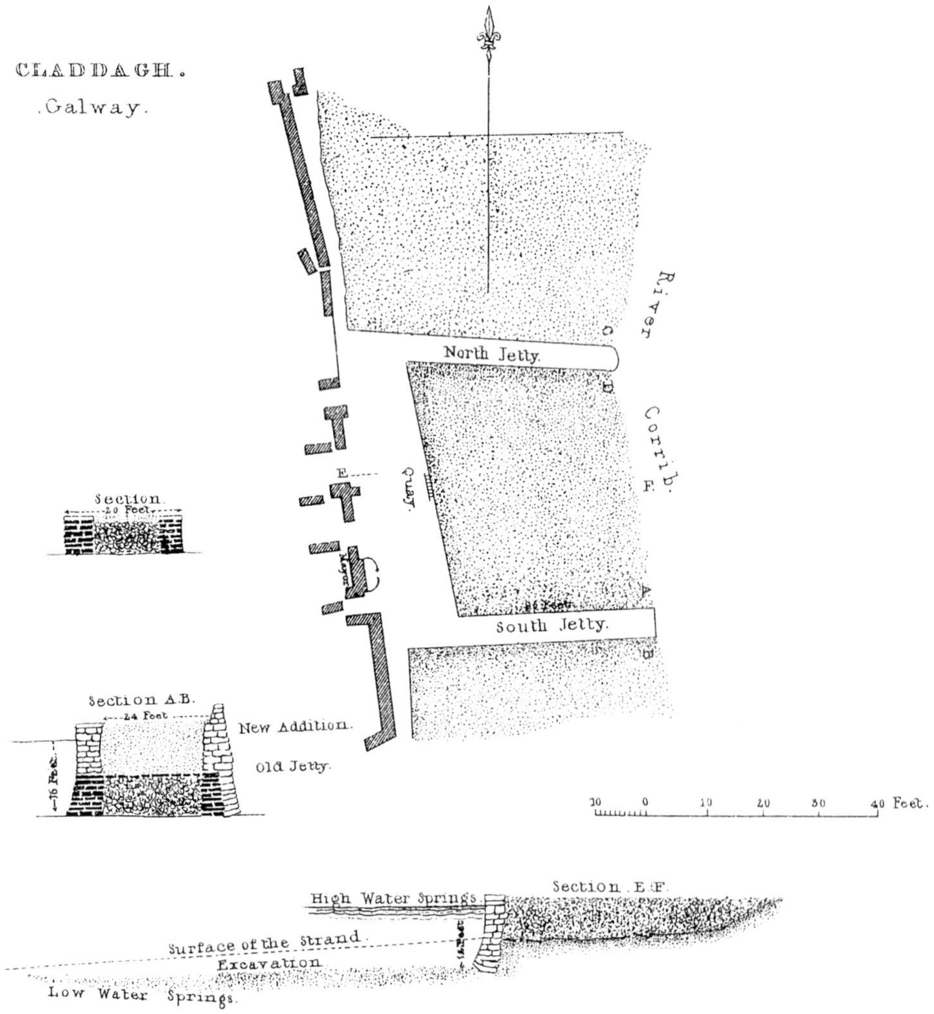

The board, under pressure from the merchants of Galway, had sent a gunboat, *The Plumper*, to try and bring order to the bay in 1820. Some of the merchants of the town had set up a fishing company and had fitted out several boats to fish the bay. This had angered the Claddagh men, who saw this as 'an infringement' of their rights and subsequently, 'attacked the company's boats, destroyed their nets, cut their sails and cables, threw their anchors overboard, and ill-treated the crews'.[24] Advised of the events taking place in the bay, the Fishery Board came to the conclusion that the Claddagh men were controlling the fishing rights and prices in the area and preventing healthy competition, 'to save their own labour'. They had stationed boats 'full of young men armed with stones and clubs to destroy the nets and sails, and assault ... industrious fishermen who might attempt' to fish more often or at times or places not specified by the 'monopolists'. This resulted in no capital being spent and no improvements being made to the fishery in Galway. The following year the board was advised that, thanks to the presence of *The Plumper*, the monopoly had been broken and the 'industrious fishermen' were now free to compete, 'which had produced a supply of fish beyond all expectation'.[25] Three hundred boats, each containing from ten to twenty thousand herrings, lined the quays each day and cheap food was available to the town and neighbourhood:

> while all the roads and avenues for forty miles around are crowded with countless hundreds of sotters and cadgers carrying the fish of Galway to the farthest extremity of Ireland, and promoting through the whole interior, a circulation of industry, money and human food, truly astonishing.[26]

When Nimmo arrived to open public works in Galway, in 1822, there were already two narrow jetties across the Claddagh beach, providing the fishermen with easy access to their boats, but these were in need of improvement.[27] The old town quays were on the opposite side of the Corrib River and the Galway Chamber of Commerce had already petitioned the lord lieutenant for aid to improve these and to cut a canal to Lough Corrib. The lord lieutenant wrote to Nimmo requesting his opinion. The need to make Lough Corrib navigable to the sea had already been suggested by Nimmo in his bog report and the merchants of Galway were at this time pressing him to adopt the project as public works. Nimmo strongly supported the application and he informed the chief secretary that the work would cost between £20,000 and £30,000 and would take several years to complete. He began immediately to prepare plans and estimates, but in the meantime he recommended that aid be granted to improve the jetties on the Claddagh beach, estimated at £925, of which the Fishery Board had agreed to pay half. He also proposed to build a rough pier at the end of the Slate, a reef in front of the quays.[28] He drew up a plan and specification for improving the jetties, but the Fishery Board, already in dispute with the Claddagh fishermen, withdrew their financial support and nothing further was done. In the mean time Nimmo began constructing a pier on the Slate rock, which would forever afterwards be known as Nimmo's Pier.[29]

Slate pier [Nimmo's Pier] was nearing completion in 1826 and was not only an asset to the fishermen, but also offered protection to the old quays against southerly and south westerly gales, which formerly wrecked vessels lying there. Any further improvement of the quays, and the construction of a wet dock, 'though highly desirable for the mercantile interest', would be of no benefit to the fishermen and should, he recommended, be left to private interests. However, he did suggest that a short cut in the form of a road or dyke be run out across the marshy creek that separated Claddagh village from the Slate pier. This could be used as an additional quay and behind it a dock could be formed, 'in which all the fishing boats of Galway could be laid up in complete security.[30] Donnell considered the piers at the Claddagh, including the Slate pier, to be 'ill built and unfinished'. He put forward plans and specifications for 'rebuilding and completing the present imperfect piers and quays' and the Fishery Board passed these in 1828, and work was expected to begin shortly thereafter.[31]

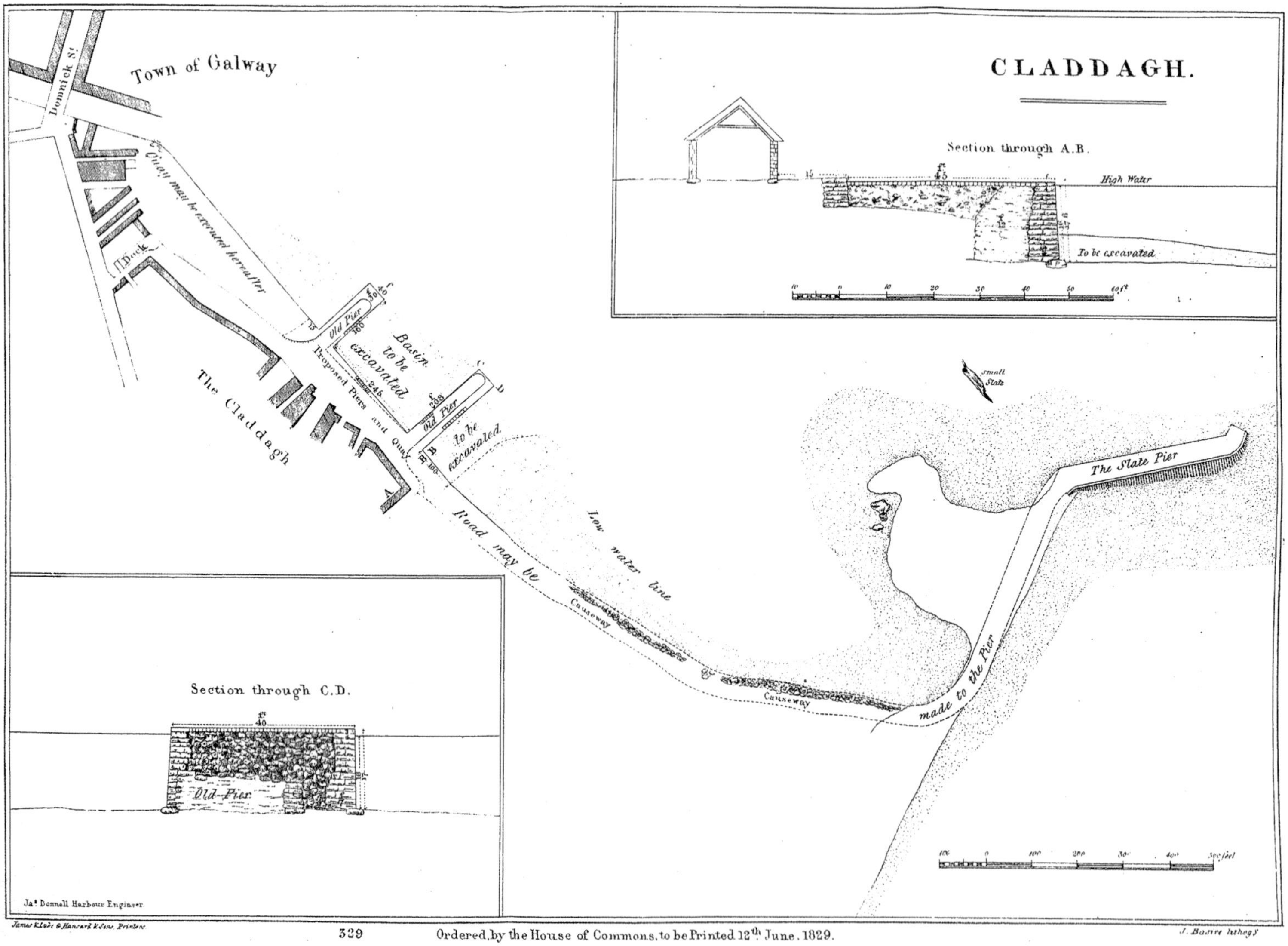

James Donnell's sketch of the Claddagh Piers, 1829
The Slate quay, (the stretch leading to the pier where the words 'made to the pier' appear on the above map) was completed at this time, the remainder of the causeway linking the Claddagh with the Slate Pier was not constructed until later.
The First Edition 1 inch Ordnance Survey map for Galway (1841) names Slate Pier as Nimmo's Pier and it has remained so ever since.
(Appendix 10, Tenth Report of the Commissioners of the Irish Fisheries)

Nimmo made further suggestions for improvements to the harbour and continued to press the importance of building the Corrib canal, but added that he did not feel his fishery report was an appropriate place to discuss such a project.[32] Elsewhere, he stated the opinion that, while other local committees were finding ways and means of acquiring funds for improving their harbours, the Galway merchants lacked the 'spirit' to take on the project themselves: 'The town of Galway, which is always crying out for improving their harbour, has never yet had the spirit to do any thing by its own means'.[33] However, as trade in Galway improved in the 1820s, the merchants and politicians of the town were actively seeking ways of implementing Nimmo's plans for the new docks and canal.

They drew up a bill to be brought before the House of Commons, seeking a loan for the works, and a Harbour Commission was formed in 1830 to promote the development. Other engineers would later alter Nimmo's plans for both the docks and canal, before their eventual construction. Hamilton Hartley Killaly was employed by the Galway Harbour Board in 1832 and he draw up the plans that were eventually used in the construction of the Commercial Docks. Work on the docks was completed in 1842.[34] Some years later, the Commissioners of Public Works under the direction of the County Surveyor, Samuel Usher Roberts, constructed the Eglington Canal, linking the Corrib with the sea. The canal was started in 1848 and officially opened by the Earl of Eglington on 28 August 1852.[35]

Rinville and Ballynacourty Pier

Donnell did not share Nimmo's optimistic assessment of the six other piers on Galway Bay. New Harbour [Rinville] did not have a village attached to it, nor a road leading from it, and it was slow to attract the traffic anticipated by Nimmo. Nimmo described it as, 'a beautiful and safe harbour for small ships', frequently used by vessels, 'especially for careening or wintering'. The works were still in progress under a contractor when Donnell visited the site in 1824. He had 'some defective parts taken down and rebuilt', but was still not satisfied when he visited again in 1826: 'there are some parts of the work shaken and disturbed, and a few of the ashler closures have been washed out'. The board eventually sanctioned funding to rectify the damage.[36]

Ballynacourty, on the estate of Walter Blake, was built to accommodate the fishermen of Ballynacourty village. It was a landing place for the sale and curing of fish and a place of shelter for boats of the bay. Begun in 1822, it was partly finished by 1824, but work was at a standstill as funds were exhausted. Funds were later provided and the work continued, but, in 1826, Donnell was recommending more repair work to make it safe and secure.[37]

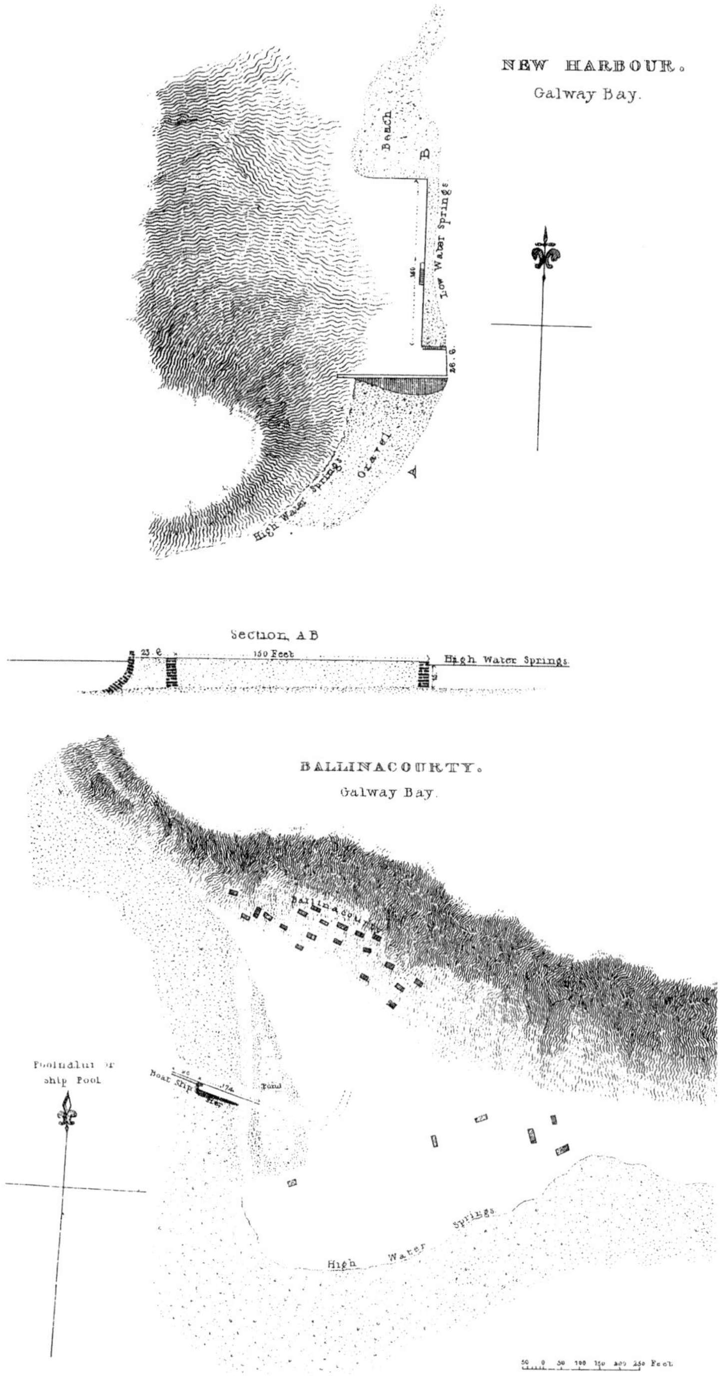

Nimmo's sketch of New Harbour [Rinville] and Ballynacourty Piers
(Nimmo's Report on the Fishing Stations of the West Coast of Ireland)

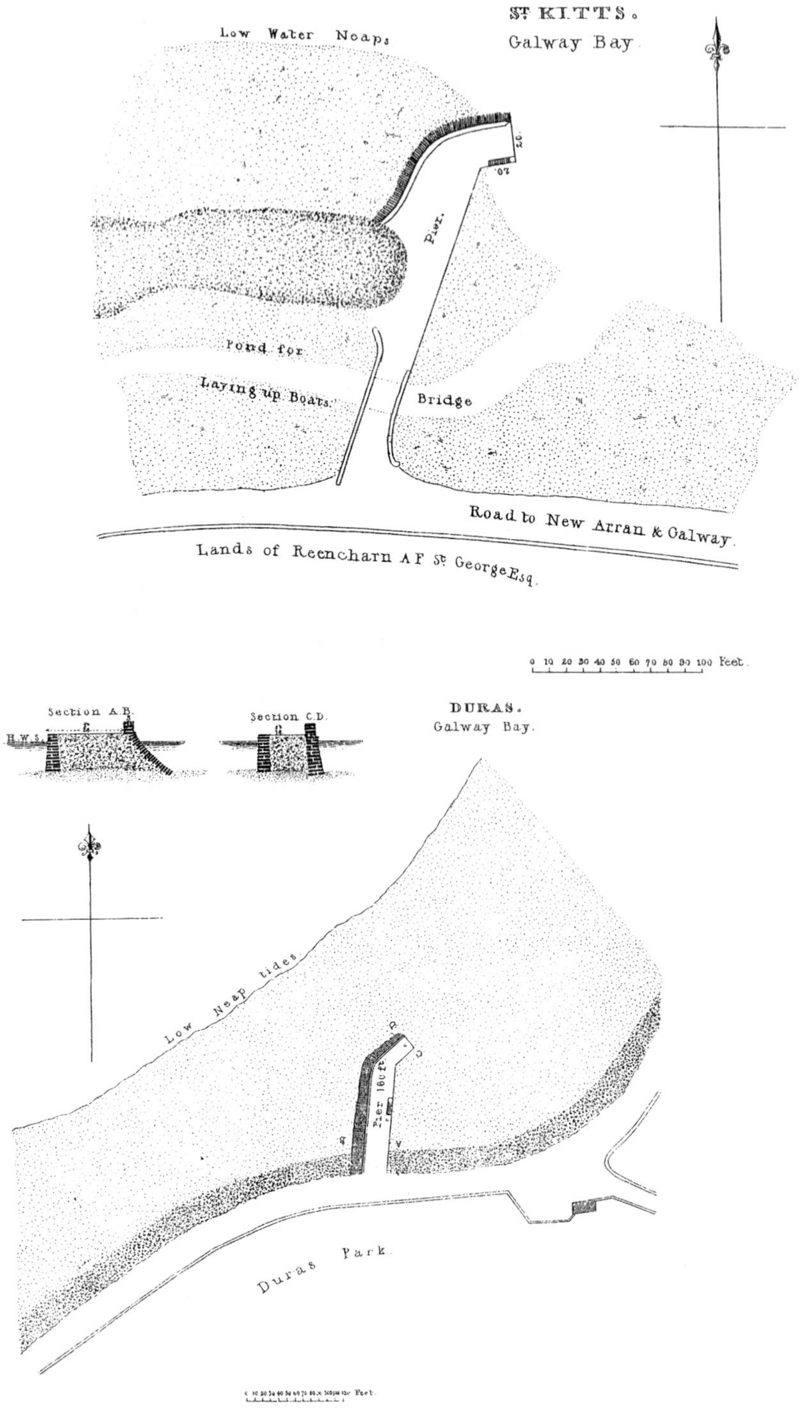

Nimmo's sketch of St Kitts and Dooras Piers
(Nimmo's Report on the Fishing Stations of the West Coast of Ireland)

St Kitts Pier, Killeenaran

Neither was there a road to St Kitts pier [Killeenaran], situated two miles beyond Tyrone House, home of Arthur French St George. Nimmo, in his original plan, designed an interesting feature for this pier, which can still be seen today: 'at the upper end is a small bridge, through which boats pass into a pool behind the quay, covered from the sea by a gravel bank, where they may be laid up in complete shelter'.[38] On his tour of inspection in 1826, Donnell found several defects in the pier. Because it had not been raised above high water level of spring tides, it was sustaining damage from the sea breaking against it and 'about fifty feet of the land end of the pier was not built at all', leaving the pier completely isolated at flood tide. Donnell was reluctant to suggest that the board supply the sum needed for the repairs, as it was regularly used for exporting turf and he was of the opinion that those who got the most use from it should supply a portion of the funds.[39]

Dooras and New Quay Piers

The pier at Dooras was contracted to the proprietor of the site, Patrick M. Lynch, who contributed one half of the cost. The work was completed by 1823 and for once both Nimmo and Donnell were in agreement, both declaring that the pier was neatly and substantially built. When Donnell visited the site in the summer of 1826 he found a tablet in the parapet with the following inscription: 'This pier was erected by Patrick M. Lynch, Esq. Of Durass Park, A.D. 1823'. The board was none too pleased as it gave the impression that it was a private enterprise, built without any assistance from the Fishery Board.[40]

At New Quay, Nimmo laid out his quay close to the ruins of two earlier quays. The work was begun as day labour but was suspended when funds ran out. It was eventually finished by contract with funds obtained from the board. A great deal of business was done here and some good houses were erected on the shore. Donnell also found the quay busy and well used: 'a substantial and useful work, but sometimes too small to contain the numerous craft seeking shelter in it'.[41]

Killeany Pier, Inishmore, Aran

Killeany pier on the east end of Inishmore was built to facilitate the fishing village of Killeany, 'on the estate of the Rev. John Digby'. Work began as day labour in 1822 to Nimmo's plan, but funds ran out before it was completed. It was later finished to Donnell's plan and by 1829 he noted a further increase in the number and quality of the boats to be found there.[42]

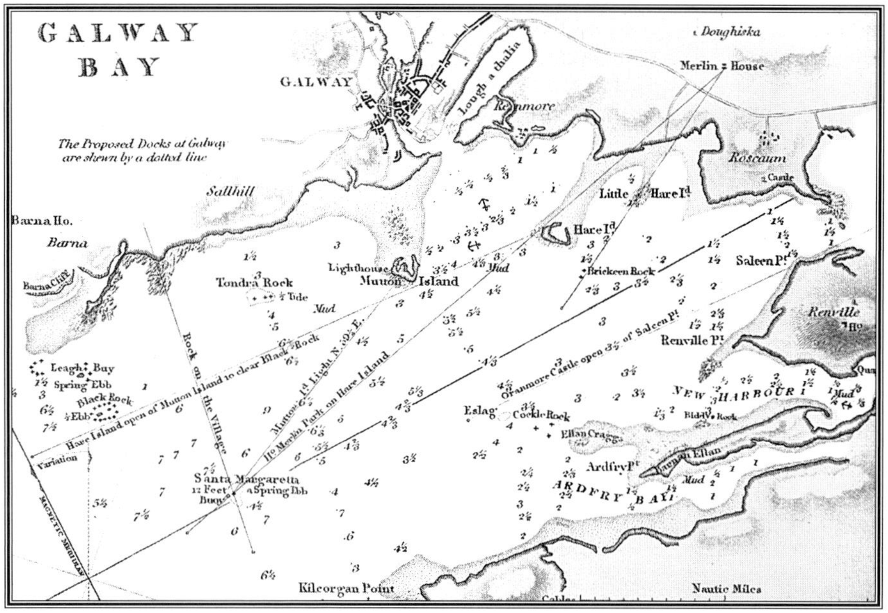

Nimmo's chart of Galway Bay circa 1831
(*Chart of St. George's Channel and the coast of Ireland*, by A. Nimmo)

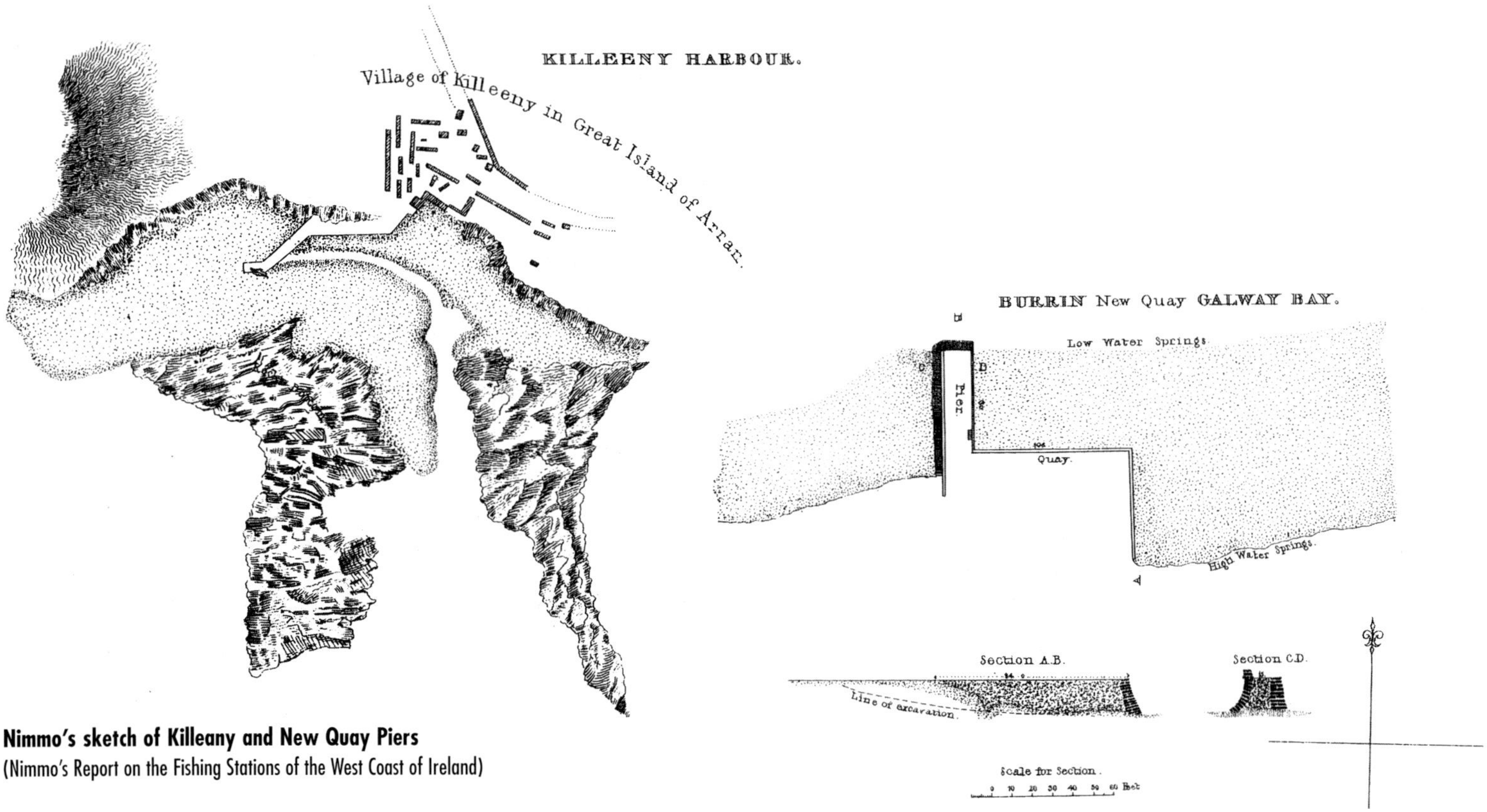

Nimmo's sketch of Killeany and New Quay Piers
(Nimmo's Report on the Fishing Stations of the West Coast of Ireland)

Assessing the new roads

Nimmo's assessment of the many government roads under construction along the coast gives an indication of the extent of the public works still ongoing in the west in the mid 1820s. In Sligo, the new road from Bundroes, near Bundoran, to Manor Hamilton, continuing on to Carrick on Shannon, was ongoing. Developments begun by the Grand Jury and landlords of Mayo were given a tremendous boost by the injection of government funds under Nimmo. Killala was now 'a town of some consequence', and exported a good deal of corn and was a place of rendezvous for the fishermen during the herring season. However, Ballina would soon surpass it, due to an increase 'in trade and population', attributed to the improved access to the town carried out under Nimmo.[43]

The road from Killala to Ballina, Foxford and Swinford would soon be completed. Work was under way on the Castlebar to Belmullet road. Work on the North Coast Road (Killala to Belmullet) was suspended and work on other coastal roads was sporadic. However, on completion these would form a ring road from Ballycastle to Broad Haven, south along the coast to Newport, and on to link with the Castlebar to Crossmolina road, which was being extended north to Ballycastle. From this road branches ran into the Mullet peninsula, down the east coast of Broad Haven and Blacksod Bay, through the peninsula of Curraun and across Achill Island. A ferry linked the island with the mainland and Nimmo recommended two small ferry piers at the place where the road from the Curraun peninsula and the road through the island were brought to the shore. He considered the Curraun site [Achill Sound] to be the natural place for establishing 'a village or sea-port town'.[44]

The improved roads along the coast of Clew Bay were expected to revitalize the fishery industry along that stretch of coast. The towns of Newport and Westport were already well provided with quays; however, the new roads meant that fish could now be easily transported to the markets of the interior. Lord Sligo of Westport House had done much to

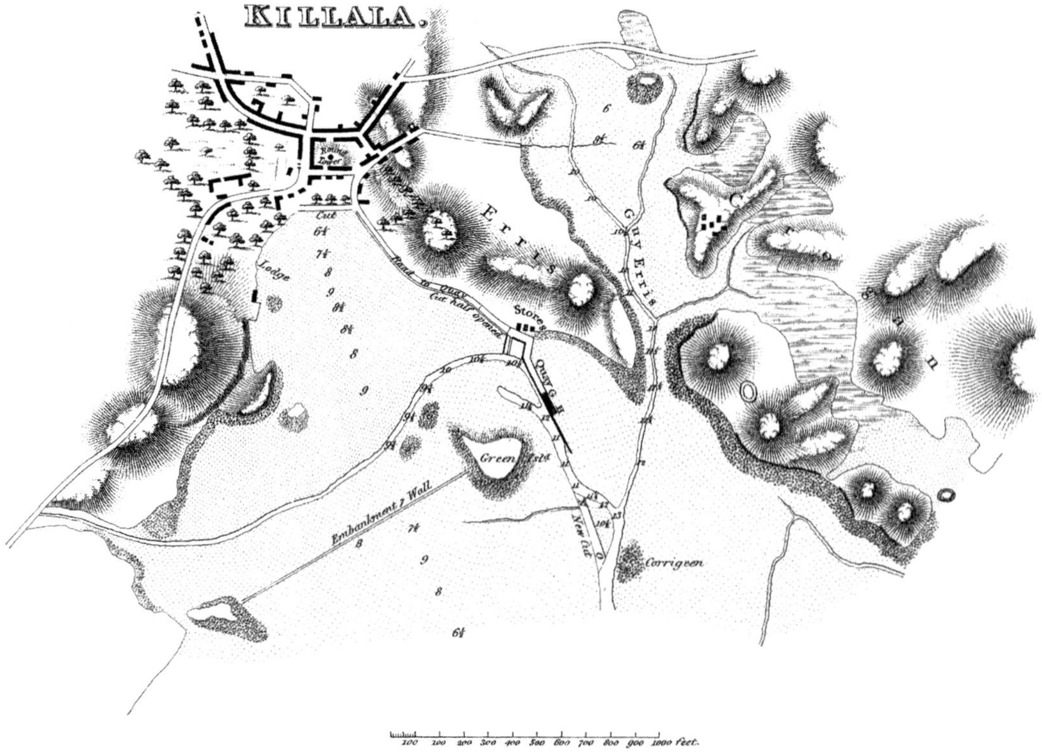

Map of Killala circa 1824
(Nimmo's Report on the Fishing Stations off the coast of Connaught)

assist the development of the district by encouraging his tenants to make small roads through their farms and by contributing towards the erection of the piers at Old Head and Bundorragha. He had also made a carriage road, through the mountains, from Westport to the village of Bundorragha on the Killary. The new government line from Westport to Leenaun was nearing completion. It was met at Leenaun by the roads from Clifden and Maum. A good deal of employment was to be had in this district with the convergence of new roads and the erection of two piers.

Road construction on the west coast of Connemara was progressing more successfully than in the east, although few were in the advanced stage that Nimmo would have us believe. According to Nimmo, a very good road from the north coast was opened, 'round Ballynakill Bay by Streamstown to Clifden', on by Ardbear to Ballinaboy, 'across the moor to the head' of Bertraghboy Bay. Here the road came to a stop as the bridge at Toombeola was not yet built. East of Toombeola, work was under way from Cloonisle all along the line to Cashla Bay and from there to Galway had been improved. But none of this was anywhere near fit for use and the two sections of road could not connect until the bridge at Toombeola was erected. At Ballynakill a branch road was taken to the quay at Derryinver and the road from Streamstown to Cleggan pier was near completion. The old road, along Mannin Bay to Bunowen, was improved. Other branch lines under construction, but as already seen not yet completed, were leading off to the piers at Roundstone, Gorumna, Rossaveel and Spiddle.[45]

Nimmo made a pier of a part of the road at Cloonisle where it came close to the shore, believing that it might be needed sometime in the future, 'especially for the shipment of the beautiful green marble from the Twelve Pins Mountains'. A marble quarry had been recently opened on the Martin estate, which 'promises to introduce a new source of employment into Conamara' (sic).[46] The crossroad from Spiddle to Moycullen was completed and taken to Knockferry on Lough Corrib. Nimmo's planned crossroad from Kilkieran Bay to the head of Lough Corrib was underway, but progress was slow due to difficulties on the central road.

The effects of the various public works on the general prosperity of the Galway coast had surprised even Nimmo himself: 'the result has far outrun what I had anticipated. Trade and commerce have been introduced into the extremity of Conamara, capital has accumulated in a surprising degree, and with it enterprise and exertion in the fisheries, in agriculture, and in foreign trade; cargoes of country goods, salt, timber etc., are now imported direct into Conamara'.[47] But not everyone was happy with the extension of roads into the heartland of Connemara. According to Hely Dutton, many of the under tenants, 'who hold immense tracts of ground at little rent', were complaining that Connemara had 'gone to the devil since good roads were made; the value of land [was] too well known. Villagers in general [did] not wish for good roads to their villages; they [said] it only encourages others to bid over them'. No stronger argument could be put forward in favour of good roads, he wrote, as it clearly showed their effect on the value of land.[48]

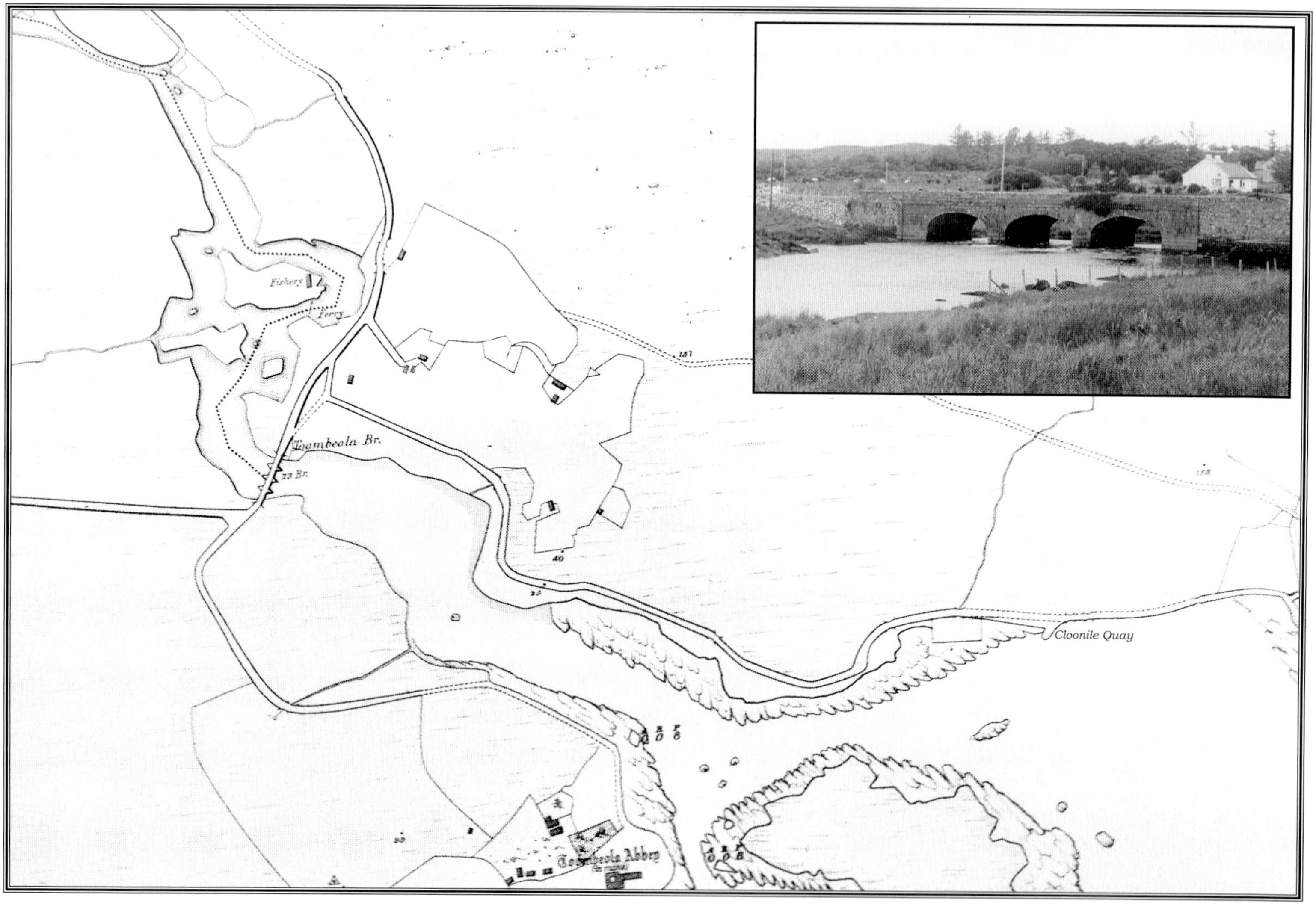

Toombeola Bridge and Cloonisle Quay
(First Edition 6 inch Ordnance Survey Map circa 1839)
Inset: Toombeola Bridge 2006

Nimmo concluded his coastal survey at Black Head on the coast of Clare, giving no indication that it would not be continued southwards to include the remainder of the coast from Clare to Dublin. In a revision of the board's funding and expenditure in 1826, the commissioners decided not to resume the survey. The termination of the coastal survey brought Nimmo's term with the Fishery Board to an end. Just two thirds of the coast had been surveyed, from Dublin northwards to Black Head in County Clare. In Nimmo's opinion the greatest benefit to be derived from the coastal survey would be for the board to publish proper charts of the west coast. A chart of Connaught was prepared, along with detailed charts of its harbours, in particular those of Sligo and Galway. These were sent to the engravers in London for publication, but copies have to date proved impossible to locate.[49] Scaled down versions of Nimmo's charts of the bays and harbours, along with

ACCOUNT of the Total Number of Decked and Half-decked VESSELS, together with Undecked or Open Boats of every description, registered and employed in the Fisheries; also their Tonnage, and Number of Men employed thereon,

STATIONS. In the year ended	Decked Vessels.			Half-decked Vessels.			Open Sail Boats.		Row Boats.		Total Number of Men employed.
	No.	Tonnage.	Men.	No.	Tonnage.	Men.	No.	Men.	No.	Men.	
5th April 1825.											
Westport	7	156	32	1	17	4	133	532	1,293	5,172	5,740
Galway -	8	236	52	10	130	50	566	2,830	1,327	5,308	8,240
5th April 1826.											
Westport	2	37	8	-	-	-	84	336	969	3,876	4,220
Galway	8	236	52	10	130	50	600	3,000	1,390	5,560	8,862
Clifden -	13	581	52	-	-	-	99	306	818	3,272	3,630
5th April 1827.											
Westport	3	72	18	-	-	-	116	530	1,061	4,621	5,169
Galway	13	304	65	4	82	20	454	2,270	1,150	4,600	6,955
Clifden	-	-	-	16	306	48	149	447	976	3,904	4,399

Table giving the number, type and tonnage of boats working in the Westport and Galway fishery stations from 1825-1827
Clifden Station was established in 1826. (Sixth, Seventh and Eighth Report of the Commissioners of Irish Fisheries)

other surveys conducted by him for the Fishery Board and others, were eventually included in his chart on St George's Channel and the coast of Ireland, prepared for the Corporation for Improving the Port of Dublin (the Dublin Ballast Board). New piloting directions for navigation of the coast, written by Nimmo in 1831 and published the following year, accompanied the chart.[50]

The coastal survey, the Fishery Board announced in 1826, was abandoned in favour of a more detailed survey to be undertaken by the Admiralty. The Board had decided instead to direct their funding towards the continuous construction of fishery piers and quays.[51] The number of boats and men employed in the fishery was increasing annually and the board was anxious to build even more piers along the west coast to further advance the industry. However, the board was still having difficulty raising the required local contribution from the impoverished districts.[52]

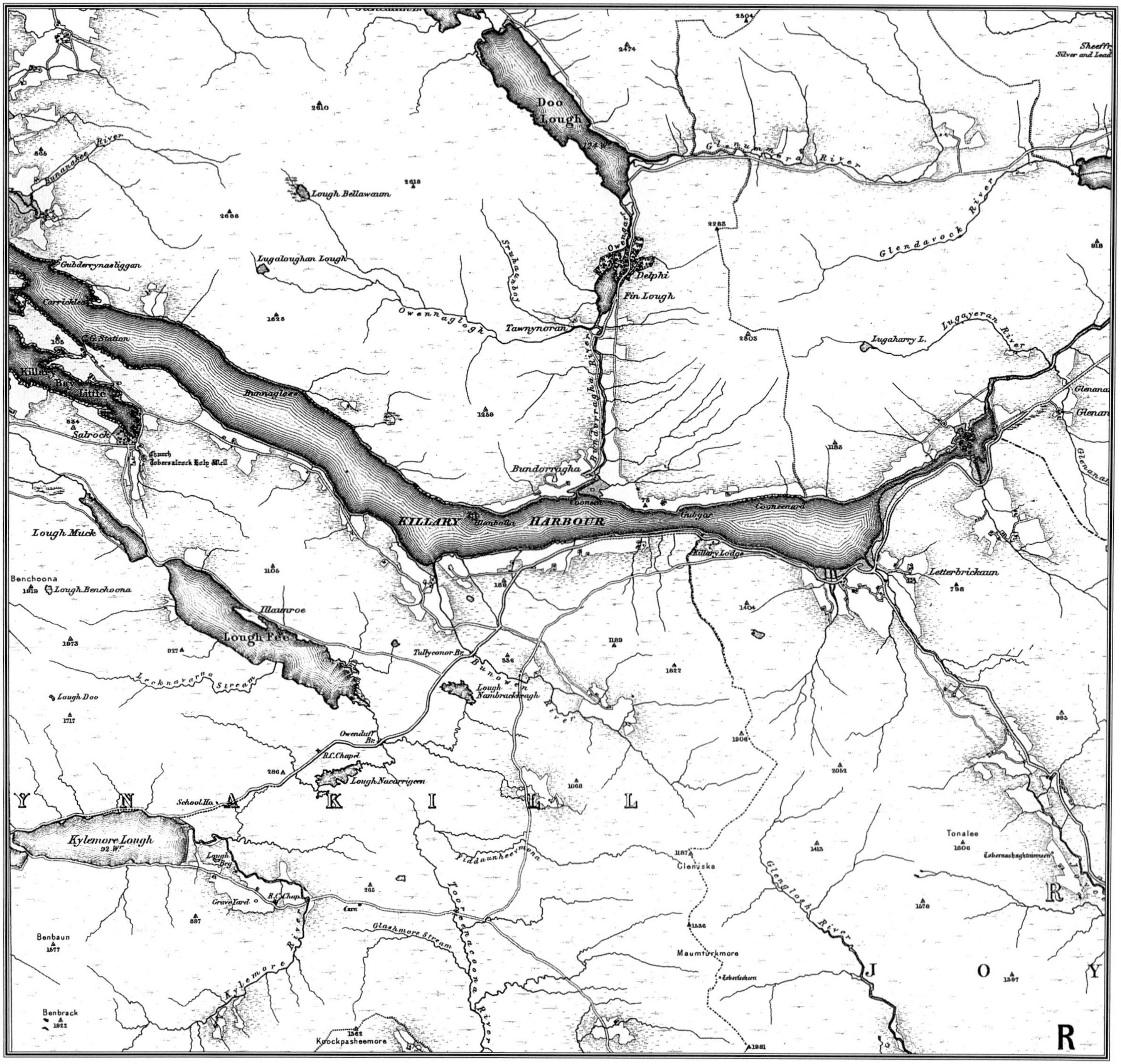

Killary Harbour roads circa 1862
(First Edition 1 inch Ordnance Survey maps)

Chapter 5

Nimmo's private enterprise and new town development

Nimmo still had his home and office at 78 Marlborough St, Dublin. He possessed no other property in Ireland or England, he told a committee in 1824. He frequently visited England and travelled to London on business but had never resided there.[1] However, in the ensuing years, Nimmo did increase his holdings in Ireland when he erected several buildings in the west and took possession of almost two hundred and forty acres of land at Roundstone Bay on which he developed a fishing village. The buildings were erected entirely at his own expense, he later told Dublin Castle. Some of these were for his own use or for the use of his staff and workforce, others he rented out to the government. Nimmo does not appear to have intended residing permanently in any of the buildings; they were a commercial venture, constructed to serve the needs of the day and to make a profit. And the way in which he rid himself of them later would indicate that he had no emotional attachment to them. Strangely enough, Nimmo would seem not to have considered his enterprise in the west as in any way smacking of jobbery and no one seems to have suggested so at the time. However, it was, to say the very least, opportunistic.

Leenaun

At Leenaun he built 'a lodge' to accommodate the workforce in that area. This was later (1831) given over to a tenant, who converted it into 'a billeting place for travellers passing to the western coast'.[2] He also built a store, office, carpentry and smithy workshops, which he rented to the government for three years, from June 1828 to June 1831, at £30 per annum.[3] The exact location of these buildings is difficult to ascertain, Maria Edgeworth, who stopped here during her tour of Connemara in 1833, makes no reference to them. Early maps show various clusters of houses or *bailes* in this district but it would be some years before the coastal village of Leenaun would develop.

Maum

At Maum, in the townland of Moneenmore, where the Bealanabrack and Failmore rivers enter Lough Corrib, Nimmo built Corrib Lodge. It is difficult to know exactly what year this building was erected; the bridge in front of the Lodge was not finished until 1829, and the first mention of the building is to be found in Nimmo's Progress Report for 1830.

In the report, Nimmo describes the building as a hotel. However, both Maria Edgeworth and William Wilde later describe it as a residence and pay office, and while it may have been used as such in the early days, Nimmo's intentions for the building were very clear from the start: 'At the head of Lough Corrib I have built and furnished at my own expense a comfortable Hotel for the accommodation of travellers who might be called into these hitherto secluded districts'.[4] Maria Edgeworth also stopped here in 1833; at the time it was in the hands of Nimmo's Scottish servant, who had accompanied him on his visits to Edgeworthstown: 'Corrib Lodge is a neat, bleak-looking house, which Mr Nimmo built for his own residence when he was overseer of public works in these parts, but dying, as usual with public works overseers in Ireland, in debt, his lodge was turned into an inn'.[5] There is no indication that Nimmo died in debt, although he does not appear to have accumulated the wealth one would expect from such a busy professional life.

Wilde, writing in the 1860s, laments the passing of the former proprietor, who was then a man named Rourke, and remarked on the suitability of the hotel, then re-named Maam Hotel, for anglers and other tourists:

> The Maam Hotel – where good accommodation and conveyances can always be obtained, and which in the days of Rourke, its original proprietor, became attractive, not merely as a summer residence, but on account of its good cheer – ... and, if well managed, would be a most attractive resting-place for tourists, as, independent of the beauty of the scenery, and the interest attaching to the neighbourhood, and its admirable trout and pike fishing, it is within four miles of the main road between Galway and Clifden, upon which several conveyances ply daily.[6]

Corrib Lodge and Nimmo's bridge at Maum circa 1842
(*Ireland: its scenery, characters &c.*, by Mr & Mrs S. C. Hall)

Developing towns and villages

In the years to come, villages evolved on some of the sites proposed by Nimmo, but not on all. Nimmo had hoped that new villages would develop at Derryinver, Cleggan and Spiddle, but the only new villages advancing at this time were Belmullet and his own village at Roundstone. It is interesting to note that the Martins, the most renowned of Connemara landlords, established no towns or villages on their Ballynahinch Estate, although Nimmo had pointed out several suitable sites.

From the outset, Nimmo was in favour of establishing towns to encourage the development of industry and commerce in backward regions. Towns, he argued, helped to win people away from their dependency on agriculture and the land:

> Generally speaking, it strikes me we should now endeavour to congregate the surplus population of Ireland into towns, where they could be useful to each other, rather than have them so much scattered over the country. The towns in Ireland bear a very small proportion, in respect of population, to the towns in England; and the industry, in like manner, is less.[7]

For this reason he welcomed the introduction into Ireland of the cotton industry over that of the linen industry. The cotton industry removed the worker from the land and brought him to reside in towns, creating an intermediate class between the proprietor and the agricultural cultivator. In contrast, the linen weaver was 'almost universally a farmer; and therefore, however prosperous his manufacture may be, it [was] only to him a means of enabling him to pay more rent than before for his land'.[8] Donnell was of the opinion that villages would develop at all piers where the site accommodated it and where there was no legal impediment. The creation of towns and villages would, he told the Fishery Board, 'render men at present but partially civilized, amenable to the operations of the laws and subservient to the better arrangements of society'.[9] In 1826 he pointed to three villages in particular where the construction of piers had greatly assisted their development:

> [Killeany, Inishmore,] is one of an existing fishing village, requiring an increase of accommodation on account of an increase in number and size of the boats. The harbour already has had the effect of increasing exertion, by giving increased confidence for venturing out to sea, and by decreasing the anxiety and labour of the fishermen when returned on shore.
>
> At Roundstone, a wild district in Conemara a fishing village had scarcely been marked out, when twenty-two lots were taken for building a good class of houses upon contiguous to the pier; and Clifden in Conemara, which six years back did not contain six houses, and was the focus of smuggling and illicit distillation, has become an extensive well built town, with commercial stores and other appendages, and actually paid some thousands of pounds excise duties in the last year.[10]

Towns helped the spread of law and order to remote regions. The recent establishment of a coastguard had done much to put down smuggling along the coast and the construction of coastguard stations had in turn contributed to the commercial viability of the emerging towns and villages. The coastguard service was established in 1819 but was not fully operational throughout the country until 1822 or 1823. District commanders were appointed to Killala, Belmullet, Westport, Clifden and Galway and coastguard stations were set up at the following locations:

	Stations:	Chief Officers:	Men:		Stations:	Chief Officers:	Men:
Sligo				Westport			
	Inniscrone	-	6		Innisturk	1	4
	Pullagheeny	1	7		Old Head	1	11
	Pollnadivva	1	11		Islandmore	1	9
	Raghly	-	6		Moynish	1	11
	Mullaghmore	1	11		Achillbeg	1	9
					Keem	1	8
Killala							
	Dunkeehan	1	9	Clifden			
	Porturlin	1	6		Inishlackan	1	8
	Belderg	1	10		Mannin Bay	1	6
	Ballycastle	1	7		Cleggan	1	11
	Lacken	1	6		Killaries	-	5
	Kilcummin	1	12				
				Galway			
Belmullet					Isle of Aran	1	8
	Dugurth	1	9		Ballyvaughan	1	8
	Doohooma	-	5		Kilcorgan	1	5
	Ballycovy	-	5		Barna	1	7
	Blacksod	1	7		Cashla Bay	1	7
	Ballyglass	1	5		Recorder's Quay	-	7
	Renown, Tr	-	6		Kilkieran	1	9[11]

Belmullet and Binghamstown [An Geata Mór]

The construction of the piers and the laying down of new roads in the Mullet peninsula prompted William Henry Carter, the proprietor of one-third of Erris, to visit his estate. It was now possible to make the trip from Castlebar to the Mullet in one day, 'about fifty English miles', and the possibilities for further developing the region were becoming increasingly apparent.[12] Carter was only the second member of his family to ever visit the Mayo estate, although it had been in their possession for almost a century. Accompanied by his agent, Rev. Cecil Crampton, and with Patrick Knight acting as his guide, Carter quickly recognized the advantages of building a town on the isthmus of Belmullet and instructed Knight to 'make the necessary plans'.[13]

Carter instructed Knight to build a pier at Belmullet. Nimmo, on viewing the pier while under construction, declared it inadequate for the needs of the town and he convinced the Fishery Board to issue funds, in addition to a similar contribution from Carter, to complete the works to a more satisfactory state.[14] Knight decided to take the task even further and contributed funds of his own (£227.1.1d) to ensure that his pier at Belmullet would be of a similar standard, if not superior to any other on the peninsula.[15] The town developed rapidly and by 1824 boasted a store, several small shops, and almost all of the building lots had been taken. The store was full with oats and barley and two cargoes had already been shipped that season.[16]

Over the years a certain amount of rivalry developed between Binghamstown and Belmullet and officials lined up behind Bingham and Carter to offer opinion on the advantage of one over the other. Knight was, needless to say, very much in Carter's camp and was inclined to downgrade Binghamstown in his book, *Erris In The Irish Highlands*, while at the same time paying tribute to Bingham as an improving landlord. The Fishery Board inspector, William King, was supportive of Bingham's efforts to establish Binghamstown and of Nimmo's choice of site for a pier at Saleen. Donnell, in 1826, praised the location of Saleen pier over that of Belmullet, for the advantage of the fisheries. He felt Binghamstown should be viewed as a supply centre for the fishery and a market for fish. But just two years later he was favouring Belmullet. He was obviously impressed by the rapid growth of Belmullet: 'a few years back there were only some scattered thatched cabins about this place, and a single boat of sixty or seventy tons was found sufficient to take away the whole agricultural produce. At present there is a regular village, which is rapidly extending; most of the houses two stories high, and slated. There are also respectable stores for merchandize imported, including all materials necessary for boat-building, fishing etc.,' and fish and corn were exported. In the previous December a large schooner had taken shelter at the unfinished pier during a storm, without sustaining any damage: 'She was taking in a cargo of corn; of which there is now a considerable export'. A large number of boats were built in the district in 1825 and these sold to fishermen on credit, 'payment being taken in fish, oil, or any other exportable commodity'. Binghamstown was not advancing to the same extent; 'a street, containing many houses [was] built; some public buildings, and a large store were also commenced, but they as yet remain unfinished'.[17]

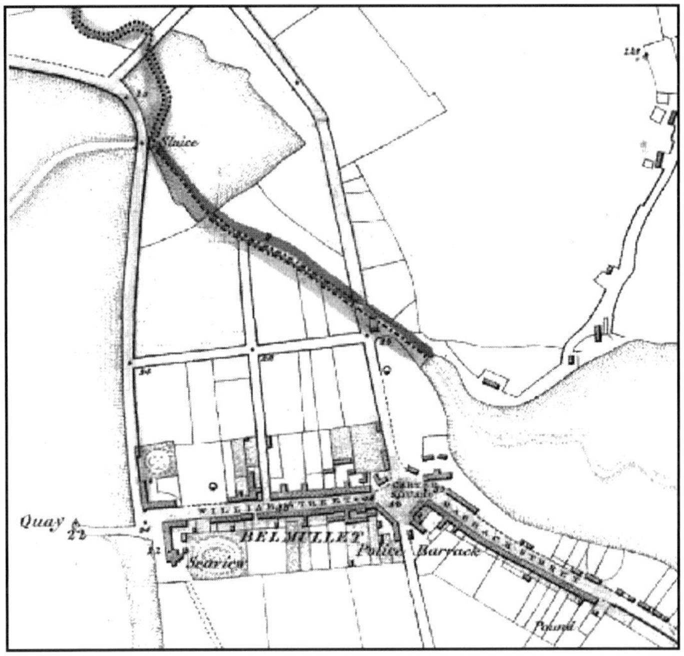

Belmullet circa 1838
(First Edition 6 inch Ordnance Survey Map)

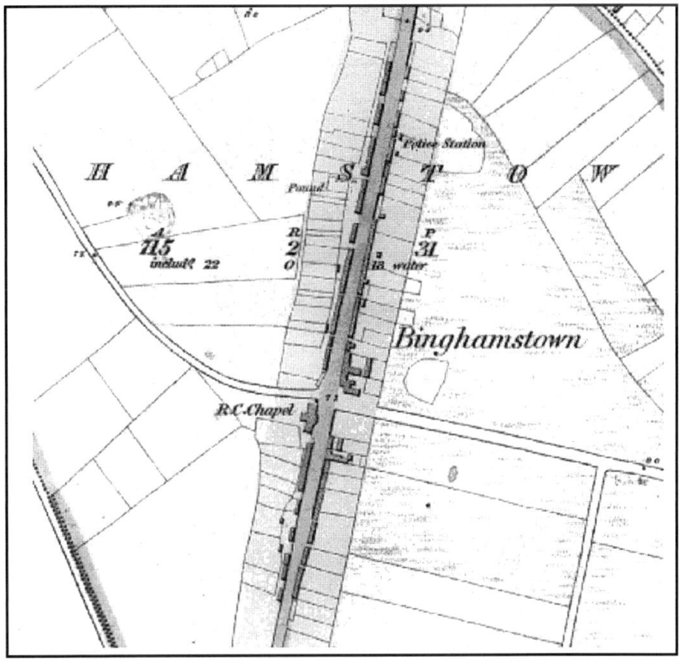

Binghamstown circa 1838
(First Edition 6 inch Ordnance Survey Map)

Nimmo supported the establishment of both towns and of both piers; indeed he suggested other piers for the west coast of the Mullet, but the Fishery Board did not sanction these. However, as developments continued he came to the conclusion that Belmullet was better placed than Binghamstown for the material development of the district. The extension by Knight of the Central Erris Road to Belmullet completed the domination of this town over the latter, and convinced Nimmo to terminate his north coast road at the same place, and not at Doonkeeghan as was previously planned.[18]

William King was the first to suggest to the Fishery Board that a channel be cut across the isthmus of Belmullet from Broad Haven to Blacksod Bay, although the notion had been tossed about for many years. In 1822 the Fishery Board considered his suggestion, but decided that they had insufficient funds to carry it through.[19] Nimmo put the suggestion forward in 1824, for the improvement of communications along the coast. The channel would, he wrote, 'permit boats to pass at high water from one haven to the other, as each of the inlets dry to a considerable distance at low water, and Broadhaven (sic) in particular is shallow at the head'. He outlined the works needed and estimated the cost at about £5,000. He attached a plan of the isthmus, and the probable direction of the cut. He indicated that the 'cut would permit the passage, in most tides, of such vessels as use the Forth and Clyde Canal, and would certainly be a great convenience in the time of the fishery'. The sound of Achill, he pointed out, 'permits vessels of the class of coasters to pass at high water, and with the proposed cut at Belmullet we would have a kind of inland navigation from Westport to the bay of Donegal.'[20] The work was later carried out under the Board of Works during the Great Famine; it was funded by government and local contribution and completed in 1851. A swivel bridge was put in place to maintain land communication with the mainland.[21] A fixed bridge more recently replaced this, once again cutting off sailing access from one bay to the other.

Clifden

John D'Arcy took up residence on his Connemara estate in 1810 and two years later he founded the town of Clifden. However, despite strenuous efforts by him, the town was slow to develop and was in danger of never fulfilling the dreams of its founder. When Nimmo came to Clifden in 1822 he found 'nothing but scenes of misery and despair'. There was 'only one slated or two-story house, and a few thatched cabins'.[22] However, with the opening of public works in the region, and the offer of written agreements and low rents by D'Arcy to his tenants, more people began to arrive. Land in the neighbourhood of the town was quickly taken up and brought under cultivation. The reclaimed land proved very productive and soon sufficient wheat, barley and oats were grown to supply the town, and the surplus was exported.[23]

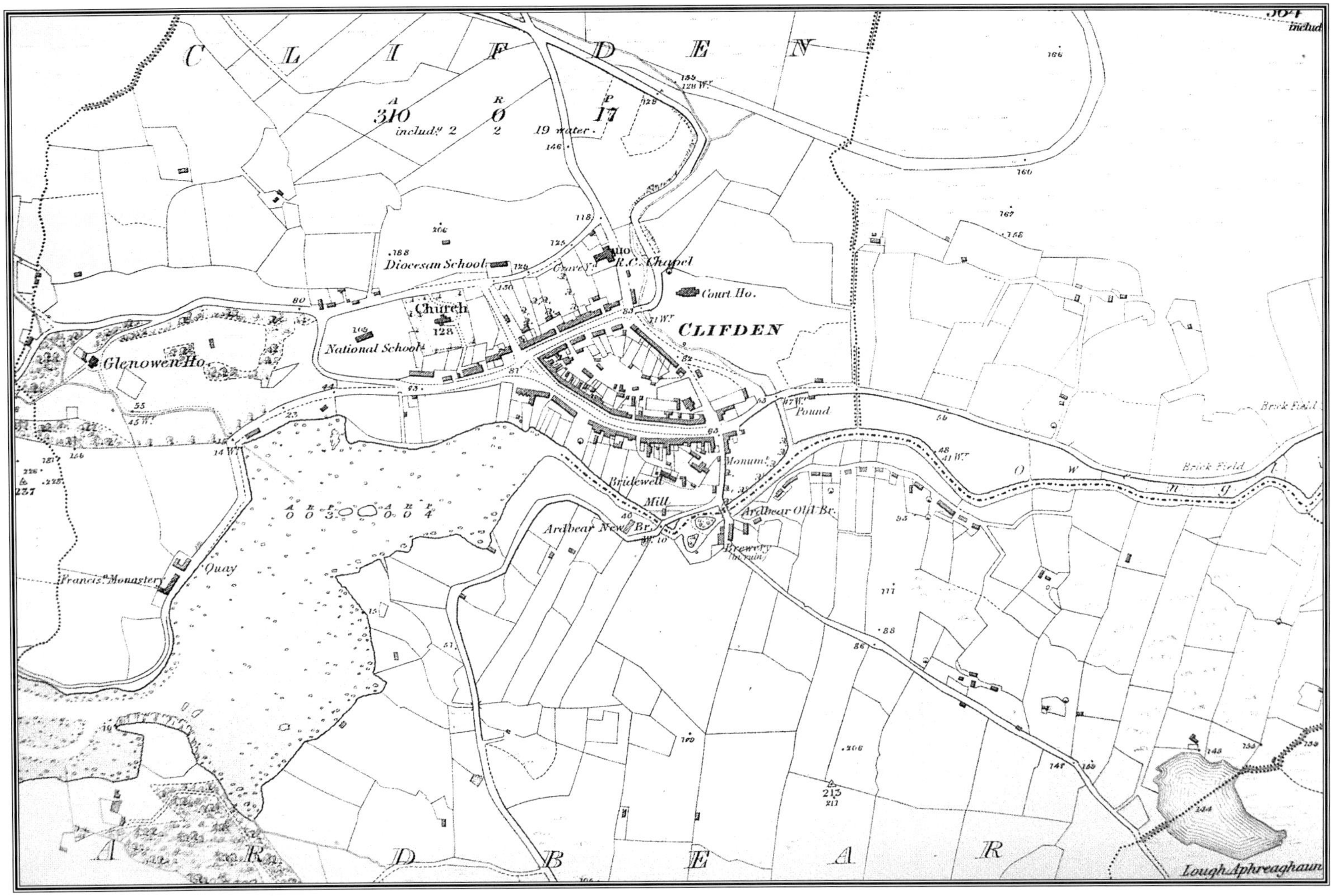

The construction of the pier offered D'Arcy the opportunity to promote the town as a supply centre for the fishery. Clifden, he pointed out, was ideally situated to service the needs of the fishery along that coast, being centrally positioned halfway between Westport and Galway. The town stood at the head of Ardbear Bay, where there was a good basin, well sheltered in all weathers, and where up to two hundred wherries could lie aground at low water, 'on a fine even strand', for repairs or to shelter in rough weather.[24] Tradesmen and merchants from other parts of Ireland came in response, attracted by 'the progressive state of this little town surrounded by all the advantages of trade and industry by sea and land'.[25]

D'Arcy made strenuous efforts to attract investors for the purpose of establishing a fishing company in the town. He outlined a plan showing the projected expenditure and profit arising out of such a company, which Hely Dutton later published in his Survey of Galway.[26] The National Fishing Company, scouting for suitable sites at which to establish a

Clifden circa 1841
Showing Ardbear Old Bridge and Ardbear New Bridge. The Galway road was by then taken across Waterloo Bridge and entered the town at the bottom of Marked Hill. Interesting to note that the brewery was already marked as a ruin at this early date.
(First Edition 6 inch Ordnance Survey Map 1841)

Ardbear Old Bridge with brewery circa 1860s
(Courtesy of National Library of Ireland, SP1133)

fishing station, felt Clifden was worthy of consideration, but took it no further.[27] It was afterwards found that attempts to establish fishing companies usually failed or proved to be unprofitable. This was attributed to the fact that fishermen employed by companies or societies rarely worked as diligently for their employers as they would for themselves.[28]

By 1826, Clifden had expanded beyond Nimmo's expectations and was now the principal town in Connemara. It contains 'about one hundred tolerable houses, mostly two stories, and roofed with Bangor slates; about thirty country shops, several of the owners worth £500 to £1,000, who were not worth £50 three years ago'.[29] This increase in housing would indicate that D'Arcy had succeeded in attracting a wealthier tenant than heretofore. It also gives some idea of the extent of construction work that went on in the town in the previous four years.

Oats, hay and straw were readily available at Clifden market, where, up until 1825, Nimmo and his men were forced to bring the oats needed for their horses in barrels from Galway. A mill was built close to the brewery and distillery, and it was fully operational in 1826. All three provided a market for the grain, which had previously been only used in 'private distillation'. Two stores were erected on the quay and both held herrings waiting for export. The removal of the salt duty had resulted in a direct import trade from Liverpool, and three cargoes of salt had arrived during the winter of 1824 and 1825. Several cargoes of various goods, for sale in the town, were imported in 1825, including a cargo of timber direct from America. With this increase in trade, Nimmo estimated that the town was annually paying £3,000 of direct revenue to the government.[30] Money was circulating among all classes and Nimmo was able to report that:

> The peasantry, who were almost entirely unacquainted with money, now pay their rents in cash, a thing heretofore unknown; they are well clad, evidently in new stuff; they bring to the markets of Clifden a variety of articles for sale, and never depart without making some purchases. A few bad years, previous to 1822, had left them in rags and nakedness; and being forty miles from the nearest market town, over bogs and bad roads, without money to buy, they were in want of most of the conveniences of civilized life.[31]

Roundstone

In his report to the Fishery Board, Nimmo explained how he came to hold the land adjacent to the Roundstone pier and how he set about establishing a village there. Construction of Roundstone pier began in 1822 and Donnell tells us that, at the time, there were, 'only three or four scattered cabins near this quay, and the small quantity of corn produced in the county adjacent was entirely used for illicit distillation'.[32] The pier was situated two-thirds up on the west side of Roundstone Bay. It was quickly brought into use, 'shipping kelp, curing herrings etc'. However, the tenant of the farm on which the pier was situated was demanding compensation for damages allegedly sustained during the progress of the work. For years Nimmo had talked of the necessity for a village along this stretch of coast and so, rather than pay the compensation, he decided to buy out the tenant at his own expense and take over his lease.[33]

On I August 1826, Nimmo renewed the agreement with Thomas Martin. He took out a ninety-nine year lease on the townlands of Roundstone and Letterdyfe, comprising almost two hundred and forty acres of land. The land was then divided into lots and offered for rent. Several of the leases contained a covenant to build 'good dwelling houses, with slated roofs and, in most instances, of two stories high'. To help promote the village, Thomas Martin obtained patents for four fairs yearly, a weekly market and for 'a court of pie powers, tolls and customs'.[34]

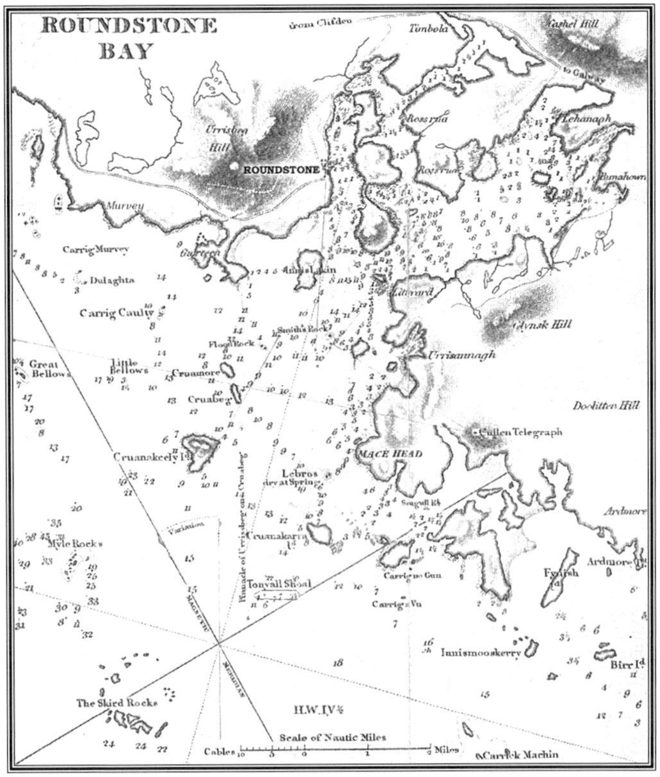

Nimmo's chart of Roundstone Bay circa 1831
(*Chart of St George's Channel and the coast of Ireland*, A Nimmo)

Nimmo then made Roundstone the centre for the public works in the Barony of Ballynahinch. He built a store, office, carpentry and smithy workshops in the village, which he leased to the government for six and a half years at £50 per annum.[35] He also built a substantial slated store on the quay for the fishery.[36] His brother John took up residence there, as did Joseph Reville, pay clerk for the South West Division, John Minty, superintendent of works in the Barony and William England, the foreman and assistant storekeeper. It would appear that John Nimmo took charge of the project for his brother and he held the lease after Nimmo's death in 1832. John continued to reside in Roundstone until his death in 1850, his wife, Marion, having predeceased him the year before. John's son, Alexander, died a bachelor in Dublin in 1839. George Nimmo died at Ballynahinch in 1851. He was

Nimmo's pier and Roundstone Village circa 1880s
© Private Collection

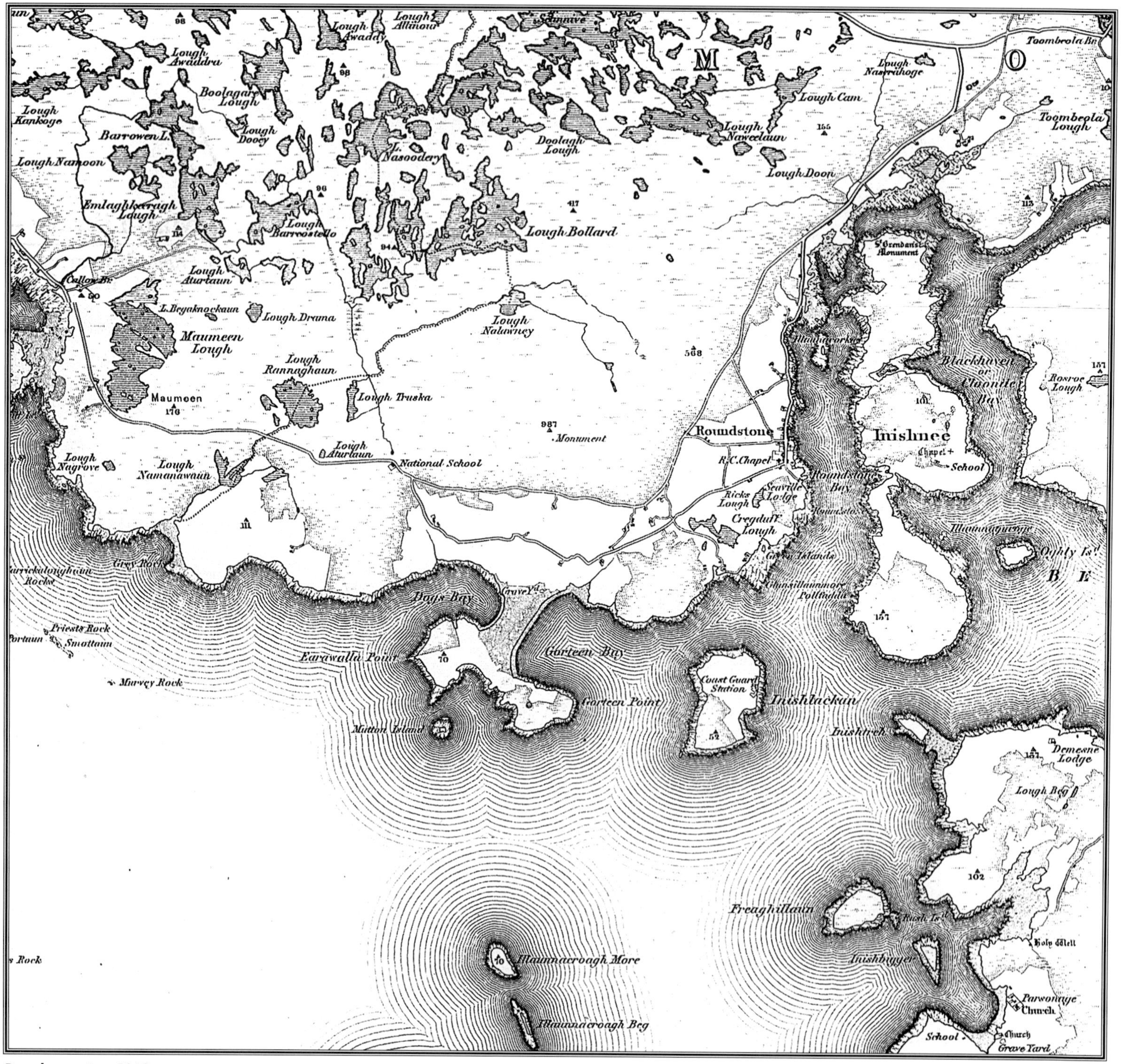

Roundstone circa 1862

(First Edition 1 inch Ordnance Survey maps)

predeceased by his two daughters, Margaret and Charlotte, both of whom died in 1836, and his wife, also named Charlotte, who died in 1839.[37] This seems to have ended the Nimmo connection with the village.

A branch road from Nimmo's coast road was brought to the village on the east side and the old road to the west was later realigned and improved by Nimmo, but it took some years to complete. Donnell, like Nimmo, was optimistic for the future of the village and he wrote in 1826: 'There can be no doubt but this hitherto sequestered and almost unknown district will shortly give extensive employment in fishery, agriculture and commercial avocations'.[38] Because Roundstone was situated on the east side of Slyne Head, it was frequently used by the people of west Connemara, preferring to make the journey to the quay by land rather than face the treacherous waters around the head.[39] Samuel Lewis tells us that in 1837 about 250 people were employed in trading and fishing at Roundstone and the pier was frequented by '30 sailing-boats, averaging 10 tons, and 40 row-boats, of 4 tons each'. As well as fishing, the sailboats were employed delivering corn, kelp and turf to Galway.[40]

Nimnmo's private commissions

Although fully occupied with the many works in progress in the Western District, Nimmo still found time to take on private commissions. Throughout the 1820s and in the first two years of the 1830s he was involved in many of the major engineering works taking place in the country. He also produced several papers for publication at this time.

Works attributed to him in the 1820s are two bridges at Pollaphuca, on the turnpike road from Dublin to Baltinglass, a bridge at Cahir in Co Tipperary, in 1825, and the Wellesley Bridge, 'a five-arch bridge of sophisticated structural form and architecture with an opening span of two iron swivel girders, quays and a wet dock' at Limerick.[41] It is considered to be 'a masterpiece in function, proportion and style'.[42] Work began on the Wellesley Bridge (Sarsfield Bridge) in 1824, but it was not completed until after Nimmo's death (under Charles Vignoles C.E., in 1835), and the docks were never finished.[43] In 1823 he submitted a plan for a second bridge over the Shannon, but this was not accepted. He published a paper in *Transactions of the Royal Irish Academy* on the advantages knowledge of marine geology could bring to marine navigation.[44] In 1825 he planned a railway from Limerick to Waterford and published a paper on railways in *The Dublin Philosophical Journal and Scientific Review*. A year later he drew up a proposal for improvements to the harbour at Drogheda and produced a plan of the River Boyne, with proposals for improving the navigation of the river.

In a court case between the Liverpool Corporation and the Mersey and Irwell Navigation Company in 1827, Nimmo's calm delivery of complicated mathematical evidence, while under severe pressure from the prosecution, impressed many. Afterwards he acted as consultant engineer to a number of railway companies in northern England. Nimmo became a member of the Institute of Civil Engineers in 1828. Also at this time, he produced several reports with Thomas Telford and Robert Stevenson recommending two new seaports and a new ship canal through the Wirral peninsula in England.[45] All of the above must have required regular trips to England in the latter years of the 1820s and must surely have taken Nimmo's focus off the works in the Western District. It would also appear that Nimmo was experiencing ill health at this time. In a letter to Robert Stevenson in April 1829, written from the Spring Garden Hotel in London, he complained of being too ill to attend to business for some time: 'I have been very poorly all this winter and have full escaped a fever and cannot attend much to business'.[46]

Chapter 6

Nimmo criticized for slow progress of works 1827–1830

For the latter years of the 1820s, Nimmo came under pressure to complete the roads in the Western District, but he would fail to comply. Hampered by insufficient funding and bad weather, the works dragged on, much to the annoyance of Dublin Castle and to the inconvenience of the commuter, with Nimmo drawing criticism from both.

The lord lieutenant, the Marquess Wellesley, was decidedly unimpressed with Nimmo's estimates for public works in the Western District for 1827. He demanded an explanation as to why Nimmo was making such a high estimate for completing the works, when, going on the previous year's figures, an estimate of a little over £6,000 was all that had been expected. In his report and estimate for 1827, Nimmo gave a figure of £31,315.7.2d for completing all works in the western district. However, it seems he had in the previous year estimated £32,771 as the sum necessary for completing the same works. On the strength of his 1826 estimates a sum of £26,000 had been voted by parliament, and the records show Nimmo eventually receiving a total of £27,229.4.6d for the year ending 5 January 1827. Under-secretary, William Gregory, wrote to Nimmo requesting details of the year's expenditure in the west for 1826. The lord lieutenant, he told him, would recommend to parliament a vote of no more than £12,000 and he added that he was considering allocating part of these funds to another engineer to complete some of the works hitherto under Nimmo, rather than run the risk of 'another miscalculation' on his part.[1] A sum of £10,099.1.6d was finally issued to Nimmo for works in 1827.[2]

This questioning of Nimmo's capability seems to have been the first of its kind and must surely have hit him hard. There is no record of his response to the lord lieutenant, but in July 1827 he complained to William Lamb, who had replaced Henry Goulburn as chief secretary three months earlier, that, because of the limited sum voted by government for that season, he felt it necessary to pay up and discharge a considerable portion of those receiving regular employment in his district.[3] This decision greatly frustrated Nimmo and left the people in need and his bank account in arrears: 'the sums at my disposal are not equal to the payment of the arrears of work executed'. Since 1825, he explained, there had been 'an arrear of £6,000 of the parliamentary grant' for works in his district, leaving him in a 'similar arrear' on all his works. He found it 'seriously distressing to have the people so very unpaid' and impossible 'to prosecute those works to a state of completion as [he] had expected'.[4]

The Board of Accounts was disallowing certain sums claimed for in Nimmo's accounts. The board pointed out that there were discrepancies in the accounts going back to 1822. These could be found in the documentation relating to many of the presentment roads

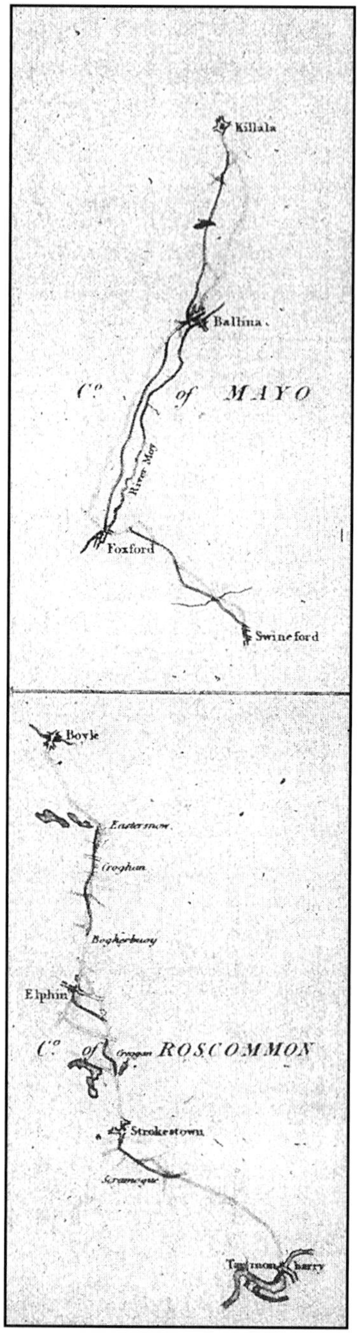

Sketch maps of road from Killala to Swinford and Boyle to Termonbarry 1826
Showing early stretches of road and realignment carried out by Nimmo's team. (Second Report on certain Roads and Bridges in Ireland, 1826 –27)

and on almost every road constructed by Nimmo as public works. In many cases the amounts were small, but when added together they came to a considerable sum, £19,368.14s.[5] Nimmo dispatched his accountant, Patrick Carney, to visit the west and knock on the doors of the overseers and 'gentlemen' of the various counties, men to whom Nimmo had issued funds in the past, and request that they produce the necessary vouchers. In some cases these were produced, in others they claimed that they had already been issued, and refused to produce duplicates. For others various explanations were offered, such as the death or emigration of the overseer to whom the funds were paid. The process of collecting the vouchers would take some time and would not be fully completed in Nimmo's lifetime.[6] In the meantime, Nimmo struggled on, with the government continuing to demand that all works be completed as quickly as possible.

The government was anxious to complete all public works started on the west coast in 1822 and to transfer these to the Grand Jury of the relevant county. In the interim, the Commission for Inland Navigation was given responsibility, in 1825, for the repair and maintenance of all completed roads and bridges financed, or partially financed, by government.[7] The maintenance of roads was always a contentious issue among landlords and Grand Juries, and now the government. No one wanted to take on the responsibility and cost. Nimmo was coming under pressure to complete the roads under his supervision and to deliver them up to the commission. His contemporary, John Killaly, had completed the roads in the Middle District, comprising Clare and Galway (sometimes referred to as Central District). He handed over ninety-three miles of road to the commission in 1825.[8] Griffith was still working in the Southern District, but would hand over one hundred miles of finished road in 1828.[9]

Killaly was next employed by the Commission for Inland Navigation to examine the government roads under construction, to determine their condition and, if satisfactory, to take them under the authority of the board.[10] In December 1825, thirteen miles of Nimmo's Boyle to Termonbarry road in Roscommon had been approved by Killaly and taken over by the commission. The road passed through Eastersnow, Croghan, Bogherbuoy, Elphin, Greggan, Strokestown, and Scramoge to Termonbarry. Killaly described this road as 'well and judiciously executed ... with a carriageway varying in breath from twenty-one to twenty-five feet. The materials used in repairing it were chiefly limestone'. This was a busy road, accommodating 'the great land-carriage trade in Wheat from Strokestown to Sligo on the north end, and Oats to Richmond Harbour on the south'.[11] The Royal Canal terminated at Richmond Harbour in Termonbarry. The road from Killala to Ballina, Foxford and Swinford was taken over in September 1826, along with roads at Ballyfarnan and Gortlownan in Sligo.[12]

By 1827 some of Nimmo's major works were nearing completion but there were still many left unfinished. The slow progress of the works was drawing criticism from the west and complaints were soon arriving on the desk of the chief secretary, William Lamb. The majority of the roads were in use but many were still unsuitable for a carriage. Some sections of road had been in use for four or five years without receiving any repair or surface dressing. Nimmo was using all funds he received to extend the new roads as far as possible and was leaving repair and maintenance work until the entire stretch of road was ready for opening. Winter floods were causing damage in the mountain districts, bringing down bridges and destroying road surfaces. The result was that it was not unusual to travel for some time along a perfectly good road only to be brought to a halt for want of a bridge, or by the fact that the link between a government road and a county road was awaiting Grand Jury funds.

Major Denis Bingham of Bingham Castle wrote a damning letter against Nimmo in September 1827, regarding the road from Castlebar to the pier at Blacksod (Central Erris Road). Bingham had employed William Bald to prepare a report on the state of the road and had sent a copy to Henry Goulburn in the previous January. He was now seeking the attention of the new chief secretary. Large sums of money had been allocated to this road, he told Lamb, and it was still unfinished. Bingham, on behalf of the inhabitants of Mayo, wanted Nimmo's accounts regarding this work investigated, in particular the engineers' charges. He requested that the authorities send down 'an engineer of the first class and unconnected with Mr Nimmo, to examine the state of those roads', for he was of the opinion that 'no work of a public nature yet carried on in Ireland requires to be more carefully examined into than this.'[13]

Corrick Bridge in 2006
Otherwise known as 'The Musical Bridge'. It produces a musical sound when a stone is drawn across the parapet.
© Private Collection

Nimmo furnished the chief secretary's office with detailed expenditure accompanied by a report from his assistant, Patrick Knight, stating that, while the road was not yet finished, it was in good repair and passable. Conditions had greatly improved in the district since public works had commenced; the mountainous regions in particular were showing obvious improvement in cultivation and housing. Good houses with chimneys and windows were erected along the line from Castlebar to Corrick Bridge and 'the foundations of a very commodious inn for general accommodation' was laid out at the bridge. Corrick Bridge was designed and constructed by William Bald for the Mayo Grand Jury and it was already in place when Nimmo took over works on this road. The road covered almost fifty miles and contained two significant bridges costing £5,000, and several smaller ones. A sum of £19,000 had already been spent on the road, £8,000 of which had come from the Mayo Grand Jury. This, Nimmo argued, was 'a very moderate charge' considering the country traversed by the road, and it would be completed as soon as funds became available: 'I admit that it still requires much improvement and wish I could get funds for that purpose, but even as it is you will [recognize that] the benefit to that district has been great and manifest.'[14]

John D'Arcy of Clifden Castle was complaining that Nimmo was concentrating on cross roads rather than completing the coastal road in that district. He too suggested that an engineer be sent to assess progress on all works, piers and roads, under Nimmo's direction. Henry Blake of Renvyle wrote requesting further roads in his district. In response, Nimmo told Dublin Castle, 'The truth is these Gentlemen [D'Arcy and Blake], though very thankful for what has been done in their district and very sensible of its great and rapid improvement, look for much more from the Government than it would be reasonable to expect'.[15]

Here again Nimmo furnished a report from his assistant in the area. This is one of the few surviving reports from Nimmo's assistants in the field. In this report, Daniel Mackin, the engineer in charge, tells Nimmo that the new line of road on either sides of Toombeola Bridge was open to the full width and was fit to drive on but the surface was not fully formed. The bridge itself was still under construction: 'The bridge of Timbola (sic) remains in the following state, one land abutment and centre pier raised to the proper height, centre of one arch up, and six rings laid on each side, another centre pier nearly to full height, and a considerable quantity of materials prepared on the spot'. The branch road leading from this line to Roundstone was in part newly formed and part improved bridleroad. The first one hundred and forty perches was a new road 'gravelled in the centre about 8 feet broad'. This connected with the improved bridleroad, 'for some distance', and then continued as a new line 'going directly to the village of Roundstone'. The road to Cleggan pier was in need of a general repair, a large pipe had been torn away near Cleggan, 'which will require a double gullet'. The road from Cleggan crossroads to Letterfrack was also in bad repair and required work, 'as the proper quantity of materials has not been laid on'. Work had started up again on the twelve-mile stretch of the central Connemara road from Halfway House to Emlaghdauroe. This was the section begun by Thomas Martin and was then contracted to Samuel Jones. It was, according to Mackin, partly formed and partly gravelled, and work on the bridges was progressing. The road from Emlaghdauroe to Clifden, the section formed by John D'Arcy some years before, was in places 'broken and bad and in other parts, particularly along the lakes, very narrow and in bad repair', but no funds were available for repairs to this section of road.[16]

Centre Connemara road

Since arriving in Connemara, Nimmo had tried to stay away from the old centre road, leading from Galway to Clifden. In 1822 and 1823 he had granted funds towards repairs on sections of the road between Galway and Oughterard, 'no better object being at hand for the employment of the people'. He was, however, reluctant to do more. The twelve-mile section running through the Martin estate had been left unfinished since 1819, when funding was withdrawn, and communication through central Connemara was almost non-existent. Many attempts were made to 'induce' Nimmo to recommend the works for free government aid and he was criticized locally for not doing so. For years he held back, waiting for the Grand Jury to do its duty, but nothing was done. He did not wish to exonerate the landlords from their share of the county cess, he told the lord lieutenant: 'I should have conceived it as consistent to recommend that the Government should pay a portion of their rents'.[17]

However, in 1825, perhaps under pressure from his friend Thomas Martin, Nimmo examined the road to ascertain the extent of the works already executed under the grants from the Consolidation Fund, and to estimate the sums needed for its completion. He estimated that £2,000 would complete the works, but the lord lieutenant turned down requests for funding, stating that government had allocated no new funds for that year; funds voted by government for Connemara were needed to complete works already begun under Nimmo.[18]

The landlords continued putting pressure on Nimmo to take the road under his control; eventually, 'with much hesitation', he capitulated. In 1828 Nimmo proposed that work on part of the coast road, between Kilkieran Bay and Toombeola, be suspended and the funds transferred to completing the works on the old centre road. In defence of his decision, he pointed out that completing this road was essential not just to improving communications between central Connemara and Galway town, but also to linking the coast road to the south with the inland road (Maum Road) to the north. The chief secretary's office sanctioned Nimmo's proposal and the Grand Jury appointed him joint overseer of the works. With Nimmo on board the commissioners of the Consolidation Fund resumed their advances.[19]

Nimmo took the opportunity to introduce further changes to the proposed works. Under Nimmo's direction the central road was redirected to follow the route first outlined by him in his bog report. At Derryneen (marked Boheeshal on Nimmo's map p136) the road was taken along the northern shore of Garroman Lake [Glendallaugh] and onwards to pass between Derryclare and Ballynahinch lakes, crossing the connecting river at the Canal Stage, to meet up with the old road at Emlaghdauroe. This new line became known as the new centre road. In realigning the road, Nimmo was removing the public road from directly in front of his friend Thomas Martin's front door and bringing it to the other side of Ballynahinch Lake. This gave Ballynahinch Castle more privacy than it had previously

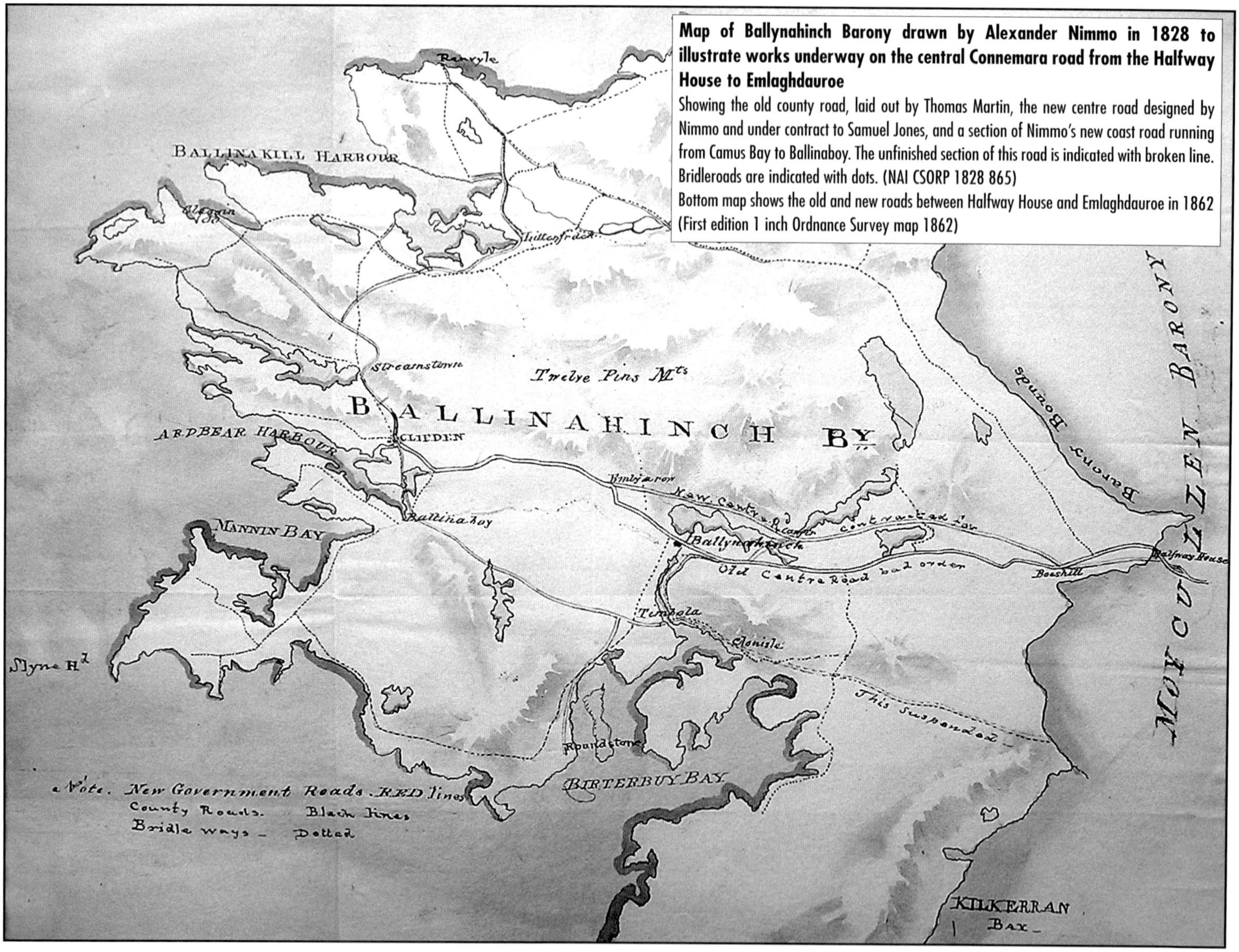

Map of Ballynahinch Barony drawn by Alexander Nimmo in 1828 to illustrate works underway on the central Connemara road from the Halfway House to Emlaghdauroe

Showing the old county road, laid out by Thomas Martin, the new centre road designed by Nimmo and under contract to Samuel Jones, and a section of Nimmo's new coast road running from Camus Bay to Ballinaboy. The unfinished section of this road is indicated with broken line. Bridleroads are indicated with dots. (NAI CSORP 1828 865)

Bottom map shows the old and new roads between Halfway House and Emlaghdauroe in 1862 (First edition 1 inch Ordnance Survey map 1862)

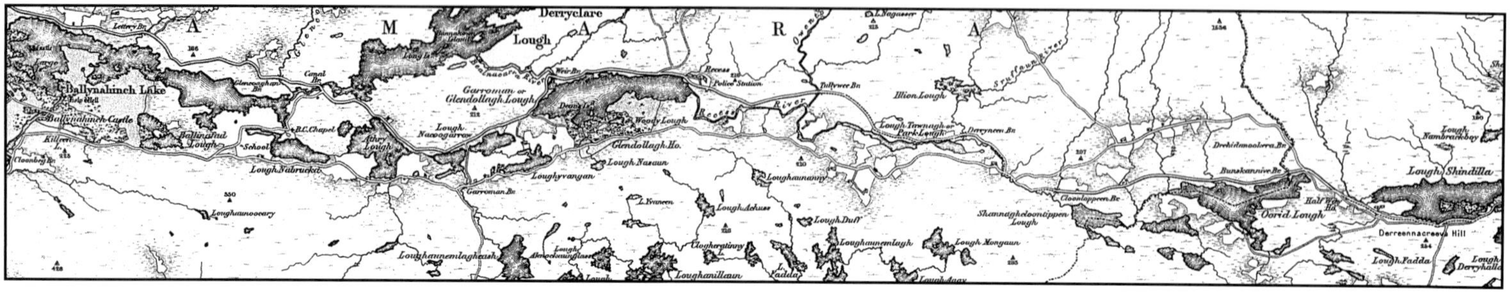

enjoyed and it was built entirely at the expense of government and the Grand Jury. Martin then adapted the old road to his private avenue.

The work was put out to tender and the lowest bidder, Samuel Jones, was awarded the contract. Half of the money had been paid out before it became obvious that Jones was unfit for the job and the works came to a standstill. The Grand Jury was making every effort to induce Jones to complete the work or abandon the contract, but without success. With Jones continuing to hold the contract, Nimmo was prevented from stepping in and completing the job, thus leaving the road, once again, unfinished and impassable.[20]

The contractor, Samuel Jones, was a surveyor and connected by marriage to John D'Arcy of Clifden Castle. Jones was born in Edgeworthstown, Co Longford, the son of a carpenter and tenant of the Edgeworth family. He was educated by William Edgeworth, Maria's brother, and came to work for Nimmo in Connemara. In the company of Nimmo he was accepted into the homes of the local landed gentry and soon formed an attachment with Elizabeth Campbell, the half-sister of Mrs John D'Arcy. Much to the surprise and the bemusement of Maria Edgeworth, Jones married Miss Campbell, and the couple went on to have a family. Sadly, Mrs Jones died young, leaving her children to be reared by Mrs D'Arcy at Clifden Castle. Jones leased 400 acres of land at Ardbear, south of Clifden, from D'Arcy and built Ardbear House, a substantial family residence.[21] In Clifden he took on the trappings of respectability, while, according to Maria, his brothers were 'in rags and whiskey at Edgeworthstown'. She had planned to visit him at Clifden, during her tour in 1833, but changed her mind when she learned of the condition of the road and the controversy Jones was causing in Connemara. Maria suspected that his hands were not 'very clean as to public money, road money etc'.[22] Any friendship that existed between Nimmo and Jones was soured by his poor performance on the central Connemara road and the row over this contract would continue long after Nimmo's death in 1832.

Roads completed in 1828

Nimmo attempted to hand over several roads in Galway, Roscommon and Leitrim to the Commission for Inland Navigation in 1828. The commission, however, rejected these, declaring them to be too short and patchy. The commission preferred to use day labour under supervision to maintain the roads under their care, but this was impossible on small stretches of roads spread over wide areas and covering several counties. Nimmo's roads would have to be given out to private contractors, which the commission knew from experience would prove costly and unsatisfactory. They suggested that Nimmo maintain the roads himself until he had completed all roads under his supervision and was ready to hand over all roads for the entire district. Nimmo refused, pointing out that funds at his disposal were for road construction, not road maintenance.[23]

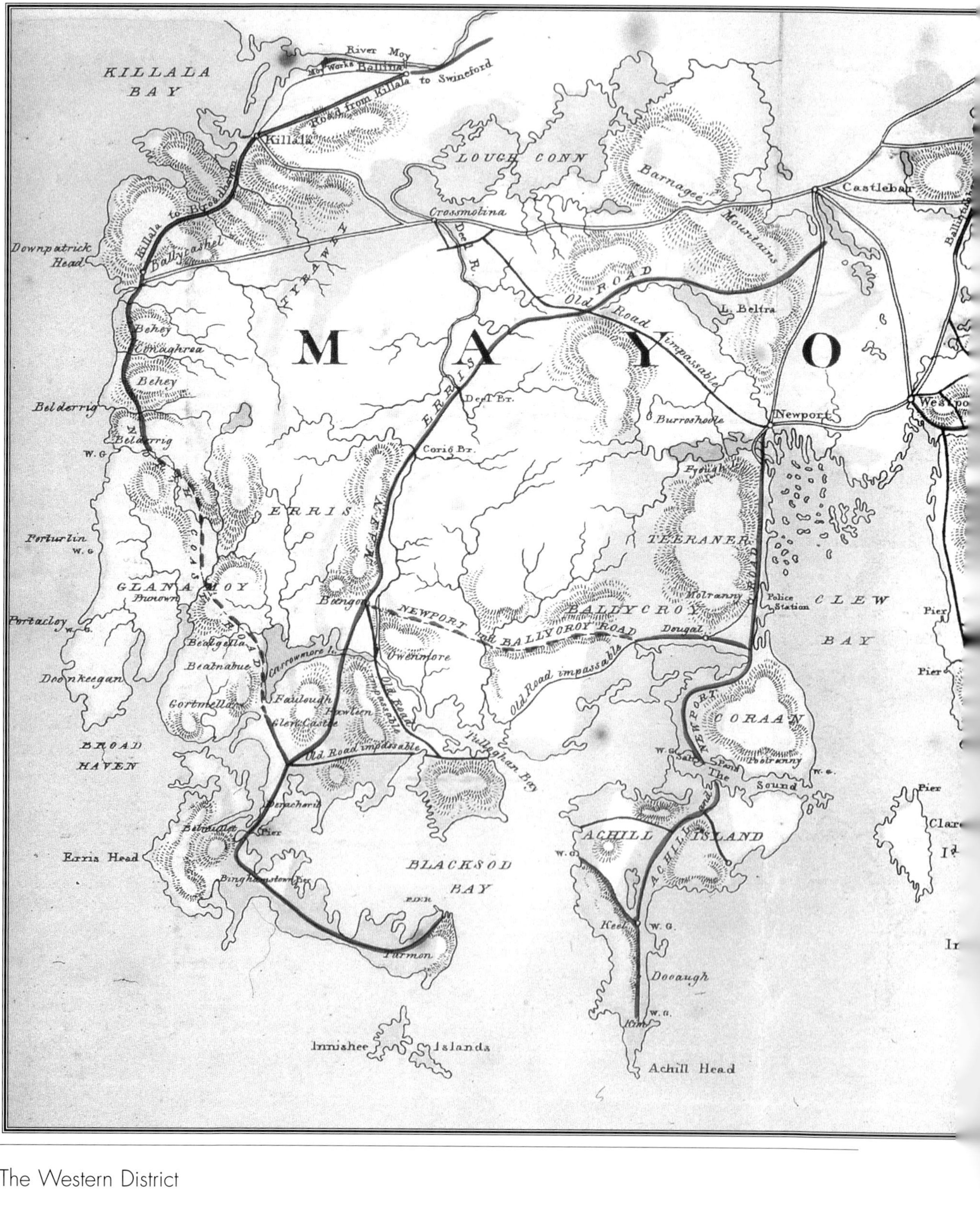

Map of counties Mayo and Galway circa 1833.
Showing roads begun by Nimmo in 1822, unfinished sections are indicated with broken lines. Also marked are several of the piers constructed under Nimmo
(BPP 1833 [285], xxxv)

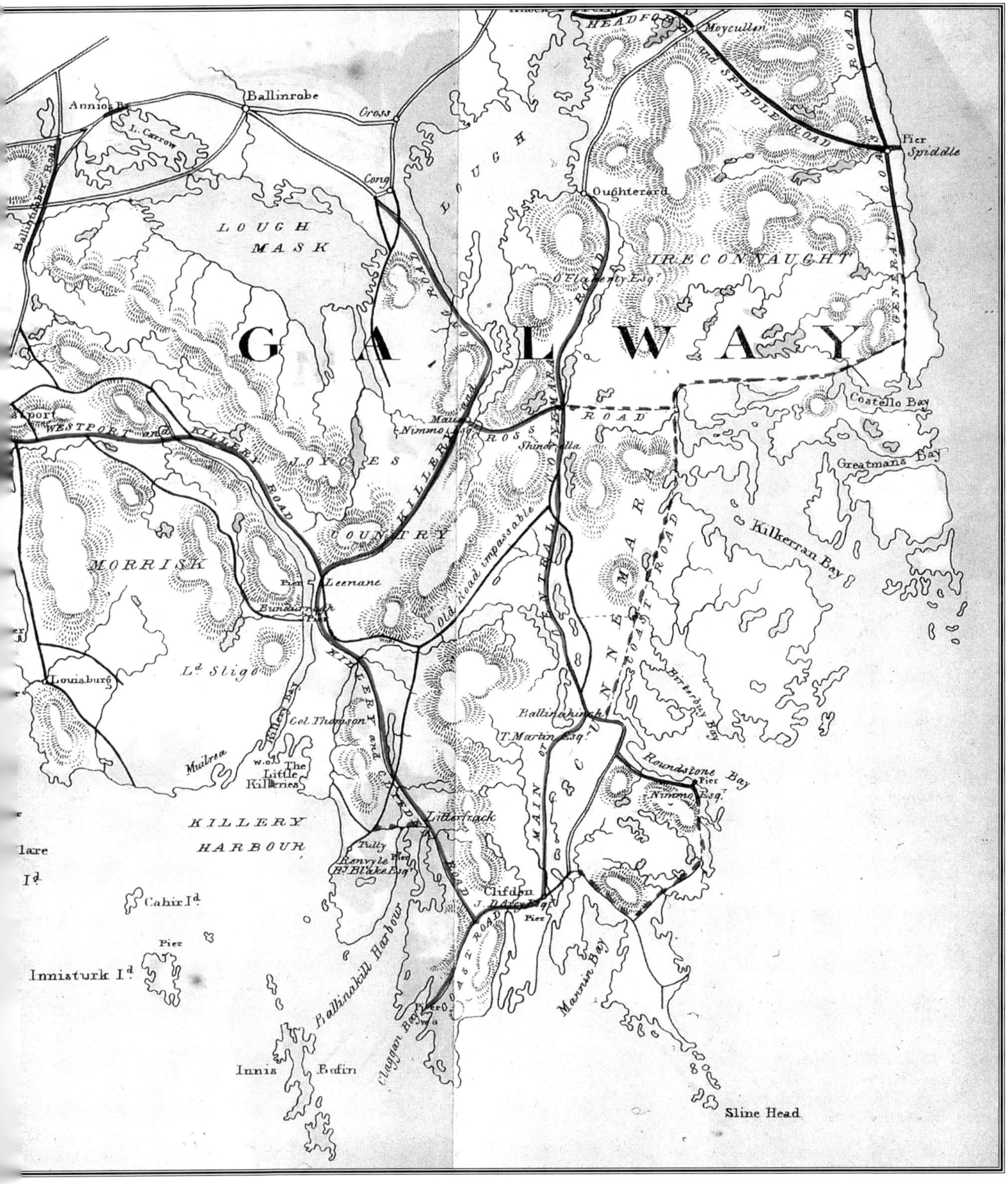
Ballinrobe
Cross
Cong
LOUGH MASK
Moycullen
HEADFORD and SPIDDLE ROAD
Pier
Spiddle
Oughterard
IRECONNAUGHT
O'Flaherty Esq.
GALWAY
Costello Bay
Greatmans Bay
Kilkerran Bay
ROSS ROAD
Nimmo Esq.
JOYCES COUNTRY
KILLERY ROAD
WESTPORT and KILLERY ROAD
MORRISK
Leenane
Old Road impassable
Ld Sligo
Louisburg
Col. Thomson
Ballinahinch
T. Martin Esq.
Roundstone Bay
Pier
Nimmo Esq.
KILLERY HARBOUR
Littlefrack
Clifden
Pier
Mannin Bay
Cahir Id.
Innisturk Id.
Pier
Innis Bofin
Claggan Bay
Ballinakill Harbour
Sline Head

In their eagerness to secure some roads from Nimmo, the government instructed the commissioners to take over any completed stretches of road. Reluctantly the board complied and took over the following roads and employed private contractors to oversee their maintenance: Barna to Spiddle, Spiddle to Moycullen and on to Knockferry on Lough Corrib; a stretch from Killroo to Cloneen on the Galway Headford Road; in Roscommon a small stretch from Castlerea to Ballymoe and in Leitrim, a small stretch just north of Mohill.[24] Nimmo still retained about two hundred miles of road under his supervision, this was in use by the public but still unfinished. The commissioners pointed out that Nimmo's inland roads in Roscommon and Leitrim were attracting heavy traffic, transporting agricultural produce to seaports and the Royal Canal, for delivery to the Dublin and Limerick markets. With this in mind, they recommended that all future road planning should take into consideration access to the river Shannon and the Canals.[25]

Bridges repaired in 1829

Nimmo was issued with £10,699.1.6d for works in his district for 1829; most of this went towards repairing and completing bridges.[26] In Mayo work was resumed on the coast road from Killala to Broad Haven, which had been suspended for some time. The surface of the road had sustained considerable damage in the interim and bridges were left unfinished, but the people had succeeded in making their own 'temporary passes, so as to take advantage of the portion done'. Following repairs the road was made passable for a carriage from Killala to Belderg, with the exception of Glenlassera bridge, where they were forced to unload and to transfer their goods across the river, before continuing on with their journey. Nimmo reported that a 'considerable increase of culture and population [had], even in its imperfect state, occurred from this opening', and a number of good stone houses had been built at Behy, Glenlassera and Belderg. He was anxious to link this coast road with the central Erris road at Bellacorick, but did not have the funds to proceed.[27]

Seven of the nine bridges damaged on the central Erris road by the 'torrants in June 1828', were repaired in 1829. The remaining two bridges at Fauleen and Bonavon (Bunnahowen), between Bangor and Belmullet, were left unfinished. Nimmo described the Bonavon Bridge as 'a respectable work'. The number of houses at Belmullet had nearly doubled by 1829 and three large corn stores, 'of three floors each', were built and held one thousand tons of corn for sale directly to the British market. The corn was grown locally in Broadhaven and Doonkeeghan and, prior to this, had been used exclusively in the manufacture of poitín.[28]

The road through Achill was also completed that year, making it possible for a light carriage, by crossing the sound at low tide, to travel from the mainland to Keem on the western side.[29] Nimmo's assistant wrote with enthusiasm, 'it is surely a new sight to see a horse and cart with 10 cwt of flour on it going through Achil (sic) Island'. It was expected that the one cart owner on the island would soon be joined by others. However, the old road from Newport to the commencement of the new road to the island was in need of realignment, having 'many steep hills quite unfit for loaded carriages'.[30] The benefits to the islanders were considerable: 'The convenience to the population is very great; they bring their corn to the shore at Saula or Cashel by the new road, and thence send it by boat to Westport or Belmullet, or take it on horesback across the sound'. But still no work had been done on the ferry piers, which Nimmo considered 'the most necessary improvement'.[31]

Erriff Bridge circa 1862
(*The West of Ireland*, by Henry Coulter)

Some levelling and fencing had been carried out on the road from Westport to Killary Harbour. Paving work was done under Erriff Bridge in 1829, but work on the parapet walls was suspended for the season. It was, however, passable and corn from the head of Lough Corrib was taken across it by cart to the market at Westport instead of Galway, 'though the distance is nearly equal', because of the inferior state of the Galway road.

Mountain floods and high tides in 1828 had also damaged the Killary end of this road and had brought down Leenaun Bridge, the collapse of which, and the many attempts to reconstruct it, is referred to by Maria Edgeworth in 1833: 'Here was an arm of the sea, across which Mr Nimmo had been employed to build a bridge, and against Big Jacky Joyce's advice he would build it in some part, where Jacky prophesied it would be swept away in the winter; but twice the bridge was built and twice was swept away. Still Nimmo said it was all the fault of the masons and the embankment, and his theory of bridges could not be wrong; and a third time he built – and there we saw his bridge or the ruins of it left in the sands, embankments swept away'.[32] However, as the road was still passable for carts, the bridge and road were at this time (1829) left in their damaged state.

The large arch had been closed on Maum Bridge, at the head of Lough Corrib, and the river diverted to its new course: 'We have added the wing walls, turned the river into its new course, and filled up the old bed, and carried an embankment across and upon the deep bog, so that it is now passable for carts'.[33]

The arches of Toombeola Bridge were closed and the line from there to Clifden was given an additional gravelling. Thomas Martin had completed the quay at Cloonisle and was expected to build stores at the mouth of the Ballynahinch River. Work was still in progress on the Roundstone road south of Errisbeg. North of Clifden the new roads were kept in repair but little progress was made on the road on the southern shore of Killary Harbour. Nimmo reported that improvements in Connemara were continuing: 'a spirit seems to have been infused into the district likely to produce important results, if not untimely checked'.[34] Nimmo was already aware of how nature had a way of 'checking' any progress in this corner of Ireland.

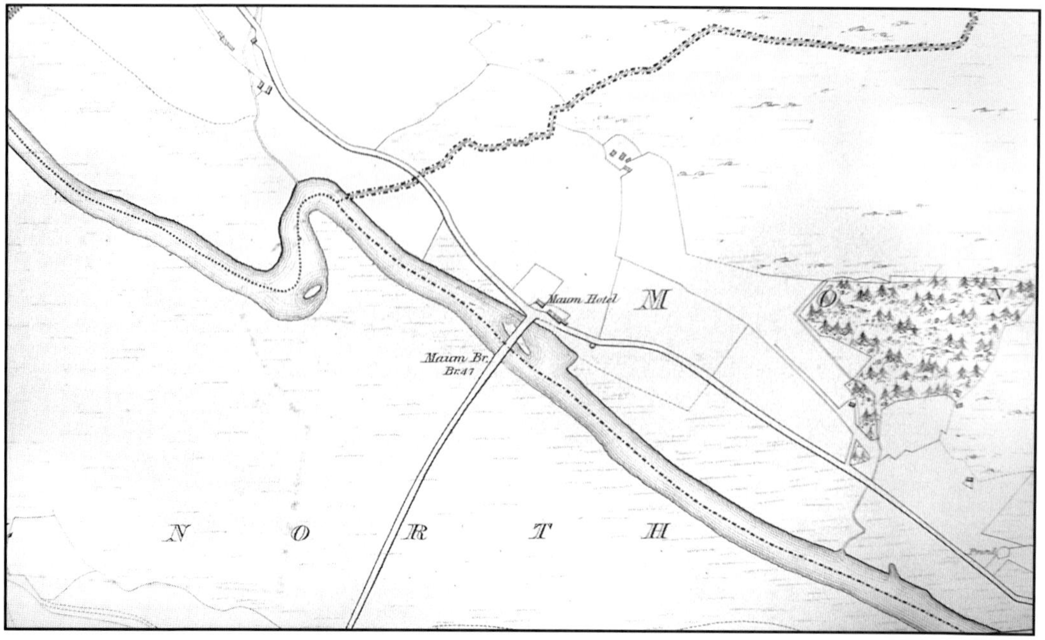

Map showing Maam Hotel and Bridge circa 1839
(First Edition 6 inch Ordnance Survey Map)

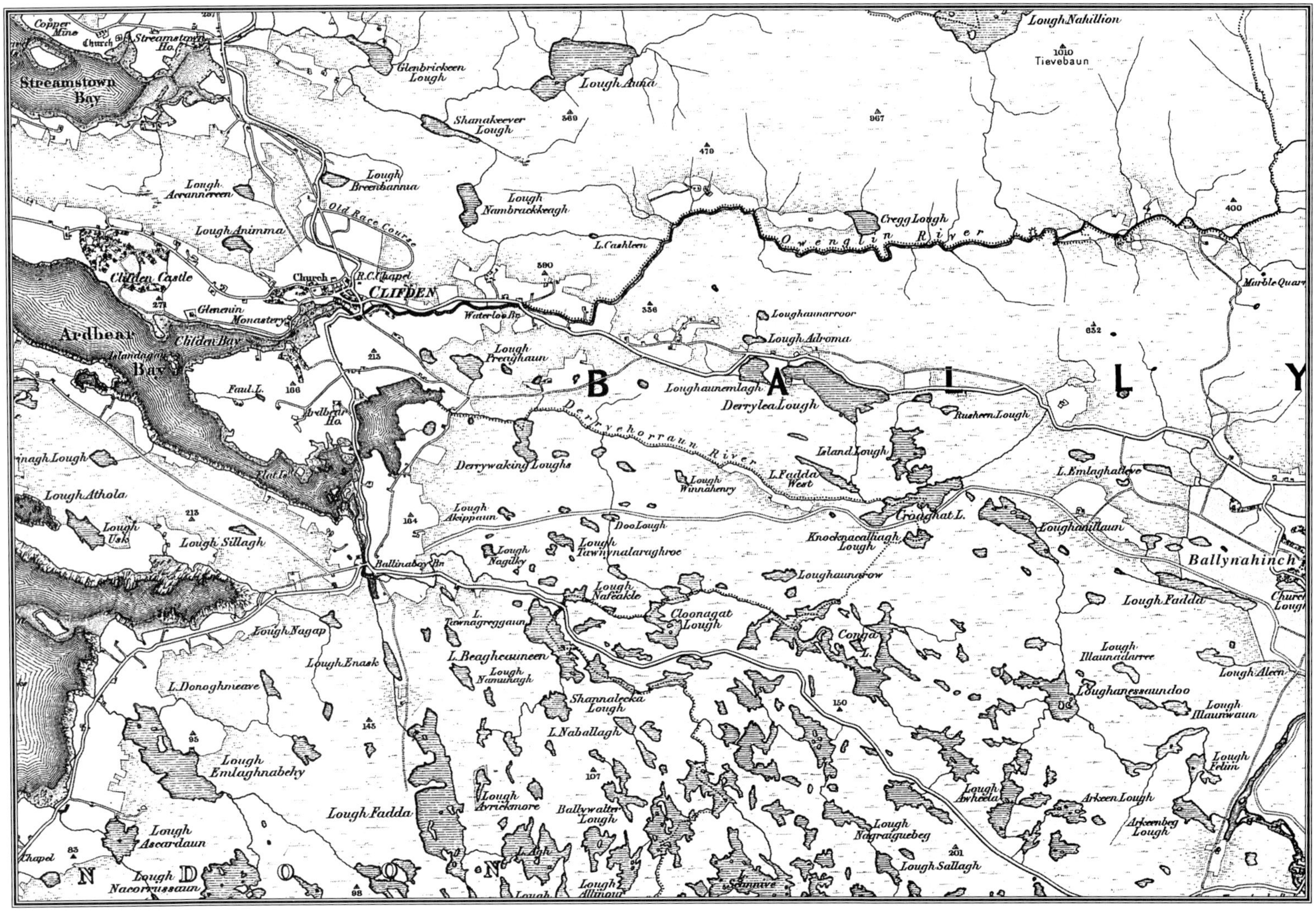

Map showing the three roads from Ballynahinch to Clifden circa 1862
(First Edition 1 inch Ordnance Survey maps)

Roads completed in 1829

In Leitrim a section of road was opened at Arigna, which had facilitated the erection of several stone houses and cultivation was extending into the mountain districts. A valuable flagstone quarry had opened and both it and the mines were offering employment. A four-mile stretch of the Celtenavina Colliery to Lough Gill road was opened from Ballinamore Bridge (Dromahair). This was used by those carrying corn and butter from the neighbouring mountain district to Sligo, with limestone been brought up as back-carriage. In Roscommon the road from Castlerea to Ballymoe was open and described as 'a great thoroughfare'. Nimmo, at the request of 'some respectable country gentlemen', had prepared a plan to be taken before the Galway Grand Jury for a presentment to carry out improvements on the road from Ballymoe, by way of Dunmore and Tuam, to Galway, 'with a view, in conjunction with the operations in other quarters, to open a great line of coach communications across the kingdom from Galway to Belfast'[35]

Nimmo concluded his 1829 progress report to Dublin Castle by reiterating the advantages the public works had brought to the west, pointing out that the resulting increase in revenue to the crown had already gone a long way towards reimbursing the government for the outlay:

> It is unnecessary for me to repeat in this Report that there is an evident spirit of improvement and increased industry visible in all this district, mainly, but perhaps not altogether, arising from the expenditure of the public money, in opening new lines of communication, new markets, and providing, to a certain extent, a supply of circulating medium in the hands of the labouring population. Capital has appeared, and been invested in buildings, agriculture and trade, where none was supposed to exist, and although the individual portions may be small, the aggregate amount is considerable. This fact is best seen in the increase in the revenue. The customs and excise of Galway have fully doubled since 1822, and, exclusive of the improvements in Mayo, Sligo etc., the surplus must have gone far to pay off all the sums Government have expended since that period in the province; while the population are now enabled to pay the same rate of taxation per head as the rest of the kingdom.[36]

However, Nimmo warned that he could not foresee the completion of the public works in his district as long as parliament continued to grant only 'limited' sums, 'for an object so evidently profitable to the State', and so much of which went into administrative charges and to repairs.[37]

Crop failure and further distress in 1830

Nimmo did not have to wait long for the 'spirit' of the west to receive its 'untimely check'. A partial failure of the potato crop in 1829, accompanied by a failure of the fishery, left the people short of food supplies and anxious for the coming months. As early as February 1830 calls went out from the people of Galway and Mayo for charitable donations, government assistance and public works, to help alleviate the distress already evident among the poor.[38] As the year unfolded it brought added misery. The newspapers reported 1830 to be the wettest year in living memory in the west: 'The gales about the period of the equinox continued longer than usual, accompanied with great rains, so that much of the oats was destroyed'. Provisions in the west were reported to be low and expensive. The price of food rose in the market place as the severe weather continued through the winter: 'The cold, wet and stormy weather continued throughout the remainder of the month of September and far into October. Great storms occurred at the end of November, and early in December, high floods – several bridges were swept away.

Great damage was done in Galway ... a great storm about the middle of November this year was most destructive to shipping along the coast. The tide was driven into the town of Sligo, leaving some vessels laden with corn high and dry on shore. The hurricane was also felt severely inland'.[39]

To prevent the poitín makers from getting their hands on what little corn was available to them, the people on the north coast of Mayo destroyed the kilns that dried the malt and pulled down the houses where illicit distillation was carried on. At Belmullet and Clifden, large crowds gathered on the quay to prevent the merchants from exporting the corn and it was only on receipt of assurances that representations would be made to the government on their behalf, that they eventually backed down and allowed the merchants to proceed. Nimmo used the incident to highlight the advances made in the west in recent years and to emphasise the improved habits of the people:

> The assemblage of the people at Clifden, Bellmullet (sic) and other parts to prevent the exportation of grain, though not productive of any disturbance is a circumstance worthy of particular observation. In 1822 when these works commenced there was no grain to export, what barley was raised was entirely for distillation, towns and markets have risen, buyers appeared, and in consequence an increased degree of agriculture throughout all the neighbourhood. The corn stores are a sure resource in the event of one of those seasons of famine recurring, which are so unavoidable with a population depending solely on so precarious a crop as the potato.[40]

Dublin Castle again came under pressure to expand its public works in the west. There was certainly plenty of scope for large-scale employment on the many works left unfinished in the Western District. Lists of suitable proposals were brought to the attention of the administration. The violent storms and heavy rains brought a further failure of the crop in 1830 and conditions deteriorated throughout the district. The distress in Geevagh and the Mount Allen Colliery on the Sligo-Leitrim border was reported to be so great that the chief secretary advised Nimmo to begin road works there immediately. Mining companies in the region were anxious to further develop their interests and the government was willing to assist them.[41]

Nimmo dispatched a man to open works on the road through the mountainous district of Geevagh to the banks of the Arigna River. The number of applicants for work was described as 'enormous' and they were immediately employed in spadework, repairing drains, fences and embankments. However, Nimmo felt that although the work was much needed, it was poorly executed, and the line was 'still imperfect'. Similar work was carried out on the Celtenavina Colliery road near Dromahair, County Leitrim, with funds diverted from the Connemara coast road. Nimmo took this opportunity to begin work on a bridge across the Shannon below Lough Allen on the Drumshambo road. Nimmo had previously pointed out the necessity for this bridge to link the new road on the Roscommon side of the river with the county road, going on under his direction, and running from Manor Hamilton to Drumsna.[42]

Given the conditions of the people, Nimmo was anxious to offer employment to large numbers wherever possible. A good deal of the funds received in 1830 was spent on repairs to keep the roads open after damage sustained by the bad weather. In Mayo, £600 was spent extending the north coast road through the vale of Glenamoy using spadework, instead of erecting bridges, which was the original purpose of the grant. The spadework would, in Nimmo's opinion, 'give more general employment than the building [of bridges], which from its nature must be confined comparatively to a few'. Four and a half miles was 'formed, fenced and drained and half a mile gravelled', bringing instant employment to a 'distressed peasantry … entirely secluded from any employment but that of illicit distillation' and 'restoring them to peace and diligent attention to earning the wages offered by honest industry'.[43]

The Central Erris Road, criticized by Bald and Bingham, received £800 for repairs and small improvements. Nimmo did not feel this was sufficient and complained that, never having received sufficient funds to complete this road, the entire route was then 'almost worn out and requiring nearly a total renewal'. The two bridges left unfinished the previous year were still in need of attention, but bad weather and the late allocation of funds prevented any further work been done on them in 1830. Instead, at Fauleen, Nimmo fell back on the method so frequently adopted by his predecessors, and so often criticized, and considered re-routeing the road so as to avoid the expense of building the bridge.[44]

In his 1830 report to the chief secretary, Nimmo outlined the condition of the roads and their usefulness to the region, pointing out that in Connemara good accommodation could be had at 'Tully, Clifden and Roundstone, so that the country [was] gradually advancing in civilization and comfort'. The Westport to Killary road, he reported, was kept in good repair but was in need of some 'short changes to avoid hills'. Erriff Bridge had been raised to the parapets and, here again, carts, 'the natural consequences of improvement', were on the increase. The Westport to Claremorris road was an extremely busy one. This was the shortest route to the port for grain from this 'extensive agricultural county'.[45]

In Galway, Leenaun Bridge had recently been rebuilt, but, as previously mentioned, we later learn that this too did not stand the test of the Connemara weather. On the south coast, the road west of Roundstone was brought down to lower ground on the west side of Errisbeg mountain and work was progressing in extending it to Ballyconneely and Ballinaboy. This line, Nimmo pointed out, was important for the protection of property and for the collection of revenue, as it was along this shore that most wrecks took place and 'the present line of communication is not practicable even for a horseman'. The bridge at Toombeola was then closed while improvement work was under way on the line from it to the old centre road. The contractor on the section from Toombeola to Clifden had to be removed due to poor workmanship. The section from Ballinaboy Bridge into Clifden, 'which is a great thoroughfare', had a repair of gravelling but improvement begun at the town of Clifden was suspended for a time. No explanation was given as to why, except to state that this was at the request of John D'Arcy.[46]

The coast road between Galway and Barna, and the works at Barna quay, had sustained severe damage from an unusually high tide in October: 'an extraordinary storm and elevation of the sea took place in the upper parts of Galway Bay … though not a spring tide, the sea rose suddenly several feet higher than was ever known'. Nimmo planned to carry out the repair work with spade labour. Nimmo's Pier [Slate pier] at the Claddagh was also affected by the same high tide, it was 'overtopped by the water and stripped of its parapet wall', and some of the stones in its front were dislodged by a vessel beating against it. As the Slate was constructed as public works it fell to Nimmo's department to provide funds for immediate repairs.[47]

Nimmo's estimates for completing the works in progress and for new works to cater for the immediate distress were £27,000. His observations at the end of this report stressed the need for more government roads to assist in the development of agriculture to replace the collapsed kelp trade:

> The entire destruction of the kelp manufacture has been productive of much distress among the inhabitants of this coast, and it is most desirable that liberal encouragement should be given by opening up roads that some substitute be found in the extension of the agriculture of the interior.[48]

In a covering letter to the chief secretary, Nimmo frankly declares he was unable to complete any individual road of sufficient extent to be delivered into the hands of the Inland Navigation Board.[49] The principal reasons for this were the 'late period at which the grants were issued, and the very small amounts placed at [his] disposal'. The government, he complained, never supplied sufficient funds in any one year to complete any works. He was further hampered by the government's policy of responding to urgent claims of distress by transferring funds from existing, well-advanced works to the opening of new works in the distressed districts. This resulted in the suspension of the earlier works and the new works being left unfinished due to insufficient funds.[50]

MONIES expended under the direction of Alexander Nimmo, Civil Engineer, for carrying on certain Public Works in the Western District of Ireland.

	£.	s.	d.
From 5 January 1826 to 5 January 1827 - - - - -	27,229	4	6
5 January 1827 to 5 January 1828 - - - - -	10,199	1	6
5 January 1828 to 5 January 1829 - - - - -	12,999	1	6
5 January 1829 to 5 January 1830 - - - - -	10,699	1	6
Total sum from His Majesty's Treasury between 5 January 1826 and 5 January 1830 - - - - - - - - £.	61,126	9	–

Table showing monies expended under the direction A. Nimmo on public works in the Western District from January 1826 to January 1830.
(BPP 1830 [624], xxvii)

Irish Fishery Board abolished

In 1830 the government published a breakdown of the money issued by the Treasury to Nimmo for public works in the Western District from 1822 to 5 January 1830. In this Nimmo is shown to have received £167,430 and this is the sum frequently quoted as the total expended by him in the Western District. However, this figure does not include funds issued to Nimmo in 1830 and 1831. Neither does it include funds issued under the Fishery Board.

The fishery laws expired on 5 April 1830, bringing the term of the Irish Fishery Board to an end. The board had warned the government against an abrupt withdrawal 'of established regulations' and financial support, provided to the fishery industry, 'which had already produced considerable advantages to the Maritime Districts of Ireland, and promise still more extensive benefits'.[51] The prosperity generated by government aid had, according to the fishery board, produced a more peaceful, industrious and grateful maritime peasantry: 'the generally peaceable and industrious habits of the Maritime Peasantry, during a period of considerable disturbance in other parts of the Country, furnish satisfactory proofs of the good effects that have resulted from aid afforded; and the Commissioners have been assured by some of their practical officers that a grateful sense of the beneficence of the Government has very generally succeeded to distress and apathy'.[52]

At the time there was no official statement on a replacement, but the government was considering ways of consolidating various works undertaken at public expense and the need to regulate these more fully. The duties of the fishery board were transferred to the Inland Navigation Board. These were limited to the completion of the piers begun prior to the 5 April 1830 and the collection of the outstanding loans issued to the fishermen.[53]

The last three reports issued by the fishery board show the increase in the number of boats and men employed in the fishery stations at Westport, Clifden and Galway in the late 1820s. They also give the numbers employed in ancillary industries such as coopers, fish curers, sail and net makers. The government later published a report on the 'state and utility' of the fishery piers, drawn up by the Board of Public Works in 1833. As previously shown, Nimmo designed many of these piers during his term with the Fishery Board and supervised their construction, although this fact is not always mentioned in the report.

Abstract of Reports on the state and utility of the Fishery Piers around Ireland; prepared by direction of the Board of Public Works, in 1833.

Fishery Piers.

Names of Piers, and where situated.	Sources from which Funds were derived for their construction.	OBSERVATIONS.
17. KILRUSH, Co. Clare.	Grant from Government, £1,340 0 0 Rt. Hon. J. O. Vandeleur, 500 0 0 £1,840 0 0	There are 20 small hookers engaged in fishing and dredging for Oysters, belonging to this port, affording employment to about 200 persons. The Pier is of much use in a mercantile and agricultural point of view. A small steam vessel plies between Kilrush and Limerick.
18. CARRIGAHOLT, Co. Clare.	Grant from Government, £379 0 0 Dublin Committee, 99 10 0 £478 10 0	A useful Pier both for the fisheries and trade. Six hookers of seven tons each, and upwards of 500 canoes or corachs, (small cots, some covered with horse-hides, others with canvass laid over with composition,) give employment to about 400 persons in fishing, and to farmers occasionally. Turf shipped from hence.
19. KILBAHA, Co. Clare.	Grant from Government, £1,148 18 0 Dublin Committee, 115 7 0 Mr. Westby, 200 0 0 £1,464 5 0	Owing to the facilities afforded by this Pier for landing sea manure, the country round it bears evident marks of agricultural improvement. It is also useful to the fisheries in which about 110 persons are exclusively employed, and it is also an asylum harbour for small craft coming in from Loop Head and other fishing stations, to the Shannon.
20. DOONBEG, Co. Clare.	Grant from Government, £386 6 0 Dublin Committee 99 10 0 £485 16 0	This Pier is not much frequented by fishing boats. It is, however, very useful for agricultural purposes, and also as a place of refuge in hard weather for small craft.
21. SEAFIELD, Co. Clare.	Grant from Government, £617 16 0 Dublin Committee, 115 7 0 £733 3 0	Can be approached at spring tides only, by a vessel of 12 tons, but is capable of much improvement, and might then be very useful.
22. LISCANOR, Co. Clare.	Grant from Government, £2309 0 0 Dublin Committee, 115 0 0 Sir Augustine Fitzgerald, 607 0 0 £3031 0 0	This Pier is situated on a very bold and dangerous shore, exposed to the force of the Atlantic, and has been several times much injured. It however affords considerable shelter to small craft, and is much used for the purposes of trade.
23. BURRIN QUAY, Co. Clare.	Grant from Government, £351 0 0 Dublin Committee, 53 7 0 £404 7 0	30 Hookers, of about 12 tons, and 150 yawls of 3 tons each, frequent this place, giving employment to about 500 men in the Fisheries. Large quantities of corn, butter, pigs, and sheep are shipped here. In hard weather, 100 sail of small craft have been known to take refuge in this harbour.
24. ST. KITTS, Co. Galway.	Grant from Government, £399 13 0 Dublin Committee, 64 17 0 £464 10 0	But little benefit is derived from this Pier, as none but small craft can approach it. No road leading to it.
25. KINVARA, Co. Galway.	Built at the expense of the County.	This is a useful Pier, both for commerce and Fisheries, besides facilitating the supply of the surrounding district with turf, which is brought in considerable quantities from Cunemara. There are about 120 men and boys employed in fishing.
26. DURAS, Co. Galway.	Grant from Government, £156 18 0 A Contribution to assist Mr. Lynch in building a Pier.	This appears to be a work more of private than of public accommodation, although a small sum was contributed towards its erection from Government funds.
27. CLORAN BRIDGE, Co. Galway.	Built by County Presentment.	Is much used for landing sea-manure and turf. 30 fishing yawls, of 3 tons each, frequent it, employing about 70 persons.
28. BALLYNACOURTY, Co. Galway.	Grant from Government, £710 0 0 Dublin Committee, 75 14 0 £785 14 0	Never properly finished, owing to which the works have been completely destroyed.

R 2

Abstract of reports on the state and utility of the fishery piers on the coast from Clare to Sligo in 1833

(First Report of Inquiry into state of Irish Fisheries, 1837)

Abstract of Reports on the state and utility of the Fishery Piers around Ireland; prepared by direction of the Board of Public Works, in 1833.

Fishery Piers.

Names of Piers, and where situated.	Sources from which Funds were derived for their construction.	OBSERVATIONS.
29. New Harbour, Ardfry, Co. Galway.	Built by Mr. Nimmo, supposed with Government Funds.	A very useful Pier for landing turf and sea-manure for a considerable district, as well as for affording shelter from the W. and S. W. winds, vessels frequently running up from the roadstead of Galway to shelter there.
30. Roscom, Three and a-half miles East of Galway.	London and Dublin Charitable Societies.	Reported to be a very useful Pier, at which large quantities of manure and turf are landed. The marble from the Menlo Park quarries, which is of a fine description, is shipped here. It is frequented also by about 45 small fishing boats.
31. Oranmore, One mile North East of Galway.	London and Dublin Charitable Societies.	Beneficial to the poor fishermen. Turf and manure in considerable quantities landed here.
32. Claddagh Pier, in the suburbs of Galway.	Grant from Government, £1597 0 0 Local Contributions, 623 0 0 £2220 0 0	A most convenient and useful Pier, for the numerous fishing boats belonging to this place, in number about 140 sail-boats, from 12 to 14 tons, and 50 row-boats, of 5 to 6 tons each, employing near 2200 persons. With all the advantages of good boats and accommodation, there are few places where the Fisheries are carried on with so little enterprise.
33. Slate Pier, Galway Harbour.	Built by Mr. Nimmo with Government Funds.	This is a work of much importance, as it affords considerable shelter to the Docks and Quays of Galway. The Pier is in a most dilapidated state; the cost of a thorough repair is estimated at £1397; should it be left without repairs for some time longer, it is probable that it would be utterly destroyed, and the materials forced into the navigable part of the river.
34. Barna, Co. Galway.	Grant from Government, £210 2 0 Dublin Committee 76 8 0 £286 10 0	The Pier was originally built by a private individual in 1799, and re-built by Mr. Nimmo in 1822; but being badly erected, in 1830 it was destroyed; and in 1831 was again partially re-built by the Officer of the Coast Guard, and Collector of Galway, with the charitable funds; it is found to be very useful for the Fisheries, and it is sheltered from the S. and S. W. gales, and the only safety harbour for small craft on 27 miles of coast.
35. Spiddle, Co. Galway.	Grant from Government, £128 15 0 Dublin Committee, 41 16 0 £170 11 0	This Pier is but of little use, and is much out of repair.
36. Costello, Co. Galway.	Grant from Government, £380 16 0 Dublin Committee, 80 15 0 £461 11 0	This is a work of but little use, and in a state of dilapidation; not worth the cost of repairing it.
37. Ballycreavey, Co. Galway.	London and Dublin Charitable Societies.	This Pier is of little use except to the Coast Guard.
38. Greatman's Bay, Co. Galway.	Originally built by Mr. Nimmo from the Government Funds.	This Pier being left incomplete, was destroyed: subsequently the materials were collected, and the Pier, for good part of its length was re-built by Lieutenant White, of the Coast Guard, and the Collector of Galway, with funds from the Charitable Societies: it is found to be very useful.
39. Killaney Pier, Isles of Arran.	Grant from Government, £564 14 0 Dublin Committee, 60 5 0 £624 19 0	There are 39 hookers frequenting this Pier, giving employment exclusively in fishing to about 350 persons. A large quantity of sea manure is shipped here for the shores of Clare and Galway. A Coast Guard station, the officers of which have made some improvements in the works with the aid of funds from the London and Dublin Charitable Societies. The inhabitants are wretchedly poor.

Abstract of Reports on the state and utility of the Fishery Piers around Ireland; prepared by direction of the Board of Public Works, in 1833.

Fishery Piers.

Names of Piers, and where situated.	Sources from which Funds were derived for their construction.	OBSERVATIONS.
40. ROUNDSTONE, Co. Galway.	Grant from Government, £393 1 0 Dublin Committee, 92 6 0 £485 7 0	30 sailing-boats, averaging 10 tons, and 40 rowing-boats of 4 tons each, frequent this Pier. The former occasionally employed in bringing corn, kelp, and turf to Galway. About 250 persons constantly employed between fishing and trading.
41. CLIFDEN, Co. Galway.	Grant from Government, £168 16 0 Mr. Nimmo it is believed also expended Government Funds on this work.	This is a Quay wall, partly unfinished. Captain Boileau, of the Coast Guard, expended some of the charitable funds in adding to its length. Corn and butter are shipped here for Liverpool, and large quantities of sea manure landed. About ninety persons employed in fishing.
42. KINGSTON, Co. Galway.	Funds of the London and Dublin Charitable Societies, under the direction of Rev. A. Thomas, Rector.	This Pier is used by the fishermen of the neighbouring creeks as a place of rendezvous. Sea manure and turf are also landed here.
43. CLEGGAN, Co. Galway.	Grant from Government, £725 9 0 Dublin Committee 115 7 0 £840 16 0	This work appears to be in a ruinous state; the materials and workmanship reported to be both bad. The pier was a useful one when in good order, and it would be worth repairing.
44. CLOONISH PIER, Co. Galway.	Built by the late Mr. Nimmo, supposed with Government Funds.	Useful for the landing of sea manure and shipping of turf for Galway. Mr. Seymour reports that the "face of the country is fast improving from the advantages derived from the sea manure, particularly so, since the manufacture of kelp was given up."
45. DERRY-INVER, Co. Galway.	Built by the late Mr. Nimmo, supposed with Government Funds.	This Pier is the only place of refuge in Ballinakil Bay, but having been badly built, is now much out of repair. The country round it is rapidly improving, and the Pier is a very useful one for the trade and Fisheries.
46. LEENANE PIER, Killenny Bay, Co. Galway.	Built by Mr. Nimmo with Government Funds.	About 30 row-boats occasionally employed in fishing, frequent this Pier. Its situation is near the junction of the Westport, Clifden and Galway Roads.
47. BUNDURRA, Co. Mayo.	Grant from Government, £344 19 0 The Marq. of Sligo, 286 18 0 £631 17 0	Of use for the landing of sea manure and sand, large quantities of which are used in improving the neighbouring mountainous country.
48. INNISTURK, Co. Mayo.	Grant from Government, £377 16 0 Dublin Committee, 99 10 0 £477 6 0	This Pier has fallen into ruins, and is a severe loss to the poor islanders, composed of about 80 families, chiefly dependant on the Fisheries for support. The Pier was built on the only landing place in the island.
49. OLD HEAD, Co. Mayo.	Grant from Government, £1388 15 0 Dublin Committee, 115 7 0 £1504 2 0 Lord Sligo, it is believed, subscribed 417 4 0 £1921 6 0	Not much frequented by fishing boats, but large quantities of sea manure are landed here.
50. CLARE ISLAND, Co. Mayo.	Grant from Government, £276 18 0	About 340 men and boys occasionally employed in the Fisheries, being also farmers; large quantities of sea manure landed, and the island improving in agriculture. Much grain shipped here for Westport. Population about 1600.
51. ACHIL SOUND, Co. Mayo.	Grant from Government, £211 15 0 Dublin Committee, 92 6 0 £304 1 0	This Pier badly executed and never completely finished. There are 50 sail-boats, averaging 8 tons, and 100 yawls averaging 2 tons each, which frequent this harbour, giving occasional employment to about 500 persons.

Abstract of Reports on the state and utility of the Fishery Piers around Ireland; prepared by direction of the Board of Public Works, in 1833.

Fishery Piers.

Names of Piers, and where situated.	Sources from which Funds were derived for their construction.	OBSERVATIONS.
52. Bull's Mouth, Co. Mayo.	London and Dublin Charitable Societies.	This small Pier was built under the superintendence of Lieut. Nugent of the Coast Guard, and affords shelter to about 40 yawls during the Herring Fishery, from N. and N. W.
53. Tarmon, Blacksod Bay, Co. Mayo.	Grant from Government, £1071 0 0 Dublin Committee, 99 10 0 £1170 10 0	This is an extensive and tolerably well-built Pier. 20 hookers of 6 tons each, and about 100 yawls frequent it. Those persons engaged in the Fisheries on this part of the coast, are, generally speaking, farmers as well as fishermen.
54. Saleen, Near Newtown Bingham, Co. Mayo.	Grant from Government, £674 11 0 Dublin Committee, 69 4 0 £743 15 0	Little used as a fishing station, but considerable quantities of corn and potatoes are shipped here for Westport, and merchandize brought back in return. A small expenditure in completing the inner dock would make this harbour much more useful.
55. Belmullet, Co. Mayo.	Grant from Government, £282 0 0 Estimate, 564 0 0 Contract taken at the Estimate by Mr. Coates, who was interested in the works.	This Pier affords good shelter from the S. E. and W. winds; at ¾ flood spring tides a vessel of a hundred tons can get alongside the Pier. Described as a good fishing station, but the fishermen not enterprising. The Pier is also of great use for mercantile purposes, the imports being timber, iron, slates, &c., in return for grain, and other agricultural produce. It is stated that from this small village about 1300 tons of oats were shipped for the English and Scotch markets last season. It has often been suggested to cut a canal across the narrow isthmus which divides Broad Haven from Blacksod Bay, by which means a vessel could put to sea with any wind from either of the harbours.
56. Killalla, Co. Mayo.	Grant from Government, £54 12 0 Dublin Committee, 41 16 0 £96 8 0	A good fishing station; about 360 persons occasionally employed. Much sea-manure landed here. The country reported to be much improved since 1827. About 7000 tons of barley and oats annually exported. Imports, timber, iron, slates, and coals (for malting.) Harbour tolls are charged. The Pier is an old one, and has been newly repaired with public funds.
57. Pulocheny, Co. Sligo.	Grant from Government, £468 9 0 Dublin Committee, 30 5 0 £498 14 0	The work was left in an unfinished state, and what had been executed is now down, the materials lying on the strand. There is a Coast-guard station here, and if the works were completed a vessel of 150 tons might run in at ¾ flood.
58. Raghley, Co. Sligo.	Grant from Government, £1,088 9 0 Dublin Committee, 191 15 0 Sir Gore Booth, 120 0 0 £1,400 4 0	About 200 persons occasionally employed in the fisheries at this Pier. Large quantities of sea-manure also landed here. A Coast-guard station. Nearest road to the pier distant ½ mile.

Public endorsement of Nimmo's work

A government committee set up to investigate the state of the poor in Ireland in 1830 focused its attention on what it viewed as the success of the public works carried out under Nimmo in the west. The committee was full of praise for the works and the benefits derived from them and it strongly recommended to the House of Commons that it extend the application of a 'principle [proven] to be advantageous in so many important particulars'. It drew heavily on evidence given by Nimmo before earlier committees and commissions, and on his reports from the Western District, to help form the opinion that the public works had contributed greatly, not just to the districts in which they were carried out, but to the country at large, in the form of increased revenue through customs and excise. Central revenue, at this time, depended entirely on customs and excises for its income. They also contributed to the suppression of smuggling and poitín making: 'That the effects produced by these public works appear to have been – extended cultivation, improved habits of industry, a better administration of justice, the re-establishment of peace and tranquillity in disturbed districts, a domestic colonization of a population in excess in certain districts, a diminution of illicit distillation, and a very considerable increase to the revenue'.[54]

They urged the government to adopt for the future 'a system of grants in aid of local funds', and took the case of Clifden to show how the increase in revenue from districts in receipt of public works could, in time, cover the outlay. Clifden had paid no revenue to the state in 1822, but that had changed with the opening up of public works in the region and by 1826, 'taxation to the amount of £2,500 was collected; and in 1828, it was augmented to £6,080.7.3', derived from an increase in the 'consumption of spirits, tobacco, tea, sugar, pepper, butter, glass, timber, and other articles; and thus proving indisputably the increased wealth and the improved habits of the people'.[55]

The committee recommended that all public works under the direction of the government engineers, along with the functions performed by the Irish Fishery Board, the mail-coach road department of the Post Office and the Board of Inland Navigation be consolidated 'under one fixed and responsible authority'. This would ensure that, in the future, all works would be properly selected and executed, and that all funds would be audited and examined, and payments made direct to the labourers.[56]

The government established the Irish Board of Public Works in the following year, 1831. The board superseded five boards that were in operation up until this time, the Board of Works for Dublin, the Board of Inland Navigation, a Board for lending Money from the Consolidated Fund, the Commissioners of Kingstown Harbour and the Commissioners of Dunmore Harbour. The Board of Public Works then became the sole body with responsibility for issuing loans in aid of public works and for issuing grants for the formation of roads, bridges, small piers and harbours along the seacoast. And in taking over the Board of Inland Navigation it also became responsible for the fisheries of Ireland and for the completion and maintenance of the roads and bridges constructed by government on the west coast.[57]

Year	Districts and Stations	Decked Vessels: No.	Decked Vessels: Tonnage	Decked Vessels: Men	Half-decked Vessels: No.	Half-decked Vessels: Tonnage	Half-decked Vessels: Men	Open Sail Boats: No.	Open Sail Boats: Men	Row Boats: No.	Row Boats: Men	Total Number of Fishermen employed
1828	Westport	6	137	33	19	167	76	115	389	940	4,196	4,694
	Clifden	- -	Nil.	- -	16	306	48	157	471	1,133	4,532	5,051
	Galway	11	267	54	4	82	20	465	2,325	1,191	4,764	7,163
1829	Westport	7	147	38	20	175	81	116	393	1,000	4,436	4,948
	Galway	12	281	59	6	111	30	484	2,420	1,199	4,796	7,305
	Clifden	- -	- -	- -	16	306	48	167	501	1,235	5,459	6,008
1830	Westport	10	170	49	21	179	84	117	397	1,085	4,79[illegible]	5,321
	Galway	12	281	59	9	150	45	499	2,495	1,219	4,876	7,475
	Clifden	- -	- -	- -	15	306	48	170	522	1,295	5,714	6,284

Year	Districts and Stations	Fish Curers: No. of Curing Houses	Fish Curers: No. of Men employed in them	Number of Persons employed in Gutting and Packing	Coopers: Number of them	Sail Makers: Number of them	Net Makers: Number of them	Total Number of Fishermen and Tradesmen employed in the Irish Fisheries
1828	Westport	Nil.	Nil.		Nil.	Nil.	Nil.	4,694
	Clifden	Nil.	Nil.		Nil.	Nil.	Nil.	5,051
	Galway	19	270		27	4	600	8,064
1829	Westport	There are no regular curing houses in this district; the herrings are cured in bulk, and ungutted. Same applies to sail and net makers, the fishermen making the latter. For larger vessels English canvas is used; for smaller boats, baidie linen-made in the country.						4,948
	Galway	23	304	460	27	4	600	8,700
	Clifden	- -	30	- -	- -	[illegible] -	Supposed 400	6,438
1830	Westport	No regular curing houses			- -	- -	- -	5,321
	Galway	23	764		28	4	600	8,871
	Clifden	30	300		12	6	- -	6,602

An account of the number of vessels, fishermen and tradesmen employed in the fisheries at Westport, Clifden and Galway Stations from 1828 to 1830

(Ninth, Tenth and Eleventh Report of the Commissioners of the Irish Fisheries)

Máimín Pier, Gorumna Island 1943
© Fáilte Ireland

Cleggan Harbour 1948
© Fáilte Ireland

Chapter 7

Nimmo's removal from the Western District 1831

The Marquess of Anglesey replaced the Marquess of Wellesley as lord lieutenant to Ireland in March 1828. Anglesey, a strong supporter of Catholic Emancipation and Reform, but against the Repeal Movement, would serve two terms as lord lieutenant, 1828–29 and 1830-1833. He had served with the Duke of Wellington at the Battle of Waterloo in 1815. During the battle he had eight or nine horses shot from under him before eventually losing a leg and being forced to withdraw from the field. Some years earlier Anglesey, a father of eight children, had eloped with the wife of Henry Wellesley, the younger brother of the Marquess of Wellesley and the Duke of Wellington. The couple eventually married and went on to have ten children together, making him the father of eighteen children in all.[1]

In the first year of his second term, Anglesey toured the distressed regions of the west to view for himself the state of the country and to ascertain the measures necessary to alleviate the conditions of the poor. Early in 1831, in response to the immediate distress so widely reported, parliament sanctioned the issue of an advance of £50,000 for temporary relief in Ireland; but the House was dissolved before an Act was passed and the money issued.[2] The act, however, was expected to go through the House in the next session and the funds would then be made available at the discretion of the lord lieutenant. In the meantime, Anglesey undertook his tour of the west so that 'from personal enquiry and observations', he would be in a position to consider which of the measures proposed would be 'most conducive to the relief of the poor, and the security of the empire'.[3]

The lord lieutenant arrived in Limerick on Sunday 3 April 1831 and spent the next eleven days travelling north along the coast. Accompanying him on his tour was Lieutenant Col. William Gosset, who had replaced William Gregory as under-secretary in the previous December, his private physician and two aides de camp. His decision to travel among the people without an armed escort greatly impressed the press and public, at a time when, we are told by the press, 'a spirit of discontent and disaffection hurries the unfortunate peasantry to acts of cruelty and bloodshed'. The press called on the people to cooperate with the lord lieutenant 'in providing for the wants of the poor, and good order of society'.[4]

Anglesey spent Wednesday night, 6 April, in Ennis, the following night in Gort, and headed out for Galway on Friday morning. At Oranmore he left his carriage and, mounting a fine chestnut horse, he rode into Galway with Valentine Blake of Menlo Castle riding next to him. The merchants, tradesmen and fishermen came to welcome the lord lieutenant, described by the *Connaught Journal* as 'the hero of Waterloo, the friend of Ireland, and the advocate of suffering humanity'. The party went to the courthouse where the lord lieutenant received the freedom of the corporation and addresses from several groups, among them the lord mayor, Colonel John Blake of Furbo, the Galway Grand Jury and the Harbour and Canal commissioners.[5] The Harbour and Canal commissioners had been set to promote the development of new docks and quays for the town, and a canal from Lough Corrib to the sea.[6] The commissioners suggested that, among the many 'benevolent objects' Anglesey was considering for 'the relief of the deep distress and the permanent amelioration of the country', he consider the improvement of the harbour and the re-construction of the docks and quay, 'presenting at the present moment a picture of ruin and dilapidation'. The condition of the docks greatly hindered the development of the port, which was 'so favourable situated for general trade, particularly that with America'. For a 'comparatively trifling expense', they suggested that an important inland navigation could be opened 'through the counties of Galway and Mayo, by means of Lough Corrib, Mask, and Carra', and united with the sea at Galway. They requested that their project be considered in the distribution of any fund for public works in Ireland. The adoption of such a project, they assured him, would perpetuate the memory of his visit 'by the noblest of monuments – a work of great, of general and of undoubted utility'.[7]

First Marques of Anglesey (1768 – 1854)
Marble bust on display in the library of the Royal Dublin Society

In reply Anglesey told the commissioners that he would be happy to consider any project that would serve the town and neighbourhood. To an address delivered by the Grand Jury, he replied in a cautionary tone, telling them not to expect too much from the government. The people, he said, expected him to have all the answers to the ills of Ireland, but such was not the case: 'I regret to say it is not in my power to relieve their distress – God knows I have not the means. We live in the age of economy – indeed, I might say, of parsimony'. The country, he advised them, could not 'bear an increase of burdens' and, 'to use a homely phrase well known to some of you. "We must cut our coat according to our cloth"'. Any distribution of public funds would go 'with the strictest impartiality to those places, which, on the fullest consideration, may appear to stand most in need of it'. But, he warned them, 'Ireland can never prosper unless there be peace, quiet, concord, and union amongst yourselves' and 'should the existing laws not be found sufficient to keep peace, and to suppress riot and outrage of every description, I shall (and it will be the most painful hour of my life), lose no time in asking for more power'. He took his leave from the gathering and went to take a look at the docks before leaving on horseback for Ross House, the home of Robert Martyn.[8]

From Ross House he travelled to Ballynahinch to spend the night at 'the hospitable mansion' of Thomas Martin. On his journey he travelled on the old county road through central Connemara and examined stretches of the new road, still lying unfinished. Maria Edgeworth, who travelled the same route two years later, described the many wide sloughs her party encountered on this road and how they required the assistance of an army of local men and boys to help lift the carriage and occupants across. One of them, Ulick Burke, a 'great giant' of a Connemara man, told her all about his encounter with the lord lieutenant on his journey to Ballynahinch Castle.[9]

Next morning Anglesey travelled on to Clifden, where he received an enthusiastic welcome from John D'Arcy and the people of the town:

> On his Excellency's approaching the town of Clifden, he was received by an immense concourse of people, who hailed his arrival with deafening cheers, which were gracefully acknowledged by the Noble Visitor. On arriving at the grand entrance to the Castle, the multitude had increased to a considerable amount; and when entering the gate, after the cheering had subsided, his Excellency was ushered in by a Royal salute from the Castle battery which was responded to by the multitude.[10]

There were several banners attached to the turrets of the Castle, one bearing the words '*Cead Milla Faltha*'. At the Castle entrance, John D'Arcy presented the lord lieutenant with an address from the inhabitants of the town. This was the first time that the region had been visited by someone of such 'high and exalted situation' and they congratulated him for making the journey. After the usual words of welcome and praise, D'Arcy concluded by pointing out that the town of Clifden was law abiding and loyal, and that, in the space of just fifteen years, 'by increasing industry and commercial integrity', it was rapidly advancing 'to affluence and independence'. In his reply Anglesey acknowledged their loyalty and referred to the fact that the proprietors and clergy spoke highly of the people. This, he told them, was a sure guarantee that their wants would 'be attended to'. Inside the Castle the dignitaries sat down to a 'splendid meal' before moving on to view the town. Outside Carr's Hotel Anglesey again spoke to the crowd and then travelled on to Renvyle and Delphi Lodge.[11]

Clifden circa 1842
Ardbear New Bridge, designed by Nimmo
(*Ireland: its Scenery, Character &c.*, by Mr & Mrs S.C. Hall)

For the next four days the lord lieutenant was the guest of the Marquess of Sligo, Sunday and Monday at Delphi Lodge, then Tuesday and Wednesday at Westport House. From Delphi he went north to Louisburgh, where a large crowd turned out to greet him, and on to Westport. Seven miles outside Westport the gentry of the town arrived in their carriages to join Anglesey and the combined parties made an impressive cavalcade as they made their way to the town. Westport was decorated with triumphal arches, and a number of banners and flags hung from the chimneys and windows of the houses. The lord lieutenant's party left Westport on Wednesday morning for Ballinrobe, Tuam and Mount Bellew, where they spent the night at the home of M.D. Bellew. On Thursday he left the west for Dublin.[12]

On his journey through Galway and Mayo the lord lieutenant travelled on roads constructed either entirely or partially with government funding, and in several places he viewed the piers erected under the auspices of the Fishery Board with the assistance of funds from Dublin Castle and various charitable organisations. All along the route he was informed of the needs of each district, the hardships endured in recent years with the failure of both the fishery and the potatoes, and the prevailing conditions of the poor and the landlords: in reply to a question put by the lord lieutenant to a gathering at Westport House, as to 'what way the portion of £50,000 could be laid out for the best advantage of Mayo on good security', the Marquess of Sligo replied that no security could be found in Mayo, as almost all of the landlords were too deeply in debt to take advantage of it.[13] This close examination of the public works, the districts in which they were carried out and the circumstances of the people and their landlords, gave Anglesey an insight not previously enjoyed by his predecessors. How this inspection of the west would influence future government spending was the question on the lips of every landed proprietor, merchant and tenant in the Western District.

Nimmo responds to criticism

Nimmo was in England during the time of the lord lieutenant's tour. But it seems Anglesey did meet Richard Griffith, and perhaps William Bald, while in Mayo.[14] Nimmo's absence left the field wide open for his critics and it appears they took every advantage of it. Nimmo himself was certainly of the opinion that the lord lieutenant, while in the west, was deliberately shown the worst stretches of road in his district. He had been taken through central Connemara on the old county road and shown a stretch of the new road lying close by, unfinished and abandoned. He was brought to Letterfrack where a large bridge had collapsed a short time before and the road from there to Killary was in bad repair.[15] While making his journey, it must have appeared to Anglesey that there was little to show after years of work and government aid. Few of the roads he travelled were complete; they were interspersed with impassable stretches, resulting from storm damage or lack of funds. Nimmo wrote to Anglesey, stating his regret that 'out of one hundred miles of New Road between Galway and Westport executed by me, and on which a carriage can run as smoothly as on any in Ireland, you should have only seen three or four miles of the North West extremity which were never completed, and which had been injured by the floods of last winter'.[16] Perhaps this was done to strengthen the case for more funds to bring the works to a speedy conclusion, but maybe also with the added intention of embarrassing Nimmo. Certainly the Marquess of Sligo, with whom the lord lieutenant spent four nights, was critical of Nimmo's work in Mayo and John D'Arcy had previously stated his criticism of works in Connemara in a letter to Dublin Castle.

The lord lieutenant was told of unpaid wages and outstanding bills going back many years. The entire neighbourhood, or at least those in his company, complained about Nimmo. In response to such criticism Nimmo wrote to the lord lieutenant on 20 April 1831, pointing out the reason he was sent to the west in 1822, outlining his brief and explaining why some works had been suspended, some unfinished, and others badly affected by the storms of the previous year. He referred to his many detailed reports down through the years, which gave the background and history of the works. He regretted the fact that someone capable of explaining these facts to him had not accompanied the lord lieutenant on his journey. In regard to the complaints of non-payment of wages, he explained; 'Allegations of this kind are not uncommon as the fact of being employed on the Public Works is a sure means of obtaining credit, but on investigation they are usually found to be barefaced attempts at imposition, the duplicate pay lists and receipts are all carefully kept in the District and any case of this kind can be easily investigated for the whole nine years past.'[17]

Nimmo's 1831 estimate for completing the various works in the Western District was waiting for the lord lieutenant on his return to Dublin. In it Nimmo estimated the expense of repairing and completing the ongoing works, as well as completing works suspended for several years; and he added estimates for new, 'useful lines of road' for Mayo and Galway. The document was very detailed and ran over five pages. In it Nimmo listed each road in the Western District that was in need of further work to bring it to completion. He outlined the work needed and gave individual costings for each section. He had, he wrote, dispatched his chief accountant to the west to audit and settle outstanding claims and on receipt of his accounts would be in a position to furnish the lord lieutenant with a return of 'the total amount expended on each work and the arrears due thereon'. He attached a lithographic map of the district to his report, pointing out the government works undertaken [this map has also proved impossible to locate]. To complete the roads in hand, and for repairs and extensions, he estimated £27,291.4s. To complete works suspended, he estimated £17,544.16s and to finance the new works proposed, £15,635. This gave a total of £60,471.[18]

The chief secretary, then Edward S. Stanley, ignored Nimmo's estimate for completing suspended works and for opening new ones; instead he turned his attention to the ongoing works. He had his staff extract the section outlining the ongoing roads in Co Mayo and forwarded this to the Marquess of Sligo seeking his opinion. Stanley had taken up office in November 1830 and he was anxious to know which works would bring immediate relief to the distressed population of that county. It seems the new chief secretary did not trust Nimmo's assessment of requirements in the west. Sligo divided the works into their respective baronies and showed Nimmo's estimated expenditure per barony. He prioritized the works and pointed out where the distress was most acutely felt at the time. He accused Nimmo of estimating twice for the same barony, when in fact, Nimmo was estimating for roads not baronies. He rejected several of Nimmo's proposals for extending roads, and also the suggestion that the Moy navigation works should be re-opened, and refused to comment on works that would benefit his own estate. In conclusion, Sligo suggested that they talk to William Bald, who was to call to the chief secretary's London office with a copy of his Mayo map.[19]

Nimmo team disbanded

Later in the same month Nimmo issued a further report to Stanley, dated 27 April. This was to be Nimmo's last report as engineer for the Western District. In it he listed the roads ready for handing over to the Director General for Inland Navigation. Many of these had been in public use for sometime and were in need of repair, mostly resurfacing and bridge building, but these defects, according to Nimmo, formed 'no obstruction to the thoroughfare upon them'. The roads covered over one hundred and fifty miles and would require about £2,500 per annum for maintenance: 'being much liable to injury from mountain floods'. But, he warned, 'a considerable addition to that sum [would] be required in the first instance', as all of the roads were 'much worn' and in need of a new coat of gravel. Nimmo was of the opinion that maintenance was not his responsibility and that any future funds granted by parliament for making roads should be used to complete 'minor improvements' still ongoing and for the repair of bridges.[20]

Nothing in Nimmo's report and covering letter indicates that he expected his term in the Western District to be brought to a close at this time. However, within a month the chief secretary was handing over the Western District to another engineer and Nimmo was instructed to cease all operations in the west, dismantle his administrative staff and hand the entire operation over to John Killaly. How Nimmo was informed of this decision is unclear. His correspondence with Dublin Castle during 1831 came from various addresses in England, (March and July he was in the Spring Gardens Hotel, London, in April he was writing from Leamington and in October and November from Liverpool); whether it was private commissions or ill health that detained him, it is difficult to say. Nimmo was engaged by the Liverpool and Leeds Railway Company in September 1830 to survey a proposed railway line and to help set up the company, and at the same time by the

Manchester, Bury and Bolton Canal Company. He was employed by the Preston and Wigan Railway Company, perhaps in early 1831, and also acted as consultant to several other railway companies in the north of England.[21] In 1830 he produced a report on the works under way in Courtown Harbour, he designed a bridge to cross the Blackwater at Youghal, which was constructed after his death by George Nimmo. November of the same year he produced a report and estimates on a proposed railway from Dublin to Kingstown; C.B. Vignoles later constructed this line.[22]

After nine years of work, Nimmo was to walk away from the west with almost his entire principal roads still unfinished and his reputation under attack. The only road completed, apart from the small stretches previously mentioned, was the Joyce Country road, running from Leenaun to Cong and this was fast falling into disrepair.[23]

(NAI)

John Killaly takes over the Western District

The chief secretary wrote to John Killaly on 28 May 1831, instructing him to travel to the west and to report on the exact state of the various works in the Western District.[24] Killaly was given the title 'directing engineer' and his son, Hamilton Hartley Killaly, joined him later as 'superintending engineer'. John was paid four guineas per day, this included all expenses; Hamilton received a salary of £500 per annum plus 10s 6d per day for accommodation and travel expenses. However, John, perhaps due to ill health, seems to have adopted a lesser role soon afterwards and his salary was later adjusted to a little over two guineas per day. Then, on 31 August (1831), Hamilton was made the principal engineer, with a salary of two guineas per day and the same allowances as before for accommodation and travel expenses.[25] The chief secretary informed John Killaly that funds for the completion of certain works would be sought at the next sitting of parliament, but he was to begin work without delay. Killaly was reminded that it was not the government's intention 'to afford gratuitous assistance' to any new works, but to expend small sums so as to bring into use those roads on which large sums had already been spent. An advance of £11,000 was obtained from the treasury, a portion of which was retained to pay off Nimmo's staff, and nine thousand pounds was put at Killaly's disposal; £6,500 was for works specified by the chief secretary, with the remaining £2,500 to go towards the completion of small works identified by Killaly himself.[26]

The funds were chiefly for the completion of any small portions of road already in progress, to bring employment to the poor in those parts of the county where the greatest distress prevailed and to repair and finish all government roads in the west so that they might be handed over to the Director General of Inland Navigation, 'in order that the repairs may hereafter be born upon the local funds'. Using the information provided in Nimmo's estimate for 1831, Stanley itemised the roads he wanted Killaly to attend too, the Central Erris Road from Castlebar to Belmullet, the Westport to Leenaun and the centre

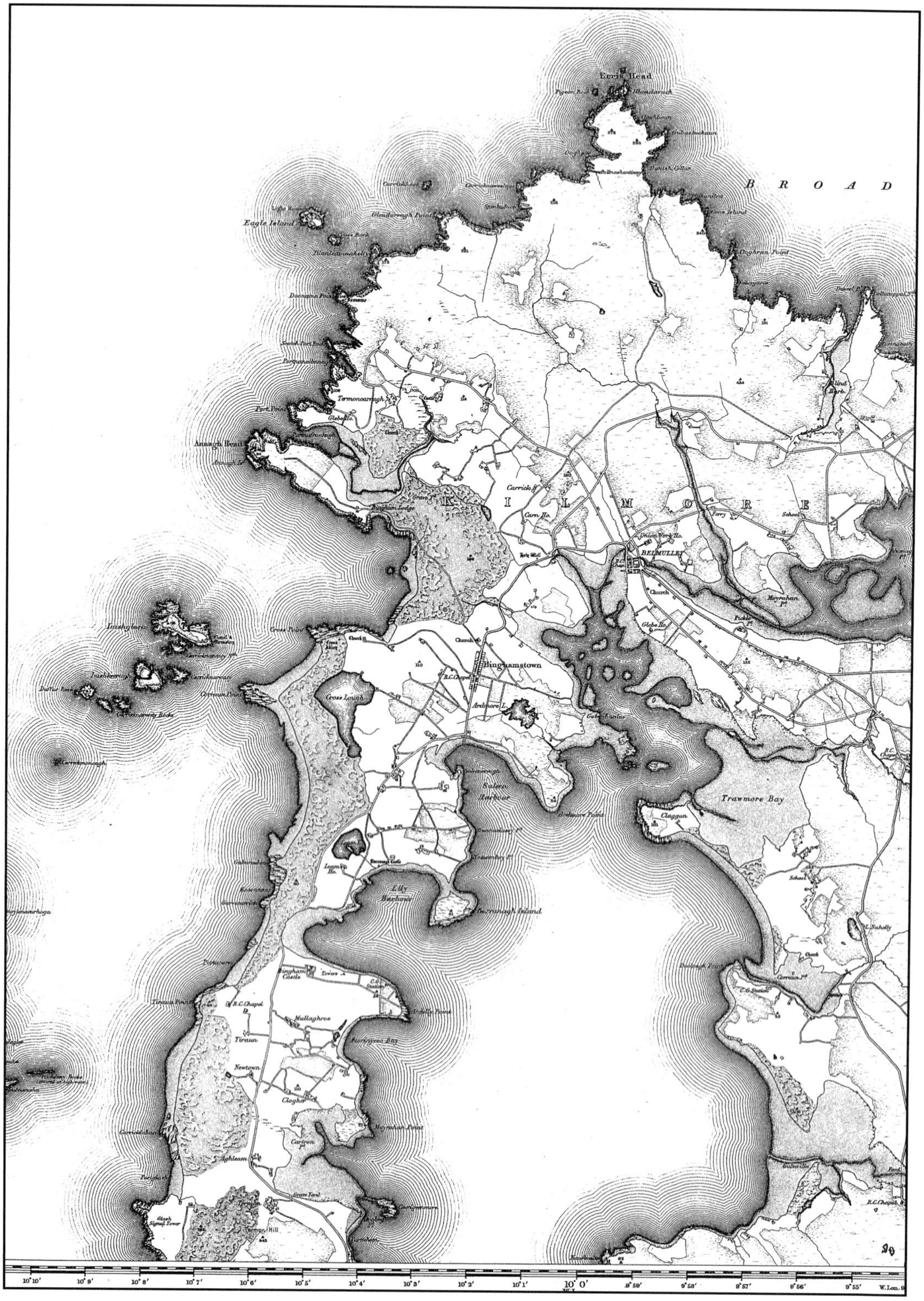

Map of Northwest Mayo circa 1862
(First edition 1 inch Ordnance Survey map 1862)

Connemara road. He was to pay particular attention to that stretch of the centre Connemara road under Samuel Jones. He was to report on Jones's progress and if it was found that he was unable or unwilling to complete his contract, then Killaly was to take the work into his own hands without delay. He was advised to spend about £2,000 on this road, directing his attention to that stretch east of Ballynahinch, which was left completely impassable. He was instructed to keep down administration costs, to make it clear to all that the government did not 'countenance any permanent establishment in that district', and to direct his immediate and constant attention to 'handing over the public roads of the county to be maintained by the county funds'.[27]

Throughout the month of June reports of distress and near famine conditions in the west filled the newspapers. Hot and dry weather causing poor growing conditions had heightened concern for the people. Meetings were held in London and other parts of England, to collect aid 'to save from absolute starvation the most destitute beings on the face of the earth'; all were calling for employment to be extended on public works.[28] In September the parliament granted £40,000 for the 'Employment of the Poor and Relief of Distress in certain Districts in Ireland'; from which Killaly would receive a further £8,000 during September and November, for work in the western parts of Mayo and Galway. Supplies to the value of £15,000 were also issued through the coastguard service and a further £3,000 was issues to Killaly in January 1832.[29]

Nimmo informed his staff that their services would not be required after 30 June 1831. When details of the changeover became known, demands for outstanding bills, some going back several years, began pouring in. Most of the staff were owed back wages and money was owed to contractors in each county in the district. Nimmo was still arguing with the administration over sums due to clear his debts, but was having little success. He was asked for a breakdown of sums due and produced a statement showing arrears of £2,382.9.6d. He was claiming £933.4.3d in back wages for his staff and £1,449.5.3d due to contractors.[30]

Nimmo's team was still in place when Killaly's men arrived in the west and friction developed between the two camps. Nimmo, writing from a London hotel, informed the chief secretary that he was unable to dismiss his men without first settling their arrears. He also complained about the attitude and behaviour of Killaly's men towards his staff: 'I am sorry also to find that some officers employed by Mr Killaly in their intercourse with mine, have been conducting themselves in a most unwarranted manner neither I should think creditable to the service nor useful to the public, which cannot I think have the sanction of that gentleman, but must greatly embarrass the final settlement of the accounts of that district'.[31]

Disagreement broke out as to which administration should pay out the arrears; Killaly wanted his staff to be responsible for paying off all debts left behind by Nimmo. Nimmo was insisting that the task fall to his officers. Killaly reported that Nimmo's officers could not be trusted and accused them of having removed materials, such as timber and iron, from the works' store, for their own use.[32] The chief secretary was in favour of accepting

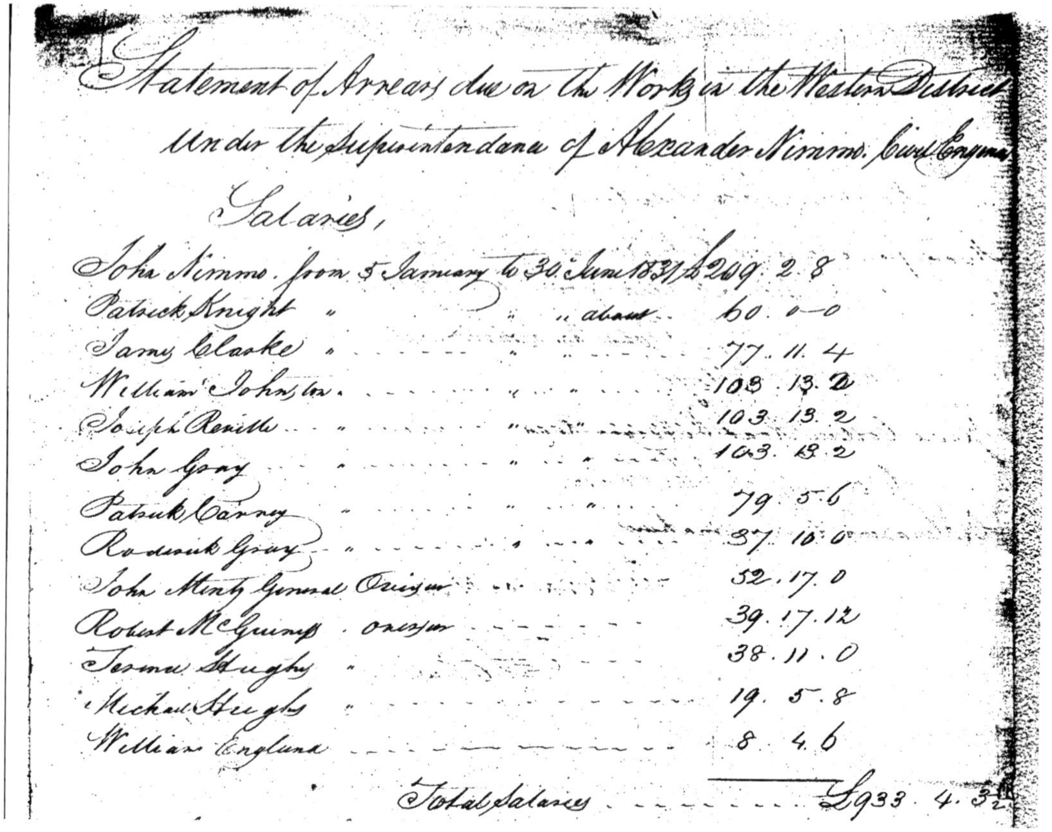
Statement of Arrears due on the Works in the Western District
Under the Superintendance of Alexander Nimmo, Civil Engineer

Salaries,

John Nimmo. from 5 January to 30 June 1831		£209. 2. 8
Patrick Knight	" " about	60. 0. 0
James Clarke	"	77. 11. 4
William Johnston	"	103. 13. 2
Joseph Reville	"	103. 13. 2
John Gray	"	103. 13. 2
Patrick Carney	"	79. 5. 6
Roderick Gray	"	37. 10. 0
John Minty General Overseer		52. 17. 0
Robert McGuiness	overseer	39. 17. 1½
Thomas Hughes	"	38. 11. 0
Michael Hughes	"	19. 5. 8
William England		8. 4. 6
Total Salaries		£933. 4. 3½

Statement of arrears due on the works in the Western District
(NAI OP 1831 973/87)

Nimmo's figures and allowing him to pay off his staff and contractors himself. The under-secretary, William Gossitt, was of the opinion that complaints made against Nimmo for non-payment of contractors were well founded and that he, or his agents, had in the past 'neglected to pay for work which had been performed, and for which he had been credited by the government'. The lord lieutenant directed that enquiries should be made before any money was paid out. And although Gossitt declared his distrust for 'what two of a trade say of each other', after reading Killaly's report, he recommended that the government pay direct whatever was owed rather than have the money go through 'Mr Nimmo's hands'. He further advised that deductions be made from the salaries of those who had received articles from the stores and not returned them, and was requesting a full account from the storekeeper.[33] Nimmo was instructed to draw up his full account of expenditure on public works in the Western District. This was to be handed in, on or before 7 December 1831 and was separate from the arrears due to staff and contractors.[34] A settlement was finally reached regarding the arrears, on 22 December 1831, and Nimmo was paid £2,312.1.5d by the treasury. In the end there was no deduction made from the salaries and £933.4.4d was issued. However, a deduction of £50 for 'expenses' was made from the figure sought to clear arrears to contractors.[35]

Killaly's progress report

In a triumphant tone, Killaly wrote to Gossitt at the end of September (1831) that, despite the bad weather, 'the works of the Western District under my direction are proceeding very satisfactory'. Killaly was critical of the standard of work carried out on the Mayo roads and a good deal of 'draining, firming, gravelling, building of bridges and gullies' was needed before they could be fully opened to the public. However, this was nothing more than what Nimmo himself had pointed out in his 1829 and 1830 reports and in his estimate for 1831.

The road from Newport through to Achill Head, which was previously impassable due to storm damage from the previous winter, was expected to be open to the public by summer 1832. Extra funds would be needed for the Central Erris Road, where the 'drainage, fence formation and gravelling' were in general 'so very defective'. Because of the severity of the mountain torrents, sections of this road were impassable, many of the gullies had to be rebuilt, two large bridges and several of the minor bridges had to be

Road in Achill circa 1860s
(Courtesy of the National Library of Ireland SP1453)

Road in Connemara circa 1860s
(Courtesy of the National Library of Ireland R923)

replaced or repaired. Perhaps in an attempt to prove that he was a better bridge builder than Nimmo, Killaly decided to go ahead with the construction of a bridge at Fauleen to replace Nimmo's bridge, brought down by the floods in 1828. After a 'minute examination' of the site, Killaly chose to locate his bridge about half a mile west of Nimmo's and to span the river with a twenty-foot arch. Nimmo's bridge had a thirty-foot arch. Another replacement bridge was built at Beltra, a little west of Beltra Lough.[36]

The Killala to Belmullet road was advancing rapidly and bridges were erected on sites previously neglected by Nimmo, bringing a considerable portion of the road into public use. On the Westport to Clifden road, 'securing and completing' of Erriff bridge was under way and repairs were carried out on a stretch of the road along by the Killeries, which had been destroyed by the sea. This must have included Leenaun Bridge, but he does not mention it specifically and, if so, neither did Killaly's work hold, as, according to Maria Edgeworth, the bridge was still down in 1833. A stretch of new road near Kylemore Lake, two miles in length, 'together with one large bridge, two minor ones, and several gullets', all of which were started in June, was nearing completion. This would shorten the road from Leenaun to Clifden by four miles; more work on the Leenaun side would eventually shorten the entire route by nine miles.[37]

Killaly replaced Nimmo's bridge at Letterfrack, which had collapsed prior to the lord lieutenant's visit, with one of more secure foundation on a more suitable site. Work was needed on the road from Leenaun to Maum, where some of the gullets were 'originally ill constructed'. Killaly was not impressed by Nimmo's bridge at Maum: 'The large and costly bridge at Maam adjoining the hotel, appears to be in a very precarious state. This bridge does not by any means afford a sufficiency of waterway, there being at the time of my visit after a few hours rain, a difference of 6 feet in the level of the water immediately above and below it, and I understand a still greater difference takes place at times, the parapets and approaches to this bridge, and to another of large dimensions a short distance beyond it are in a very unfinished state'.[38] Time would show that Killaly's fears were justified and heavy flooding eventually brought down Nimmo's bridge. Sir William Wilde in *Lough Corrib, Its Shores and Islands*, published in 1867, described how it happened:

Maam Hotel with the iron bridge circa 1880s.
It was replaced in the 1980s by the present bridge. A section of Nimmo's limestone abutment is retained in the present structure. (Courtesy of the National Library of Ireland SP1153)

> A very handsome limestone arch spanned the river at this point, but a slight settlement having taken place, some of the top stones were never put in their places. It remained, however, without any further injury, until a few years ago, when an immense flood coming down suddenly, and filling the valley on either side, rose above the arch, and immediately the whole fabric was swept away. Possibly side arches might have prevented this catastrophe. An iron bridge now occupies its place.

Wild had crossed the bridge with a friend on the previous day and noted that it had sustained damage from an earlier flood, it had moved two inches. However, 'it was only when the water rose to the battlement that the whole gave way'.[39]

Killaly, however, had also failed to bring Samuel Jones to a 'fair settlement of his account'. He had instead confined his men to those sections of the road not included in Jones's contract: the stretch of road leading to and from Letterfone Bridge, at Lough Bofin, a bridge and about two miles of new road at Bunskannive, at Lough Oorid, and some preparatory work along the shore of Derrylea Lake.[40] Two years later (1833) Hamilton Killaly would report that the road from Oughterard to Clifden was still unfinished and the old county road was 'completely broken up'.[41] The Galway Grand Jury eventually issued funds in 1834, when a presentment of £1,750 was granted for the completion of the roads from Galway to Clifden and from Clifden to Westport.[42]

Nimmo's limestone abutment in 2006
© Private collection

Death of Nimmo

Nimmo died at his residence, 78 Marlborough Street, Dublin, on the evening of the 20 January 1832; he was aged forty-nine. According to *The Times* he had, for some time, been an invalid from rheumatic pains and had died from 'dropsy'. Dropsy is an archaic medical term now called oedema, which implies retention of fluid, evidenced by swelling of feet, legs and in severe instances the abdomen. It usually resulted from an attack of rheumatic fever, which may well have been contacted several decades before. The condition would gradually become more severe over a period of time, resulting in winter bronchitis, shortness of breath, swelling of legs and feet and eventually death, due to heart failure. It is impossible to say for how long Nimmo had suffered from this complaint or when he may have contracted rheumatic fever. He had complained of being fatigued and unwell on his first tour of the Western District in 1822. And he would appear to have been very ill in the winter 1828-29. His visits to Leamington, a spa town popular with rheumatic patients in the nineteenth century, would indicate that he was in poor health in the latter years of his life and may have found it beneficial to remove himself from the damp Irish climate. A week after Nimmo's death the *Galway Weekly Advertiser* marked his passing with the following glowing eulogy:

> Alexander Nimmo, Esq., F.R.S.E, M.R.I.A., Etc., Etc., These honorary distinctions are but as dust weighted in the balance when compared with the sterling talent and [intrinsic] merit of this excellent and lamented individual. Eulogium is unnecessary, as the word *Ireland* alone will be both his most merited monument and suitable epitaph. No man so well understood the remedies required for its practical evils, and the effects will be felt long after the very remembrance of his name will have passed away. As a theorist and scientific member of his profession he has left *no equal*, and in conclusion it may safely be said, "the British empire in general has sustained an almost irreparable loss".[43]

Nimmo was in the process of drawing up his final account of the receipts and disbursements for the Western District at the time of his death. His account was still in arrears and his accountant, Patrick Carney, had only the month before succeeded in assembling the vouchers and explanations for many of the queries raised by the auditors. In their capacity as administrators of his estate, Nimmo's brothers, John and George, continued the task and submitted the account. The account claimed the total expenditure for the Western District under Nimmo's team, from 6 January to 11 December 1832, to be £10,449.5.1d. Some of this was for day labourers' wages on almost all of the major works between 5 June 1830 and 11 July 1832 and some contract payments outstanding from earlier years. The account also claimed for the arrears disallowed in previous accounts, then totalling £8,972.10.4d.[44] Prior to his death, Nimmo, perhaps aggrieved by the administration's treatment of him in the end, had instructed the brothers to include claims for back wages due to him for time spent working for the commissioners of the Consolidated Fund. They also claimed £1,026.18s in wages for three hundred and twenty six days worked by Nimmo, at 3 guineas per day, from 5 January 1831 to 20 January 1832. This would indicate that, although removed from the Western District, Nimmo was in the employment of the government to the end. Nimmo, it would seem, had worked on Western District business for almost every day of the last year of his life, excluding Sundays, right up to, and including, the day he died. Also included in the account were charges for work carried out by pupil surveyors working in the Dublin office and conducting fieldwork in the west; wages for their work had not previously been sought. The account was forwarded to the Commissioners of Imprest Accounts of Ireland, with a duplicate sent to the lord lieutenant.[45]

The commissioners, however, were still contesting the disallowances and were demanded more documentation, some going back as early as 1822; their report on the account showed a balance of just £562.11.3d due to Nimmo.[46] The brothers complained of the difficulty in trying to meet such a request and tried to explain the conditions that prevailed in the west when these works first began. They pointed out that given the general confusion and panic that prevailed at the time, 'from the distressed state of the peasantry in the west of Ireland' and their brother's 'anxious desire to carry into effect the benevolent intentions of government', it was not surprising that correct procedure were not always adhered to. As works in the west got under way, they explained, funds were issued by Nimmo, in accordance with government policy, to 'several respectable gentlemen in the respective neighbourhoods' who were acting as overseers. The overseers issued vouchers [receipts] in return. Several of these vouchers were disallowed by the commissioners because they were not written on the correct paper, although the brothers argued that they were 'the best vouchers that could be procured (from the general panic and confusion that existed)' at the time.[47]

In their report on the account, the auditors exonerated Nimmo and his team of any attempt to defraud the treasury. Although still insisting that the vouchers in question were imperfect, they accepted that the documentation produced by the brothers had provided a clear 'chain of evidence, connecting the party receiving the money with the execution of Public Works under the late Mr. Nimmo's superintendence' and they were in favour of allowing the payment to go through: 'We feel fully justified in the declaration of Our opinion, that no moral doubt exists of the faithful appropriation of the monies stated to have been advanced for the execution of the Public Works; these advances have, in many instances, been made under the express orders of the Irish Government to country gentlemen, residing in the neighbourhood of those districts where the pressure of distress was heaviest, and whence the loudest calls were heard for immediate relief, which was as instantly as benevolently yielded by the Government'.[48]

In 1833, the Board of Public Works, who had then taken over responsibility for the Western District, received a request from the Treasury to supply a detailed account of the expenditure on public roads in the counties of Mayo and Galway from 1822 to 1833. Piecing together various bits of information received from the Auditors of Accounts and

RETURN, showing the GROSS EXPENDITURE in the Counties of *Mayo* and *Galway*, under the Directions of Messrs. *Nimmo* and *Killaly*, Civil Engineers, from the Year 1821 to 1833.

	£.	s.	d.
COUNTY OF MAYO:			
Amount of expenditure by the late A. Nimmo, civil engineer, from the year 1822 to the year 1832 - - - - - - - - -	45,076	1	11
Proportion of the general account for management, contingencies, Dublin office, &c. during the above period, chargeable to Mayo - -	10,910	16	10
	55,976	18	10
Amount of expenditure under the direction of Messrs. J. and H. Killaly, from April 1831 to April 1833 - - - - - - -	15,111	9	9
Total - - - £.	71,088	8	7
COUNTY OF GALWAY:			
Amount of expenditure by the late A. Nimmo, civil engineer, from the year 1822 to 1832 - - - - - - - - - -	68,262	16	6
Proportion of the general account for management, contingencies, Dublin office, &c., for the above period, chargeable to Galway, as nearly as can be ascertained - - - - - - - - - -	16,523	2	3
Expenditure under Mr. J. Killaly, civil engineer, in 1822, 1823, 1824 and 1825 - - - - - - - - - - -	6,272	12	5
	91,058	11	2
Amount of expenditure under Messrs. J. and H. Killaly, from April 1831 to April 1833 - - - - - - - - - -	11,098	5	7
Monies due by the late Mr. Nimmo, and paid out of the funds lodged to Mr. Killaly's credit - - - - - - - - -	714	1	4
Total - - - £.	102,870	18	1

Return, showing the gross expenditure in counties Mayo and Galway under Nimmo and Killaly from 1821 to 1833.
(BPP 1833 [285], xxxv)

Hamilton Killaly, the board produced the following tables showing the funds received by Nimmo for both counties, for the years 1822 to 1831, and by John and Hamilton Killaly for 1832 to 33. Also shown is the figure received by John Killaly for his earlier work in that part of Galway included in the Middle District from 1822 to 1825. In addition, the board furnished the treasury with the breakdown of the sums expended on individual roads and with a report from Killaly on the state of the roads in 1833. The figures, they wrote, were not entirely accurate and they did not include all of the roads. However, the figures would show that there was very little difference between the sum sought by Nimmo to complete the works in the Western District in 1831 and that expended by the Killalys in the two years that followed.

In 1831, Nimmo had requested £27,291.4s to complete the works in hand and for repairs and extensions to the roads in the Western District. Instead of issuing the funds,

RETURN, showing the detailed Expenditure on the principal Works in the Counties of *Mayo* and *Galway*, under Directions of Messrs. *Nimmo* and *Killaly*, from the Year 1821 to 1833.

DESCRIPTION OF WORK, and WHERE SITUATED.	AMOUNT EXPENDED THEREON.						TOTAL.		
	By Mr. Nimmo.			By Mr. Killaly.					
COUNTY OF MAYO:	£.	s.	d.	£.	s.	d.	£.	s.	d.
Road from Westport to Killery Harbour -	9,779	2	–	1,497	8	5	11,276	10	5
Ballintubber Road - - - - - - -	2,165	2	–	50	–	–	2,215	2	–
From Ballinrobe to Claremorris - - - -	965	11	–	–	–	–	965	11	–
Achil Road - - - - - - - -	4,435	17	–	6,097	1	11	10,532	18	11
Erris Road and Branches - - - - -	14,714	18	–	2,692	3	7	17,407	1	7
River Moy Works - - - - - -	2,901	10	–	–	–	–	2,901	10	–
Road from Killala to Swineford - - - -	7,370	10	–	–	–	–	7,370	10	–
From Bonyconlan to Foxford - - - -	787	12	–	–	–	–	787	12	–
Road to Killala Pier - - - - - -	348	15	–	–	–	–	348	15	–
North Coast Road - - - - - - -	3,206	9	–	4,774	5	10	7,980	14	10
COUNTY OF GALWAY:									
Conemara Coast Road - - - - -	20,711	10	–	7,697	4	3	28,408	14	3
Piers and Quays - - - - - - -	3,992	13	–	–	–	–	3,992	13	–
Woodstock Road - - - - - - -	550	13	–	–	–	–	550	13	–
Road from Kilkerran Bay to join the Killery Road - - - - - - - - -	8,620	12	–	–	–	–	8,620	12	–
Road from Spiddle to Moycullen - -	1,836	7	–	–	–	–	1,836	7	–
Road from Cong to the Killeries (Joyce's Country) - - - - - - -	10,552	17	–	–	–	–	10,552	17	–
Road from Kinvara to Kilcolgan - -	596	6	–	–	–	–	596	6	–
Road from Creggs to Galway - - -	463	13	–	–	–	–	463	13	–
Road from Killala to Kilconnell - -	122	13	–	–	–	–	122	13	–
Menlo Road - - - - - - -	497	18	–	–	–	–	497	18	–
Road from Killogh to New Forest - -	889	10	–	–	–	–	889	10	–
From Headfort to Galway - - - -	690	–	–	–	–	–	690	–	–
Several short Roads - - - - - -	997	–	–	–	–	–	997	–	–
Road from Tuam to Galway - - - -	315	12	–	–	–	–	315	12	–
Ballinderrig Bridge - - - - - -	909	–	–	–	–	–	909	–	–
Approaches to Piers - - - - -	1,071	–	–	–	–	–	1,071	–	–
Contract Conemara Road - - - - -	2,291	12	–	3,728	7	10	6,019	19	10
Road from Roundstone to Bunowen Bay and Slime Head - - - - - - - -	593	13	–	–	–	–	593	13	–
COUNTY OF TOWN OF GALWAY:									
Slate Breakwater at Galway, and Murrough Quarry - - - - - - - -	3,271	7	–	–	–	–	3,271	7	–
Galway Market-house - - - - -	1,000	–	–	–	–	–	1,000	–	–

Return, showing the detailed expenditure on the principal works in counties Mayo and Galway under directions of Nimmo and Killaly from 1821 to 1833.
(BPP 1833 [285], xxxv)

Dublin Castle saw fit to remove Nimmo and to replace him with Killaly. However, between April 1831 and April 1833, John and Hamilton Killaly (John is believed to have died in April 1832[49]), spent a total of £26,209.15.4d on the Mayo and Galway roads; £15,111.9.9d in County Mayo and £11,098.5.7d in County Galway. They had, however, received just a little over £20,674 from the Treasury. This, complained the Office of Public Works, left over six thousand pounds 'owing to labourers and others, and to meet which there are no funds'[50] It seems that little had changed since the removal of Nimmo from the Western District.

Hamilton Killaly's report dated April 1833, on the state of the roads in Galway and Mayo, reads similar to those sent in previous years by Nimmo. In Mayo, the Newport to Achill road was fully open, but the Killala to Belmullet and Castlebar to Belmullet roads were still unfinished. As previously mentioned, the Oughterard to Clifden road was still closed and funds for its completion were released the following year (1834). Sections of the south coast road were under construction. According to a map submitted by Killaly, that section of the road from Cashla to Cloonisle, along with the road from Camus Bay [Screeb] to Maam Cross, was still unfinished. This road had been abandoned when funds were transferred from it to works on the old centre road in 1828 and again in 1830, when funds were transferred to roads in Sligo.[51] Other roads in Connemara were open but required a 'moderate expenditure' for their completion and repair.[52] The Killalys had achieved no more than Nimmo had outlined in 1831, and may well have completed, had he been given the time and resources to do so. Nimmo, however, had died before the figures, accompanied by Killaly's report, were published and he could be vindicated.

Board of Works vindicates Nimmo's work in the west

The Board of Public Work, in their efforts to complete and maintain the roads, were struggling with the same difficulties encountered by Nimmo, lack of funding and an unwillingness by Dublin Castle to understand the need for more government roads for the west. The early reports from the board pay tribute to the standard of road construction carried out under Nimmo in Mayo and Galway. They found their condition to be 'generally good; in all cases much better than that of other roads in the same district'. They fully endorsed Nimmo's decision to build more roads than the funds allocated allowed for: 'Numerous judicious lines of road have been commenced, by means of which the whole face of the country has undergone the most rapid improvement. These roads are not by any means more extensive than was required under the circumstances, though more so than was authorized by the amount of the funds provided'.[53]

They, perhaps more than any other, appreciated the difficulties Nimmo and his team faced in the day-to-day task of trying to establish road communications in the remote mountainous regions on the west coast. The benefits, they reported, were best appreciated by those who witnessed the effects they produced: 'In traversing a country covered with farms and in a high state of cultivation, showing every sign of a good soil and of amply remunerating produce, it becomes difficult to credit the fact, that 10 or 12 years since the whole was a barren waste, the asylum of a miserable and lawless peasantry, who were calculated to be a burden rather than a benefit to the nation; and that this improvement may entirely be attributed to the expenditure of a few thousand pounds in carrying a good road of communication through the district'.[54] The construction of roads in remote districts had 'a miserable pauper population converted into a class of consumers', thereby contributing greatly to the revenue of the country, and had brought peace and prosperity to the region: 'These roads have been the means of fertilizing the deserts, and of depriving the lawless disturbers of the public peace of their place of refuge, affording them at the same time resourced for an active, honest industry, of which we must do them the justice to observe, they have not shown any indisposition to avail themselves'.[55]

The new piers erected under the Fishery Board had brought great advantages to their neighbouring districts, although no funds had been provided by government for their maintenance and several were in need of repair: 'The shelter these small piers afford to fishery vessels, and those employed in the distribution of manure and fuel, together with the facility they offer for the transmission of agricultural produce, has created in their immediate vicinity a striking contrast to the wild uncultivated wastes by which they are in many instances surrounded'.[56]

The board was anxious that the government grant more funds for the extension of the roads in the western counties. They declared their 'firm conviction of the good policy of promoting these eminently useful' public works, pointing out that, while it would be natural to assume that the Grand Juries would carry out the works needed to improve these districts, the reality was somewhat different: 'these are naturally the poorest counties in which these unproductive districts abound, and it is with difficulty they can maintain the communications already established. They have the greatest wants and the smallest means, and the remote wastes attract little of their attention'.[57] In this the board simply echoed the sentiment delivered by Nimmo in his bog report twenty years earlier.

Epilogue

The methods used to fund, manage and regulate the works started in the west and south of the country in 1822 were experimental, although the government of the day did not conceive or sanction them as such. Prior to this time government had not involved itself in long-term public works projects and, accordingly, did not have reliable experience or written procedures to draw on. The evidence would suggest that the government was ignorant of the high costs involved in the construction of good quality roads in remote regions. It was also ignorant of the sums needed to maintain such roads. The result was that the continuous management of such projects as those undertaken by Nimmo, Killaly and Griffith was left virtually entirely to the engineers in charge, with, it would seem, very little interference or supervision from Dublin Castle. So long as the progress reports kept coming in, Dublin Castle seemed happy to place its trust in its engineers. It was later recommended that the engineers responsible for such works should not have any pecuniary interests in them, and should not be made to act as public accountants.[1]

At the outset this early experiment was considered a success. The setting up of public works in 1822 brought immediate assistance to the distressed regions and saved the people from famine. The benefits the schemes brought to the remote, underdeveloped districts were generally seen as examples of good government practice, providing the means whereby agriculture and industry could be advanced. However, at a time when the dominant view among the thinking classes was that central government should not become too deeply or extensively involved in 'local government', the Irish Administration of the 1820s was anxious to curtail its involvement in road building in the west, and sought to hand the responsibility back to the local Grand Juries as early as possible. In this, it found itself hampered by Nimmo's slowness in finishing his roads in the Western District. In the Middle District Killaly expeditiously completed ninety-three miles of road in three years.[2] This would prompt one to ask whether the roads in Clare were not as deficient as those in the west or south. Griffith completed one hundred miles in seven years, at a cost of £60,000, in the Southern District.[3] Griffith continued to work in the south of the country until 1839, and he was eventually responsible for two hundred and forty miles of road in the counties of Cork, Kerry, Limerick and Tipperary, constructed at a cost of almost £150,000.[4]

Nimmo's slowness in completing his works caused obvious frustrations for the Irish Administration towards the end of the 1820s. The longer the government's presence (as effective employer) was maintained in the west, in the form of Nimmo, the stronger and more insistent were the arguments for continuing, open-ended aid to support the development of the impoverished districts. The condition of the roads, as witnessed by the lord lieutenant during his visit, eventually caused Nimmo's removal from the Western

District in 1831. Killaly's report of impassable roads, fallen bridges and damaged gullies, was no more than Nimmo himself had reported in his 1829 and 1830 reports to Dublin Castle. But, whereas Nimmo was pleading bad weather and lack of funds, Killaly was indicating poor workmanship. (On the other hand, it is only fair to point out that two years later, Hamilton Killaly was citing bad weather and lack of funds as his reasons for not completing these same roads).

Why, then, did Nimmo 'fail', while the others would seem to have succeeded? Why were so few of his roads in fit condition to be handed over to the Inland Navigation Board in 1828? Was the Western District more remote, backward and poorer than the Southern District? Was it the terrain that Nimmo's team encountered that caused the difficulties? Could it have been the extreme weather conditions experienced in 1828 and 1830? Or was it simply that Nimmo had overstretched himself, distracted, perhaps, by private commissions that kept him away from the west, and also perhaps hampered by poor health in later years?

Certainly there is nothing to suggest that Nimmo was deficient or neglectful in any way during his early years in the west. He seems to have adhered to the guidelines given to him by Dublin Castle. He was effective in identifying necessary works to bring immediate employment to the destitute poor. He was efficient in getting these under way and, despite later criticism of accounting practices adopted in the early years, would seem to have been honest in his handling of all government money. In proposing so many new roads, he could not fairly be accused of merely making work for himself; the dire need for roads in the west was acknowledged by all. There is, of course, every indication that Nimmo had his own agenda regarding the development needs of the west, and he seemed determined to extract from the government, wherever possible, the funds to pursue this agenda. His annual reports and estimates, however, always gave his reasons for suggested new works, and these were usually associated with extending existing works to make a more complete route, to bring employment to the poor, and therefore stability to the region, or to assist the economic development of a town, port or district. However, he was hindered in his efforts to generate major development works by the lack of any coherent or integrated government plan for the district – such 'integrated development' was not within the conceptual range of early nineteenth century governments.

However, Nimmo's failure to complete the works, and the condition of his roads when he did hand them over, raises legitimate questions regarding his ability as an engineer. Recent analyses of Nimmo's engineering ability rate him very highly. Certainly, his bridge design and construction has been admired and praised.[5] Was it the case, then, that he choose bad sites for his bridges in the west? Or that he simply didn't allow sufficiently for the western weather? Was his surface work defective, or was it poor maintenance that left his roads in such bad repair? Nimmo certainly used the funds allocated to him largely for road building; he would seem to have allocated very little for road maintenance. This strategy may have left him open to criticism. But there were probably other factors at play.

His public criticism of the government's tardiness in planning and funding developments in the poorer regions can scarcely have endeared him to Dublin Castle and may well have hastened the termination of his employment. Furthermore, Nimmo's use of his own relatives and some close friends in some of his schemes (though by no means unusual by the standards of his day) may have excited resentment and amplified the criticism of him and of the standard of his work during the later years of his term in the west. There is also the question of his failing health.

Yet, for all that, the importance of his work was valued by many of his contemporaries, and its impact could not be denied. The Board of Public Works paid tribute to the standard of his roads in Mayo and Galway. The board was of the opinion that the extent of road undertaken by Nimmo was fully justified, although inadiquately funded. Hardiman quotes extensively from Nimmo's Connemara Bog Report in his notes to Roderick O'Flaherty's *H-Iar Connaught* (1846). He pays tribute to Nimmo's contribution to the west: 'That great engineer was employed by Government in this district, and he did more towards the ultimate improvement of the place than any other man that ever lived'.[6] Sir William Wilde, writing in 1867, felt that Nimmo deserved more recognition for his contribution to the west and his work deserved closer study: 'Alexander Nimmo – a man whose memory should be revered in the West ... and whose well engineered and admirably constructed roads, and valuable reports upon the capabilities and resources of this great district can never be sufficiently appreciated'.[7]

The questions raised by Nimmo's activities in the west of Ireland did not disappear with his death. On the contrary, the issue of the proper role of government in alleviating exceptional distress and systemic poverty, and in facilitating economic development and improvements in the living standards of people in underdeveloped regions, was to assume a terrible relevance during the crisis of the great famine in the later 1840s. The role and responsibility of government in that social catastrophe in Ireland prompted intense debate among economists and other commentators on the relevance of political economy, and of 'English' assumptions in particular, to the social and economic realities of Ireland. The debate thus generated would have a vital importance for such later legislation as the Irish land acts. But the specific challenge of Irish regional underdevelopment would ultimately produce its own 'Irish solution', in the form of the Congested Districts Board of 1891, arguably the first regional development board established in Britain or Ireland. In reading the baseline and annual reports of the Congested Districts Board, one has the distinct feeling of re-opening files on many of the key questions first raised by Alexander Nimmo in the west of Ireland in the years 1811–1831. Perhaps this is the fairest measure of the significance of the man and his work.

Appendix

Reports on the Fishery Stations from Sligo Bay to Galway Bay, by Alexander Nimmo.

Nimmo's survey of the west coast for the Fishery Board was conducted between 1822 and 1826. His findings were sent to the commissioners in two separate reports and these were later published as appendices to the board's annual reports. Nimmo's report on the Fishery Stations from Leitrim to Erris Head, entitled, Report on the Fishery Stations of the Coast of Connaught, was first published with the Commissioners Sixth Report in 1825.

His report on the Fishery Stations from Erris Head to Black Head, entitled, Report on the Fishing Stations of the West Coast of Ireland, was first published with the Commissioners Eight Report in 1827.

REPORT

ON THE FISHERY STATIONS OF THE

COAST OF CONNAUGHT.

IN my first report on the fishing stations of the north coast, I commenced, as directed, with the bay of Sligo, proceeding round to Belfast; but as some of the works then recommended near Sligo are now executed, or in progress, it may he useful to give a more particular account of them. I shall therefore commence this report at the borders of Leitrim and Donegal, so as to make a complete description of the coast of the province of Conuaught.

My official situation as engineer of the public works of the western district, undertaken for the employment of the poor, having also enabled me to combine these, in some cases, with the objects of the Commissioners of Fisheries, I embrace this opportunity to give also some description of such of these works as are especially connected with the general improvement of the western coast.

For the purpose of this inquiry, a very careful maritime survey of this coast has been carried on, under my direction, at the same time. The result of it is laid down in a general chart of Connaught, and in detailed charts of its harbours, executed under my superintendence, especially of the shores of Sligo and Galway. Those of Mayo have been chiefly supplied from the grand jury map and survey of that county by Mr. William Bald, the accuracy of which, in general position, may, I believe, be very safely relied on, though not sufficiently particular in some parts of the maritime detail; the soundings on his map being only laid in from Mackenzie's charts. The soundings of Clew Bay have been since, however, taken more minutely, first in a survey for the improvement of Westport harbour, begun for the Marquis of Sligo, and afterwards extended over all the rest of the creeks, at the expense of the Commissioners for Fisheries. The soundings and detailed survey of Killala Bay, and the creeks near it, have been carefully taken, under my superintendence, by my late assistant, Mr. William Grant, from a separate base and operations independent of the county maps. Some of the other harbours of Mayo, especially that of Boffin island, have also been carefully surveyed by him; Enniskea, by my assistant, Mr. Knight.

The survey of Sligo Bay was originally made for the commissioners for the improvement of the port of Sligo; but not having been done trigonometrically, it has since been reexamined, and such additional information inserted as appeared necessary. The engraving of the reduced map has been done at the expense- of the Commissioners of Fisheries.

The river Moy, also within the bars, and as far as the bridge of Ballina, having been surveyed by me, for the use of the committee of merchants of that town, the result has been reduced, and inserted on the fishery chart of Killala Bay.

The coast of Galway has been minutely surveyed from Cleggan to Blackhead, in the county of Clare, with the exception of Kilkerran Bay; and from this I have been enabled to make out charts of most of this intricate and highly important part of the coast.

The completion of this part of the survey, especially the trigonometrical part, has been suspended by the orders of the Board; but I have derived considerable benefit in this particular from the former survey of the interior made by me, under the authority of the Commissioners of Bogs in Ireland, and which survey is still in manuscript. Considerable progress was also made in the delineation of the banks so productive of cod and ling, situated on the western coast of Erris and Connamara, and especially that called the Sunfish Bank; much, however, remains to be done ere the nature of the ground within soundings, on this side, can be at all laid down with accuracy.

The examination of the banks situated so far in the offing would be more appropriately intrusted to nautical men than to an engineer; and it would, in my opinion, be of the greatest importance to the fisheries and to navigation, if the Admiralty were to direct a sounding survey to be made of that part of the ocean situated outside of the north channel, as far at least as Aitkin's Rock, for I am fully convinced our present information of that tract is of a very vague and imperfect nature.

It is evident that the bank of soundings, which near Mizen and Brandon Heads, in the south of Ireland, extends only a few miles from the land, gradually expands after passing the Shannon, is of considerable breadth between Connaught and the Western Isles of Scotland, and some say even connects with the banks of Newfoundland. Some of our soundings off Achel head extend as far as ninety-seven fathoms, being the deepest yet marked on the western coast of Ireland. Some of the other banks, however, especially those, within the bay of Donegal, we only know from the description of the fishermen.

COAST OF LEITRIM.

The county of Leitrim includes a small portion of coast, about two miles in extent, along the eastern shore of the bay of Donegal ; it has no seaport town or fishing village. The shore is for the most part a bluff of conglomerate sand stone rock, with a rough skare along the foot of it; totally inaccessible to any kind of craft, and exposed to the whole swell of the Atlantic; it is however, much frequented for the purpose of sea bathing, and the villages of Bundoran in Donegal, and Bundroes in Leitrim, are principally supported in this way; being in summer occupied by persons of a respectable class, from Fermanagh and Leitrim, whose consumption of fish would be considerable.

A few yawls are kept at these places, and there is a salmon fishery at the mouths of each of the rivers, Bundoran, Bundroes and Bunduff.

Since my first visit, a proposal has been transmitted to the Board of Fisheries for the construction of a pier at Bundoran ; and I may here observe, although not strictly belonging to this place, being in the county of Donegal, that I have again examined the situation of Bundoran with the view to the construction of a harbour. As it could only be effected by a pier of great strength and extent, carried along the rocks opposite the village and as the beach extends so far, and the surf in west winds is so violent, it is my opinion that nothing of any consequence could, at a reasonable expense, be obtained.

BUNDROES.

At Bundroes, just on the northern edge of Leitrim, and about a mile from Bundoran, there is a more favourable situation for something of the kind; a rocky point runs out from Colonel Caulfield's lodge to the northward, covering a shallow bay into which the river of Bundroes empties itself. The materials are abundant on the spot, and if a pier was constructed on this point, a tolerably dry harbour could be formed within it, sufficiently extensive to permit the surf to expend itself quietly. This would answer all the purposes of a harbour at Bundoran, and, if effected, would have the advantage of the new road which is at present in progress through the county of Leitrim, from the sea at this place to Manor Hamilton, and by Lough Allen to Carrick and Drumsna.

This line lays open the whole extent of that county to wheel carriages, which can hardly pass along the present roads. It also opens the highly improvable valley of Glennead abounding in freestone and limestone of an excellent description, and beyond it, is the district near the banks of Lough Allen.

This harbour may also derive some importance from the neighbourhood of Lough Melvin, which discharges into it, and is navigable for seven miles into the county of Fermanagh; and the level of which might, I doubt not, be continued, by a canal of about three miles, as far as the sea at Bundroes.

I am also inclined to think that a canal could be easily carried into that lake from Lough Erne. But the opinion I gave in a former report, of the possibility of having a canal from this last lake to Mullinasoal and the harbour of Donegal, I am, on further consideration, inclined to retract, as upon re-examination I find the difficulties are greater than I had anticipated; and I have no hesitation in saying, that it would be much easier to construct a harbour at this place of Bundores, than to obviate the defects of the bar of Ballyshannon, whih is the natural port of Lough Erne.

It is to be observed however, that even with a pier such as I propose, the harbour of Bundroes would be shut in westerly winds; the surf sent in by this wind on this coast is so heavy as to cause a breaker to a great extent from the shore.

In this particular any harbour that could be made on the Leitrim side must be always inferior as a fishing station, and place to shelter, to that now in progress at Mullaghmore, four miles to the westward, when the in-shore winds or ground-swell made access to it dangerous, and as vessels will use that roadstead with confidence when the pier of Mullaghmore is completed, I am inclined to think the construction of the work I speak of at Bundroes is not unworthy of serious attention from the proprietors.

As the exposure is great, and the space to be quayed is extensive, the pier of Bundroes could not be executed under 3,000 *l.*

SLIGO.

THE coast of the county of Sligo from the Bunduff river to beyond Milkhaven, is the property of Lord Viscount Palmerston.

The country as far as the inlet of Drumclief, like the neighbouring part of Leitrim, is first a narrow band of conglomerate sandstone rock, thickly peopled and tolerably cultivated; behind that a zone of bog, nearly a mile broad, to the foot of the hills Truskmore, Beenweesken and Bilbulben, which rise rapidly with a bluff and precipitous front to about 900 or 1,000 feet.

The summit of these hills is flat limestone rock, mostly covered with bog, but there are three intervening and pretty fertile valleys.

The peninsula of Raughly is limestone; the hills of Mullaglimore, Stridagh, and the isle of Derrineh, connected with the last by a stony beach, form a kind of outwork or exterior band of sandstone along the coast; separated by a shallow valley; the extremities of which at Stridagh and Mullaghmore have been filled up by blowing sand, and which has covered a great extent of country near this. The intermediate portion of the valley is a lagoon called Milkhaven; which preserves an opening along the north side of Derrinehby its great extent of back-water; and forms a safe harbour within, although the entrance channel is uncertain and barred all across in westerly winds.

The bights about Stridagh are too much exposed to the ocean to afford any shelter for craft; but within Mullaghmore the north end of the valley forms a good roadstead for small vessels in two and a half to three fathoms water, the upper part being a long shallow beach of flat sand.

MULLAGHMORE.

OF this bay the mainland side has no creek of any kind, the Bunduff river falls into the sea over conglomerate rocks, and is inaccessible even to a boat.

The upper end of the bay is a flat sand as above stated, with a considerable extent of salt marsh; south of it is an extensive flat overblown with sand, extending to Milkhaven.

On the Mullaghmore side, Lord Palmerston has contracted with the Board of Fisheries to execute the pier and harbour proposed by me in a former report ; and this harbour is now in considerable forwardness. The stone is particularly well adapted to this purpose and the work will go far towards clearing the beach up to the sand.

I have, at his lordship's request, examined the possibility of continuing this pier into deeper water, and find that at a disstance of about 150 feet from four to six feet may be obtained at low water springs, which would render this place accessible to ordinary fishing craft at all times, and at high water would admit any vessel that could take the ground.

The expense of prolonging the quay I find to be about 3 *l.* per foot ; filling and pavement behind about 22 *l.* per yard; running the end wall would cost about 150 *l.*; so that an addition of about 150 feet may probably be obtained for little more than 1,500 *l.*

At the upper end of the bay a small river falls in, but spreading over the sand, is hardly able to preserve a channel. By a causeway road, near high water, over to Mullaghmore, this river may be sent towards the harbour, and by some management, may even be made navigable for boats: this would be an excellent supply of back-water for Mullaghmore harbour in the event of its threatening to choak up with sand; and the navigation may be useful for other purposes in promoting the reclaim of the land above.

If, however, an extensive work below low water at Mullaghmore be not contemplated, the drains collected, along the hill itself, in the manner I have explained on the spot, and brought to discharge into the head of the harbour, will go far to answer every purpose required as to scouring the harbour.

It is of the utmost importance for this harbour, and generally for the prosperity of this tract of country (the improvement of which must very much depend on it), to prevent the cutting, pulling, or otherwise destroying the bent grass, which is so valuable a means of preventing the blowing of the sand.

Several hundred acres of productive land are already covered by this destructive practice. and it goes far entirely to obliterate the existing havens of that coast.

Lord Palmerston has entered zealously into the improvement of this part of his property, intending to reclaim, by his own means the larger tracts of bog, and embank the salt marsh; he also contemplates the transport of turf from Mullaghmore to supply Tyreragh and the Sligo market, which would be a valuable occupation for the people of that place in the slack time of the fishery.

The road already described as in progress through the county of

Leitrim, will also be of great service to the prosperity of Mullaghmore, which affording the only roadstead between Sligo and Donegal, is certain to become a place of consequence in the coasting trade.

MILKHAVEN.

LEAVING Mullaghmore, and passing the rocky point of Ross Keeragh, a sandy shore of one mile brings us to the bar of Milkhaven, within which there is an extensive and well sheltered harbour, as above mentioned, with two fathoms at low water in the channel.

At Mount Temple, in this creek, are salt-pans, and a pretty good landing quay, with a road down to it; nothing, therefore, seems wanted from the Fishery Board within the haven.

In the mouth of the haven or bar is a stony shoal of considerable extent, and to judge from Mackenzie's description, the entrance to the haven was, in his time, on the north of that shoal. The accumulating sand has now, however, sent it to the south side, which is not so clean as the other has formerly been. This haven may therefore be considered as having deteriorated in the last half century; and as any attempt to recover the northern entrance would be out of the question, the best thing to be done, in my opinion, would be to plant a beacon on the south end of the shoal, which would point out the best entrance in blowing weather, when the breakers run all across.

In other respects this channel is also preferable, as it avoids the outer breakers, and is better sheltered by the rock of Carricknespanagh.

The exterior side of the Milkhaven is an exposed sandy beach; nor is there any kind of harbour between it and the point of Knocklane.

The village of Cloone has twenty to thirty yawls, which are employed in fishing in fine weather; but in bad weather must be hauled up. The same is the case at Rathnoran, where there are thirty to forty yawls employed during the herring fishery; no other fishing village occurs until we arrive at Raughly.

GENERAL FISHING NOTES OF DONEGAL BAY.

MULLAGHMORE is the principal fishing village on this coast. It supplies the markets of Leitrim and Fermanagh: it also sends much fish to Sligo.

The fishing ground is between Mullaghmore and Ballyshannon Bar, outside this the bottom is foul. This is good ground for haddock with spilliards, also for flat-fish, viz. Plaice, turbot and sole.

On the Bullocks Rocks, between Mullaghmore and Milkhaven, there is good lobster fishing, also at Ennismurry. Milkhaven affords slugs, muscles and flat-fish for bait. Whelks are unknown here.

The cod fishing is chiefly from hence to Ennismurry Island, with foul ground called Bankrea, but inside is an excellent turbot bank at Stridagh, in two or three fathoms water: the best is on the east side of Stridagh.

ENNISMURRY.

THIS little island is surrounded by rough ground, but which affords plenty of lobsters, pollock "bollauns" or bream, and other, rock fish. It possesses a small natural boat harbour on the side next the main, for which the only improvement would be to set some iron rings in the rock. Like all small coves of the kind the run becomes too great for a boat to lay in bad weather, they are then hauled over a bank of rock into an inner pond; it would not be worth the expense to make any cut into this pond.

SLIGO BAY.

SLIGO Bay is situated between the point of Knocklane, improperly named Gessigo on the charts, on the north side and Aughris Head on the south.

The exterior bay is nearly a square of four Irish miles, and there are three arms, each extending about three miles farther inland, but which are barred across the mouth, and mostly dry at low water. These are the harbours of Lissadill or Drumcliff, Sligo and Ballysadere.

Across the mouth of Sligo Bay from Aughris to Raughly and Knocklane runs a reef of rocks below water, called the ledge, on which there is a heavy breaking sea, with westerly winds; but it forms no obstruction to the navigation, as there are seven fathoms water across it. Outside there are twenty to fifty fathoms, and within not more than ten.

The north end of this ledge appears above water in the islets called Hawlboulin and Little Island, in the peninsula of Raughly and point of Knocklane, and also on the Seal Rock, one mile south-west of that point. The south extremity of the ledge appears in the middle of the strand east of Aughris Point; forming the broad reef called Carrickfadda, or the Long Rock. This rock is sandstone, that of Aughris Head is limestone, but beyond it is slate, and those within it along the coast to Ballysadere are limestone.

As this is the usual position of the sandstone rock, viz, between the limestone which lies over it, and the slate or other primary rock which is under it, as may be seen about the lower part of the mountain of Benbulleen, we may, probably, in the breaking up of this sandstone, account in some way for the great mass of sand that has filled so much of these inlets, and occupied most of the eastern shores of the bays of Sligo and Donegal.

On the north side Brown's Bay, between Raughly Head and the isle of Haulbowling, though quite open to westerly winds, has a bottom of clean firm sand, and in easterly winds is used by vessels to stop in for a favourable tide to go over Sligo Bar. The usual anchorage is a breast of Raughly rock; in ten fathoms.

The Seal Rock is always above water, within this is the small peninsula of Raughly, where there is a fishing village in which pilots for Sligo Bar reside . Between it and the bar is a good roadstead where a vessel may stop on three fathoms clean fine sand; in making for which observe the following dangers :–South-west of Raughly Head one cable length is a ledge called the Wheaten Rock, only seen at low spring ebbs; and south of the point the Bird Rock, or Carriganeean, the north-west end of which dries at half high tide; but a sunk ledge runs off to S. E. near a quarter of a mile.

These rocks are laid down on the fishery chart in their proper places, with the necessary directions for avoiding them. They have been incorrectly named by Mackenzie in his survey, and are so in the common charts.

In rounding Raughly Head you avoid the Wheaten Rock by keeping the tower of the Black Rock S. E. by S. on the high house in Sligo (Mr. King's), or to the eastward until you open the old castle of Ardtarmon N. E. well out of the point of Raughly; and to avoid the tail of Carriganean, keep the Seal Rock in view outside of Raughly, until you bring the same castle N.E. by N. you may then stand to the eastward and anchor in three fathoms, half a mile S. E. of Raughly Point.

From the Wheaten Rock to Raughly is all foul; there is a boat sound between Carriganean and the land about six feet at low water.

This anchorage will now be of much greater importance to the port of Sligo since the pier has been built on the shore of Raughly near the point; as this pier will not only shelter and afford a safe landing place for the pilot and fishing boats, but small vessels may run in and take the ground witin it when the bar appears too much broken to go on to Oyster Island.

South by east from Raughly Point, and half way to the Coney Island, lies the Black Rock, on which there is a round tower of stone.

This rock dries for a cable length east and west of the tower, and is in fact the extremity of a rocky shoal called Bank Gar, which runs from hence all the way to the north end of Coney Island, on which at low spring ebbs there are only three or four feet water.

On the north side of this shoal is the bar of Sligo, a broad flat sand, with ten feet at low water; when over it we have deep water up to Oyster Island.

The only other rock in the bay is that called the Blind Rock, off the south end of Coney Island; it has two fathoms water, but must be avoided by a vessel waiting for tide on south of Black Rock. The marks for this rock are also shown in the chart.

FISHING GROUNDS OF SLIGO BAY.

Inside of the ledge of Sligo Bay, in nine fathoms water, S. W. of the Bird Rock is the Raughly Cod Bank. Turbot, whiting, &c. are also found here in harvest.

The Raughly men fish also often in Brown's Bay, about the small isles, running their yawls across the isthmus, according to the weather.

Off Ballysadere Bar in five to ten fathoms is the south turbot banks. A little lower off Dirk, in eight fathoms, is the Ruary Bradogh's Bank, producing whiting, cod and glassen. Lower down off Dunnoran, on the south side of the bay, is the whiting bank.

Farther to the eastward near Easky river, about three miles off shore, is the Kellenduff cod bank, in about thirty-five fathoms, clean ground; it runs along shore to Kamsharrock Head. To the northward off Telan is the Church Bank, in nearly the same water. The mark for it is the Minister's house, called the Church of Telan, on the east point of Glan Mountain. This is a productive cod bank, but as this fish abounds near the steep shores of Telan and the Mayo side, the fishermen seldom resort to the offing.

With large vessels the white fishery is pursued on Bank Ouil, between the stags of Broadhaven and Arranmore. This as well as Telan is a valuable ground for ling, for which the best bait is sea eels found in Mukres behind Telan, and caught by hooks of softened wire.

This fishery in the offing is, however, not much pursued of late. A Mr. Gerardean, of Rutland, formerly fished this bank with good success, by long lines, bringing people from, Youghal and Waterford for that purpose accustomed to the Newfoundland fishery.

The turbot banks being near the land are seldom visited by the cod, but frequently by scate and dogfish, which destroy and drive off the flat fish.

The turbot comes on these grounds in August, and seems to follow the sand-eel, which are there found in great abundance in the strands and loose sand of the bars.

Mackarel are found in great quantities at Telen, in the months of June and July; they come first in shoals, and do not take any bait, but are caught by seines of one hundred to two hundred fathoms. In August they separate and betake themselves to the shores, when they are caught by spillers and hand-lines, with sand-eel bait, &c.

Pollock are also pretty regular in all years, and are caught in July, August and September, if weather admits.

Sole and plaice are principally got by spiller lines about Mullaghmore, Ballysadore and Killala; drafting nets are also used about the mouths of rivers where not protected.

The sailing boats mostly go to the white fishery, about Telen and Killenduff, in June. It would be better pursued in April and May; but in this early part of the season there is a scarcity of small fish bait. Whelks are not known. Muscles are found in some of the inlets, as Milkhaven, Sligo and Ballysadore.

In May small sprats are got, sand-eels in June and to August; at which time is the best turbot fishery, as above mentioned.

The white fishery continues to the summer herrings, and throughout that season to the winter fishery.

The herring fishery in Sligo Bay begins in August, at Martinmas, first appearing on the south shore about Pouldivy; it is thought, therefore, they breed in the bed outside the ledge, among the sea-weed. Further out, about Telen and Trakel, they are found at all times of the year, and heavier fish than come up the bays.

They fall into the shoal water after N .W.gales.

This early fishery continues until October, and the fish are then full of roe. It is probable that the time of spawning is not uniform. In September they are often so sick as not to mesh, but merely to lie on the nets, on which also the spawn is often found in handfuls.

In October this fishery ceases, on account of the equinoctial gales, there being no harbours on that coast fit to shelter sailing boats.

At Hollandtide (1st November) the winter fishery commences about Raughly, and is then followed up the harbours; there are as yet no craft on the coast fit to pursue the deep-sea fishery, but the yawls are very numerous, several hundreds kept between Connaught and Donegal: this fishery continues to Christmas.

In spring the Herrings fall down again towards Telen, where they are round until March. The same fish and seasons are found about Killala Bay, the fish being apparently driven according to the wind.

In heavy gales-they fall to the bottom, and nestle in the tangle.

In spring the shore fishermen betake themselves to the cultivation of the land; sea manure is busily collected about May, much of it carried by boat up the inlets to Sligo, Ballysadore and the Moy, where it is sold and carried thence up the country. It fetches, wet, abont 4 *s.* or 5 *s.* per ton. This trade continues until the latter end of May.

Kelp is made in summer from Polnadivy to the Moy. Beyond Downpatrick, until we come to Erris; there is little, the coast being chiefly a barren cliff. Northward a good deal is made about Rathnoran, Innishmurry, and Derrinish; also about Bundoran, St. John's Point, and along the Donegal shore.

The sailing boats in this bay are mostly from Teilen, about six to twelve tons, with a bumkin bowsprit and smack-rig. Some wherry-rigged boats are also found in Donegal.

About Sligo they have a running bowsprit and smack-rig, and similar about Killala; but none of these boats exceed three to four tons, as for want of harbours they must always be hauled on the shore.

RAUGHLEY PIER AND HARBOUR.

In a former report I have explained the peculiar advantages to be expected from the formation of a boat harbour at Raughley Point; and, as the work is now completed, I shall here give a brief detail of its construction.

This work, like the other fishery piers lately constructed on the west coast was begun by the funds of the Commissioners of Fisheries, aided to an equal extent by the contribution from the charitable associations in England and Ireland in 1822; but as the estimated expense, viz. 1,160 *l.* was considerably greater than what had been assumed in the other cases as the limits to such grants, viz. 500 *l.* it could not be completed from that fund.

In 1823 I was authorized by the government to finish the work by means of the funds placed at the disposal of his excellency the lord lieutenant, for the employment of the poor, in addition to the sum granted by the Board of Fisheries. I have had the pier executed in an economical and substantial way, to the full extent proposed in my original report and estimate for this place, with the addition of a dock excavated in the land of nearly an acre of extent, and having a depth at high water sufficient for the largest class of boats, while the pier will admit any trading vessel that goes to Sligo quay.

In the month of August 1822, I received instructions from the Board of Fisheries, to commence the construction of this harbour without delay, in order to afford employment to the poor of the vicinity, and that they had granted 800 *l.* for the purpose, being only two-thirds of the estimate. I accordingly, when in Sligo, waited on Mr. Jones, the lessee of Raughley, marked out the situation for the harbour, and obtained his consent to give up so much of the bog meadow or salt marsh as would be required for the harbour, without any charge; Mr. Jones, however, stipulated that the damages should be paid to him for so much land as would form the road along the shore from the village of Raughley to the intended harbour.

As this was all arable land, and partly a lawn, I could not deem the demand unreasonable, but rather proposed (as engineer for the western district) to pay Mr. Jones the sum of 100 *l.* the proceeds to be forthwith expended by him in building a fence wall along the intended road. This offer appeared to him to be too little, and I did not feel myself authorized to go further without having the authority of the government.

Mr. Jones' conduct throughout the whole progress of this work has been so very liberal, that I think him fully entitled to receive the remuneration above mentioned for the loss of his land; but as yet, I have not obtained the necessary authority for proceeding with the road. In the meantime no interruption has been given by Mr. Jones to the labourers crossing his ground.

The work pending this negotiation having been suspended for some time, I again received pressing orders from the Board of Fisheries to commence operations at this pier, and, accordingly, as the simplest proceeding, and the best calculated to give extensive employment, I sent down an overseer to begin the excavation of the dock and quarrying stone, having provided implements for the purpose in Sligo.

The various public works carried on near the town at that time occupied all the masons and stone cutters of any skill; I had therefore to send across the kingdom for persons accustomed to harbour work, to whom I could intrust the management of the several piers, of which no less than forty were ordered to be carried on at one time along the western coast.

These persons were of necessityof very various character and qualifications, so that it is not surprising if a considerable expenditure was incurred over and above the usual rate of such works.

The works generally of that neighbourhood were conducted by one of my assistants, in whose probity I had the utmost reliance; and the weekly payments at Raughley were made at first by Mr. Alexander Cochrane of Sligo, who had been recommended to me by the commissioners of Sligo harbour, to oversee the various proceedings connected with the improvement of the port. Mr. Cochrane's claims for remuneration so far exceeded the rates at which other operations of the same kind were managed, that I am still unable to come to a satisfactory agreement with him on the subject.

Early in November this work was stopped, after 4,000 to 5,000 yards of the dock had been excavated.

In March it became necessary to employ a few men to prevent the sea from breaking over the dam into the excavation. And in June I obtained permission from his Excellency to add a sum in aid of the grant of the Board of Fisheries, for the completion of this harbour; and on receiving the orders of the Board for recommencing in June, this harbour was prosecuted as one of the public works of the western district.

At this time, besides the progress aforesaid of the excavation within the drained lake or marsh, a quantity of stone had been raised at a quarry on the west side of the point, a narrow road made from thence to the proposed pier, and some of the stone drawn there.

Finding, from the progress made in 1822, that it would be necessary to have a person of experience in such work, to complete the excavation in an economical manner, I made an agreement with Thomas Rourke, a canal excavator, of respectable character,to take out the remainder at 5 *d.* per cubic yard (what was previously done by day labour had cost 7 *d.* per yard), and subsequently also for quarrying the stone by the measured perch.

On receiving the Board's orders to proceed, I sent Mr. Alexander Macgill to commence the building of the pier, and with his assistance, contracts were entered into with sundry masons to build the quay wall and pavement, &c. by the perch; with Mr. Jones for drawing the stone from the quarry and for keeping the road in repair for that purpose. The other matters, such as loading and unloading carts, filling the pier, digging foundations, clearing the beach, &c. have been done by day labour under the management of William England, a respectable mason from Dunmore harbour, who had also conducted the pier at Clonakilty.

The payments were made and accounts transmitted to my office by Mr. J. Macleod, paymaster for that part of the western district, and he has performed the duty without any separate remuneration.

Although by these measures I considered the work to have been conducted in the most economical, though at the same time in a very substantial manner, yet the total expenditure has considerably exceeded the original estimate (24th May 1822), which may be thus accounted for, the following additional works having been executed:–

First,– Instead of confining the excavation to the depth of six feet along the inner boat quay, as originally proposed, from an idea that the bottom would be hard, I have found it practicable to deepen the whole to the level of low water, having only found rock in one place at the S. W. Corner where the scouring sluice is placed.

Secondly,– This has increased the height and expense of the inner boat quay, which, instead of averaging six feet high as proposed, is now about twelve to fourteen feet high; a part of this quay also having subsided from the softness of the foundation, had to be taken down and rebuilt at an additional expense.

Thirdly,– The same circumstance has increased the expense of the outer quays, which instead of running across the beach at a sloping direction, are now all founded at the level of the low water.

Fourthly,– The excavation was extended to the northward, and to secure the other sides of the dock thus formed from being worn down by the tide, it has been thought adviseable to line the whole of it round with a pavement, (though I should have preferred a sloping beach of gravel), and to quay regularly along the north side of the entrance from the groin into the dock.

Fifthly,– At the upper end of the dock a double wall of masonry, with puddle between, cuts off a small reservoir for back-water, which is provided with a stout wooden sluice, raised by a rack and pinion.

This reservoir is yet rather small; but Mr. Jones, the lessee of Raughley, has expressed his willingness to give up more of the bog for its extension, if necessary. It has not been lined.

The excavation from the dock has formed an embankment around the original salt marsh, which is now by the sluice above mentioned kept in an effectual state of drainage.

Some saving has been effected by restricting the kant of the head of the pier to forty feet instead of sixty feet long, the former being found quite sufficient; but this was more than compensated by the filling the whole quay with quarried stone, instead of the rolled stone of the beach, and by the additional expense of carriage from the distant quarry.

And there has also been some loss from the delays of materials and other matter, causing the detention of the workmen.

The work now consists of a very neat stone pier, extending from the top of the beach to low water, nearly 200 feet, with a kant or return at the head of forty feet, all substantially built in hammered lime-stone, paved on the sea-side, and end with a cordon course and parapet of sound masonry.

And an inner harbour, or dock, excavated within the land of an English acre in extent, lined with stone, and with a small groin pier on the side of the entrance, opposite the main pier, of seventy-five feet in length; and there is also a small reservoir, provided with a stone dam and sluice, to retain back-water for scouring the harbour.

The quay wall is in hammered limestone, of good scantling, fourteen feet high, and forty feet long along the head or kant; thence 300 feet long along the pier and dock, with two stairs, the exterior one accessible by small boats at low water neap tides: this quay proceeds 200 feet further in a more sloping form to the dam of the reservoir, making the whole extent of quay 540 feet.

The opposite side of the entrance is also quayed for seventy-five feet along the groin, and 160 feet more along the entrance.

The dock is of a triangular form, 142 feet on either side, lined with a sloping stone pavement, and with a slip for hauling up boats on the land at the northern verticle angle.

The beach on the north side of the groin has also been cleared and levelled for the purpose of hauling up boats.

The whole of the piers and quays are substantially and neatly executed, and provided with stone mooring posts.

The whole expenditure at Raughley has been 1,606 *l.* 8 *s.*

LISSADIL, OR DRUMCLIFF BAY.

WITHIN the point of Raughley the bay of Lissadil first occurs: it is nearly blocked with sand, and though the extensive back-water preserves in the upper part near Lissiadil a deep channel, in which vessels may ride at low water ; below that it divides into two, each of them barred across, and nearly dry at low water at the mouth. From this cause the inlet of Lissadil can be of little value in the sea fishery, though frequently visited by the herrings in winter.

At Ballygilgan and Johnsport a few yawls are kept; above that are salt works on the north shore. There is a productive oyster bed in the channel of Lissadil, the usual sand fish; and a good salmon fishery at Drumcliff.

SLIGO.

I HAVE formerly stated that Sligo was the principal market for this neighbourhood. For the accommodation of the fish trade the commissioners for the improvement of the town have constructed a new quay and dock in the upper part of the tide-way near the bridge, funds were derived from the grants and donations for the employment of the poor in 1822; and it is neatly executed ; but owing to some misunderstanding with the proprietor of the mill, the tail-water from that concern has not been permitted to be carried through the dock as at first intended, and which would certainly have added to its salubrity and cleanliness.

The broad strand of Sligo affords an extensive supply of muscles, cockles, and other shell-fish.

Outside of Coney Island, between this and Ballysadore Bay, is the turbot bank; it appears to be nearly the prolongation of the blowing sand, which occupies that shore, and is, probably, frequented by these fish for the sand-eels and other prey which are found there in abundance in summer, as afore described.

BALLYSADORE BAY.

THIS inlet is about six miles long, and a mile or more in breadth, being more than half covered at the mouth by the sand hills from the north side, and the remainder a sand bar, exposed to the whole run of tile Atlantic, is ill adapted to the purpose of the fishery; within the bar it forms an extensive and safe harbour, six miles deep, but mostly dry at low water, and little used for navigation.

The channel up to the falls near the village is intricate, and wants perches. Mr. Cooper, the proprietor of Ballysadore, is taking measures to form a salmon fishery on this river, and will probably do something to benefit the navigation.

TYRERAGH BARONY.

THIS district is of a triangular form between the sea, the Ox Mountains, and the River Moy; about eighteen miles base by ten perpendicular. The lower part of it is limestone country, on which for about two miles broad there is a belt of arable land along the shore and the Moy River. Behind this an extensive tract, chiefly bog and moor, gradually rising to the Ox Mountains. These hills are composed of gneiss, or granite rock; seldom above 500 feet elevation, and range from Ballysadore to Foxford, at the first of which places the Arrow, and at the second the Moy Rivers cross through from the interior.

To the westward the primitive rock of this range is traceable towards the mountains of Erris, and to the eastward, by Sleish and Benbo to Manor Hamilton, where it is lost; but probably communicates subterraneously with the granite of Donegal.

South of the Ox Mountains is a valley which may be said to run from Manor Hamilton to Castlebar, or Clew Bay, the extreme parts of which are chiefly a rich limestone country; but the centre sand, or sandstone rocks, with extensive bogs and little cultivation, rising to the sandstone hills, which range from the Curlews to Sleeve Carne in Mayo.

The land communication through Tyreragh has been by one principal road running within half a mile of the coast, lately another has been formed from Ballina to Dromore, cutting off a portion of the vertex of the triangle, and then proceeding nearly parallel to the first, to the east end of the barony; but the great tract of moorland above mentioned is hitherto unopened, and there is properly only one road across the Ox Mountains, viz, from Ballina towards Boyle, and through the hollow by Lough Talt. This is too hilly for heavy carriages.

An important branch from this pass to the sea has been lately effected by Government.

The shores of Tyreragh being level and fertile, are well peopled, several good fishing grounds lie near it, especially the bank of Killenduff or Easky before mentioned, but as there are no harbours fit for receiving a sailing-boat this source of industry cannot be prosecuted to any extent. A great many yawls are, however, kept along the shore; in fine weather in summer they go off to the bank for haddock, cod and whiting.

During the herring season they collect to the seat of the fishery, and are at times during the spring employed in cutting and carrying-weed to the rivers Ballysadore and Moy, or in the kelp trade, which is pursued to a considerable extent along the west of Tyreragh.

Many of these boats appear to have been built for the purposes of smuggling; this part of the coast, until the institution of the waterguard, having been extensively concerned in that illicit trade.

PORTAVAD.

THE coast of Tyreragh is all a limestone rock, the debris of which has been washed by the Western Ocean within the bar, and forms a spit that runs along the south shore of Ballysadore Harbour, of about half a mile in length ; behind it is the fine natural creek of Portavad, the principal boat harbour within Ballysadore, and here are kept most of the yawls and several sailing boats.

The whole number in Ballysadore is about fifty.

It would be possible to make a cut across the root of this beach or spit so as to give it a more immediate access to the bar at the bight near Stafort, which bight could be covered by a pier; but on the whole, as the harbour would still lie within the breaker, in such weather as makes it dangerous, I do not consider this an improvement worthy of the expense. Should it ever be resolved on, abundance of materials for the pier may be obtained on the spot.

The most necessary improvement at Ballysadore Bay is to have perches on the two cargin shoals, near the mouth of the channel.

Beyond this the shore is a flag of limestone as above mentioned, of 100 to 200 feet broad from the low cliff to the abrupt edge at low-water mark; the deep water being near. In this flag there are some coves, but too small to afford any kind of shelter or even to permit landing, but in quiet weather, with off-shore winds, at which time some fine hewn limestone blocks are shipped in small craft for the use of the neighbourhood. Of these Dirk and Tubberpatrick quarry coves have rings fixed in the rocks for fastening boats. One of the most remarkable coves has an ancient rath or entrenchment of great strength over it, but there is nothing to make it eligible as a boat harbour even of the smallest description.

Off this in eight fathoms is the Rowry Braddagh Bank, affording good whiting, cod, and glassen.

DUNMORAN.

BEYOND Tubberpatrick the limestone ceases, and is succeeded by a flat sandy shore backed with sand hills half a mile. Here falls in a small stream; running from the Ox Mountains by the Church of Screen, it is barred at the mouth, and some boats are kept there although it forms no harbour; but on the N. W. a reef of sandstone rock, named Carrickfadda, runs off for near half a mile, giving shelter in N. W. winds to a roadstead within, where small vessels may ride pretty safely in three fathoms or more, out of the tide.

This would be a favourable situation for a pier from the rock in a S. E. direction, previously filling up the space between it and the main over the level of high water; but as the expense would materially exceed the funds of the Board of Fisheries, I have not entered into the necessary details for an estimate, but I have made a plan of what appears the most advisable and economical way of obtaining this object.

AUGHRIS.

BEYOND Carrickfadda, and east of the point of Aughris, the sandy beach continues across a bight called Aughris or Pullaghurry Harbour, about half a mile square, alluded to in my first report on the coast; the western side runs out in a limestone cliff forty or fifty feet high; here, in all southerly winds, is a good roadstead in five fathoms, frequented much during the herring fishery; but with N. W. winds the ground-swell is such as to prevent any vessel from holding on, and the upper part of the bight being a shallow strand, the construction of a harbour is a matter of some difficulty; I therefore refrained in my former report from proposing any improvement at Aughris, nevertheless, as it is the most favourable place on the south side of Sligo Bay for a thing of the kind, which is much wanted on some part of the shore of Tyreragh, I have again considered what might be the most advisable mode of proceeding to obtain this object.

Immediately on the outside of the little gap or ravine of the rock, where some yawls are now hauled up, I would propose to run a rough breakwater or roadstead for 200 feet at least.

The materials for this break water to be obtained along the cliff within it, by quarrying the same down, so as to form a passage for vessels, defended from the sea by a rough embankment, until we arrive at the bog meadow, at the head of the bay, in which meadow a dock may be excavated to any extent that may be thought necessary and the streams that now fall in near that place being led into this dock, will preserve the channel from thence to the breakwater from being obstructed with sand.

This harbour will cost on the whole about 10,000 *l.*, but its benefit to this part of the country would, in my opinion, amply justify such an expense; and the breakwater pier aforesaid, if ever carried forward into deeper water, would afford shelter at all times to vessels forced in upon this dangerous coast.

If this work be thought too expensive, a tide harbour, sufficient for the coasting trade, may be made in this manner :–from the beach between the two streams east of the road, run a pier across to low water, with a kant directed to the middle of the cliff of Aughris; let the quay wall be founded at a considerable depth under the strand, the stream from the west side led within it, and a small breakwater built from the point within Pullaghhurry cove; the backwater might be still further increased by a cut or main drain through the marsh of Dunmoran, to lead in the stream which falls in beyond Carrigfadda.

Estimate for this Harbour:

600 feet quay wall, averaging 15 feet high, 6 feet thick, at 8 *d.* per perch	£. 600
Filling behind ditto, averaging 10 yards x 3, 6,000 yards, at 1 *s.* per yard	300
Paving back, 8oo square yards, at 5 *s.*-------	200
200 feet wall in the kant, filled and paved, 5 yards high, at 3 *l.* per foot, complete	600
Breakwater, as at Port Ballintrae, 150 feet, rough work only---	600
	2,300
Contingencies	460
	£. 2,760

At the head of the point of Aughris are two narrow caves, the shore of which is a precipitous cliff, capped with the remains of a triple entrenchment called Toole's Castle; but although it appears in former times to have attracted this degree of attention, it is too much exposed to the ground-swell of the ocean to admit of being converted into any kind of harbour.

POULNADIFFY.

THE limestone cliff, continuing nearly two miles, returns back to the general direction of the shore of Tyreragh, and is succeeded for one mile by a shallow strand named Trabaan, in which there is no place of shelter. Beyond this the limestone cliff and a low shore again form the coast.

In this skare we find one remarkable bight or opening, named Poulnadiffy, which, being covered in some degree by the flat rock Lackbin, affords a tolerable quiet landing-place for yawls in off-shore winds; but without great expense it would be impracticable to complete the shelter by means of piers. A chain and mooring buoy to haul off shore would be the most useful improvement.

Rounding a low projecting point of Skerry Rock, we enter a bight called Donaghantraa, or Slate Quarry Harbour, from the thin beds of limestone which are quarried here for roofing.

A small stream falls in here, with rocky banks, the mouth of which seems not ill calculated for the purposes of a harbour, were the stony beach in front of it cut through and the passage protected by groins. A considerable straggling village, named Ballykilcash, lies along this stream, as far as the bridge upon the coast road.

Ballyminney, a yawl creek west of this, admits of no improvement.

The coast now rises in the hill of Rathlee, and continues with a bold precipitous cliff for nearly three miles; one stream falls in, but makes no creek; but beyond this cliff to Killala Bay, the coast is all low, with projecting edges of limestone, and intervening coves that would appear to afford some situations for a boat harbour: the first is

KILLENDOOF.

THIS place has been pointed out to the Board in a memorial from Colonel Irwin, but as the limestone ledge forms a bar across the mouth of it, it becomes quite impracticable in bad weather from the north, and the bay within the beach being above the level of the sea, does not admit the excavation of a dock as he expected; nevertheless a pier along the ledge would certainly be a useful work in this neighbourhood, it might be executed for about 700 *l.* to 800 *l.*

Several other coves of this kind occur along the remaining shore of Tyreragh, and as there is no want of limestone rock for building materials, it seems very practicable at some of them to form a harbour for sailing craft.

They are however all liable to the same objection, viz. that the exterior edge of the limestone shore in which they are situated being precipitous below water, whenever the wind sets in from the Atlantic, or when there is bad weather at sea, the ground-swell brought up against this cliff makes a complete breaker along the whole coast, across which no kind of vessel dare pass. The shallow cove east of Easky River has too little water for any kind of harbour.

EASKY.

ONE however, might be formed in front of the old castle of Easky, by an extensive breakwater on the low point of Ceach, west of that river, so as to cover the bay and the river might then be applied to preserve the entrance to an inner harbour; but this must be considered as too expensive an undertaking for the present circumstances of the country.

This castle seems at one time to have been provided with a landing-place, or natural quay, on the ledge of rock below it, which, if the breakwater aforesaid were built, might be again restored and extended ; without this it would be too much exposed to the action of the ground-swell.

CASTLETOWN OR TRAMEEN.

HALF a mile west of Easky, Mr. Fenton of Castletown has attempted a breakwater on one of the ledges of limestone rock, covering a cove of the description aforesaid; and on the whole it appears the most favourable situation on the north shore of Tyreragh.

I reported briefly on this place in 1823, but since then it has been carefully surveyed, though the weather at the time did not admit of having soundings taken off the bar, which is said to have nine feet at low water.

Mr. Fenton began a breakwater along the western ledge, 430 feet in length which would require to be more effectually secured and raised, but so far will answer well enough for filling; beyond this it would be requisite to continue the same kind of structure to the end of the ledge, about 600 feet, which might be averaged at about 4 *l.* per running yard ; and as it would be quite unnecessary to quay all this length of breakwater, I would propose to have it firmly paved on each side in a sloping form, and provided near low water with a jetty pier of say 100 feet in length, where craft might lay to load and discharge.

The original breakwater, which may be valued at about 200 *l.* would require nearly an equal sum to complete it; and say for the jetty 300 *l.* would be in all 700 *l.*; for which sum a substantial amid highly useful work might be obtained.

Specification and Estimate for EASKY or CASTLETOWN PIER.

The portion of A B is to be built with dry ashlers found on the spot, packed on edge, in all respects similar to the work done at Pullagheny; from B to C the inside wall is to be brought up with a batter of two inches to a foot, and the coping firmly packed on edge.

Estimate:

		£	s.	d.
430 feet long of rough facing, old pier at	2 s. 6 d. – – – –	53	15	–
Paving back of ditto – –	5 s. – – – – –	107	10	–
840 yards additional filling, at ditto	1 s. – – – – –	42	–	–
3,000 yards filling in the new Kant	– – – – – –	150	–	–
600 yards rough masonry in front wall	3 s. – – – – –	180	–	–
840 yards super of paving in ditto	2 s. 6 d. – – – –	105	–	–
317 yards of parapet – –	4 s. – – – – –	63	8	–
Posts – – – –	– – – – –	8	–	–
20 per cent. contingencies –	– – – – –	141	18	–
Exclusive of materials already laid	£.	851	11	–

One or two other yawl coves occur between this and Kenasharrock Point; none of them are so favourable as Castletown, but the neighbourhood of the valuable cod bank makes the inhabitants of this shore keep some yawls in most of these coves; the mouth of Tennet River is the deepest and best of them.

The limestone skerry of this shore affords abundance of kelp, which is mostly shipped at Pullagheny

Kinnisharogh Point runs with a long ledge towards the N.W. when round it we are in

KILLALA BAY.

THIS bay is nearly a square of five miles each way; the east side is a low limestone shore, the west more varied and elevated, with cliffs of sandstone alternately with limestone, and has various inlets on that side of the bay; there are also several extensive rocky shoals; the bottom is a broad shallow strand, cut off by a range of sand hills from an interior lagoon, mostly dry at low water, between Killala and the Moy River.

There are two openings across this range of sand hills, viz, the bars of Killala and of the Moy. The former has four feet at ordinary ebb, and on the whole is considered the safest, being protected in the worst winds by the land too the westward, and the rock of Carrig Patricke, &c. The bar of the Moy after great floods has deeper water, but being more liable to shift is not to be depended on; and in droughts of summer is often fordable.

It is only a few years since the bar of the Moy was attempted by vessels of any burthen. Sloops now usually ascend to within a mile of Ballina, and measures are in progress to bring them up to the town, by cutting a passage across the limestone ledges that obstruct the channel of the Moy.

This work has been carried on by the funds for the employment of the poor, and promises to be of much benefit by bringing the export market so far into the interior.

Heretofore vessels for Ballina entered the Killala Bar, and passed by the channel behind Berha Island to the pool of Moyne, near the abbey of that name, whence lighters conveyed the cargoes up to the town of Ballina; since the direct navigation was opened that town has rapidly increased in trade and population, and is now the best on this coast.

Killala is also a town of some consequence, and exports a good deal of corn from the Barony of Tyrawley. During the herring season it is much frequented by the fishing-boats, and the herrings frequently get within the bar.

A material improvement was effected on the harbour of Killala by the construction of a dam to increase the back-water, and a direct cut through the strand below, which has increased the depth of water at the pier.

There are no other towns in this neighbourhood; and these were also of late confined almost entirely to the supply of the barony of Tyrawley in Mayo, and the adjoining part of Tyreagh in Sligo; the roads to the interior, across the line of the Ox Mountains, being so excessively hilly and inconvenient as to be nearly impassable for carriages of any burthen.

This defect is now in a fair way of removal, as a new and level line of road is nearly completed from Killala to Ballina, Foxford and Swineford, by which the interior country may convey its produce to these seaports; and the road is on a scale to admit a mail-coach to communicate towards the Royal Canal and Dublin.

Another line of road has been opened along the north shore of Mayo, as far as Balderig and the bounds of Erris, which it is intended to connect with Broadhaven.

A line of carriage road has also been executed on the east of the Moy, from the coast road at Inniscroan on the east side of the bay to the foot of the pass of Lough Falt. This line is of great importance to the fisheries, as giving access to the country near these mountains and the interior from the fishing places of Pullagheny and Enniscroan.

A branch from the same road is also in progress towards Foxford through the district of Coolcarney, which is also opened by another line to Ballina. The roads from the town of Ballina to the quays below it have also been levelled, and similar improvements made at Killala.

These works have been made at the expense of Government, from the funds for the employment of the poor; their importance in giving access to the seaports of Killala Bay is evident.

In addition to these, the grand jury of Mayo had, some years ago, opened a road from Castlebar to Ballina, across the narrow part of Lough Con; had begun another from Ballina towards Swineford, on the east of the Moy; and had also, by means of a loan from the Consolidated Fund, made considerable progress in opening a carriage road from Castlebar through the upper part of Tyrawley into the extremity of Erris. This road is now nearly completed by means of the Government funds, and is of great consequence to the towns of Tyrawley; which, from their vicinity, are the natural market of Erris, and are equally accessible by this road. Towns, however, are now rising in Erris itself as will be seen when we come to the description of the coast of that district.

The whole Bay of Killala has been carefully and minutely surveyed and sounded, and proper sailing marks adopted for the bars and shoals; this chart is now ready for the engraver.

The merchants of Ballina having employed me professionally to examine the river Moy, and suggest the means of improving the navigation, I am enabled to lay down the result on the chart of the Bay of Killala, constructed for the use of the fisheries, though the survey of the river within the bar was made without any expense to the Commissioners. This chart, from the active spirit of improvement afloat in this place, promises to be of great, importance to navigation; and I trust the Board will lose no time in having it made public.

The fish, and seasons for fishing, are the same in Killala as already described in Sligo Bay.

Most of the bay is a good turbot bank, especially between the bars; also off Trakel about August, following the sand-eels, which abound in those strands. Rockfish are found at Carrickpadrig and the shoals near Rathfran; cod and ling abound under Kilcummin and the deep cliffs to the westward.

PULLAGHENY.

On rounding Kenniscarrough Point into Killala Bay, we find two coves backed by a high beach of limestone shingle, and separated from each other by a small islet of limestone rock, about 300 fathoms from the shore. This islet is bluff towards the sea, which end is rarely covered; and declining towards the land, is connected with the main at low water by a spit of beach, covered five or six feet at high water.

The cove on the north side of this spit is exposed, shallow and rocky; but on the south side there is deep water, making a kind of anchorage tolerably sheltered, except at high water, where vessels, come in summer to load kelp.

The entrance to this anchorage has four fathoms at low water, and within these are two fathoms, so that, if properly covered, this would be a deep harbour, fit for any craft, and being outside the bars of Killala, or the Moy, would form a place of shelter to the trade of these places in all weather. There is, however, a violent breaker in the entrance to Pulocheny in N. W. winds.

A great number of fishing yawls are kept in the place, but must all be hauled up on the beach.

Within the beach is a marsh of several acres, over which spring tides frequently flow, and the stream passing through, it has five or six feet water, but barred to high water by the beach at its mouth.

The harbour itself is about 200 fathoms square, about half of it dry at low water, the south side a skerry of limestone flag.

The Board, in 1822, appropriated 300 *l.* to be laid out in improving this place, which was, however, very far from what would be required to make it complete. I employed Mr. Ham, of Ballina, who had built the Black-rock beacon of Sligo, to superintend this work, confining the operation at first to raising the spit, or beach, which connects the rock with the land.

The persons employed here were not accustomed to works of this kind, and ere I could visit their operations, they had built a face-wall in the usual way, backed with stuff from the spit itself, and the heavy gales of the latter end of the season had demolished the upper courses, and washed part of the filling over the wall, so as to cover it. The spit, however, had been raised a little, and gave better access to the isle; but I found it necessary to suspend this work for the winter, and until I could send one or two workmen of more experience.

In 1823 I accordingly began a breakwater across the inside of the rock, made of the blocks on the spot, packed on edge, which was prosecuted vigorously, as far as the funds would permit, and has withstood the heavy gales of the winter and last summer with little comparative damage. Something further was also done on the spit as I had obtained the authority of Government to make a road from

Lochtalt to Pullagheny, comprehending some aid to this work. Some little expense was also incurred last season in securing the ends of the breakwater, which having, from the deficiency of funds, been left imperfect, had begun to be loosened by the action of the sea.

The breakwater is now 400 feet long, forty-five feet wide, and as high as ordinary tides; high spring tides in westerly gales still come over it, but the shelter within is greatly improved, and were it continued a little further, especially at the angle, a vessel might lie behind it with confidence in any weather. The additional works for completing this harbour might cost about 1,600 *l.*

INNISCROAN.

INSIDE of Pullagheny the shore continues a flat limestone rock to the extensive strand at the bottom of the bay. Here there is a fishing village named Inniscroan, with a good many yawls, much employed in the herring season, as there is a kind of landing-place where the boats may be hauled up without much danger from the breaker.

I examined this place, in consequence of a memorial from the fishermen, and found that it would be quite impracticable to found any pier, at a reasonable expense, on the naked limestone flag: but that by means of some excavation, the landing-place could be materially improved.

This work was afterwards effected out of the funds for the employment of the poor, at the moderate expense of about 30 *l.*; and the fishermen have since expressed great thankfulness, considering it to have been the means of preserving many lives.

Inniscroan is the best fishing place on this coast.

THE MOY.

WITHIN the bar of the Moy there are a good many small boats employed during the herring fishery, and there are saltworks near Colonel Wingfield's on the Sligo side, and also at Belleck near Ballina, and at Killala.

There is nothing however within the Moy particularly to interest the Commissioners of Fisheries in the way of improvement.

I proceed therefore to the county of Mayo.

KILLALA.

THIS town is situated on the edge of a broad strand, within the inner bay or lagoon, and is not very accessible by shipping of great draught; there is, however, an excellent pool in the channel of Moyne where vessels may lay afloat near the town, and a quay has been built about half way between it and the town, at a place covered from the sea along the strand at high water, by a small green island.

The pier or quay has been principally built by a small fund raised by the merchants, of 1 *d.* for each sack of oats exported, and admits vessels of eight or ten feet draught.

A presentment was obtained spring assizes 1822, in Mayo, for making a better road to this quay, with a face-wall on the sea side, which in some degree answered as a boat quay, and the foundation had been laid of an extension of the pier.

When the Board allowed 150 *l.* to the pier of this place, it did not appear that it could be better applied than to the completion of the works above. mentioned, which has since been accordingly done in the most economical way the smallness of the grant required; by day labour first, under the overseer for the county work; and, secondly, by contract with Mr. Rutledge. But, in addition to this, a material improvement has been effected by means of the charitable funds of 1822.

As it was necessary to provide a field for the labour of the numerous poor of this place, I directed the following experiment to be made, with a view to increase the depth of water at the present quays, which, having a strand between them and the pool of half a mile in breadth, are evidently incapable of obtaining such an additional depth by any moderate extension. I had therefore an embankment wall run across the strand, from the green island over to the point of the salt work, thereby enclosing about fifty acres at high water.

This bank rises only two and a half feet over the strand, and consists of a dry wall of stone, three feet thick on each side, with three feet of puddle between. It is covered five feet at high water; but as it causes the whole of the early flood and latter ebb of the fifty acres to pass along the quay, the consequence has been, that it has lowered the channel there about a foot, and though not very skilfully made, has stood well.

As the channel from the pier took a winding course through the strand, (which, although only two feet higher, caused vessels often to ground in making or leaving the pier), I directed a straight channel, to be cut for about 150 yards, which was to have also had a bank of gravel, by way of wear, on the west side; and in order to have the flood to run fairly along the quay, I directed a bank to be made in continuation of the pier, to overshoot the green island, and of same height as the first-mentioned embankment. These directions have been rather imperfectly understood, and were incomplete when the necessity for employment ceased, but have on the whole been of material benefit.

The stuff for filling the road to the quay was to have been taken from the front of its sea wall, so as to make a boat channel all the way up to the town, which would be very useful to the fishermen. This also is yet imperfect. The expenditure on these works was 350 *l.* chiefly managed by the Rev. Archdeacon Verschoyle.

RATHFRAN.

BEYOND Killala we have a peninsula called Ross, between the inlets of Killala and Rathfran. it is curiously intersected by the sea at high water, and the creeks behind it offer a good subject for embankment, as the soil of them is valuable. There is no harbour or fishery town on it, the inlet of Rathfran on that side being an extensive sand; but the backwater being extensive, there is a good channel kept open across this sand along the western shore, which, though narrow, is rather safer and less broken on the bar than Killala. A good many yawls belong to Rathfran, which has no quay or pier, though an excellent situation for one is found at the old salt works on the west side. The rock here is sandstone, which continues on a steep cliff, about half way to Kilcummin Head, where the limestone returns.

CLOGHER.

NEAR Kilcummin Head the cliff is precipitous, and appears to have been undermined by the sea; the (*illegible*) is washed in along the shore, and forms a bank of beach round a low alluvial flat at the village of Clogher. The inner angle is a landing-place for yawls, which are hauled over the beach, and are rather of a larger construction than most others in the bay. It is remarkable as the place where the French expedition landed under Humbert, in 1798. The village is extensively engaged in the fishery, and supplies most of the neighbourhood with white fish, which they find near the shore. A small harbour would therefore be of great use; the neighbouring country is fertile, and a good deal of produce is exported to Killala. I would propose to run out a small pier from the S.E. point, in the direction of Ross, to make a cut across the beach within it, and to excavate a dock in the alluvial flat, which is composed of sand and bog stuff. The work might cost about 1,000 *l.* being pretty similar to Raughly, and being, like Pullagheny, situated without the bars, and much better sheltered against the dangerous winds, would be of great utility even to large vessels.

This is the extremity of the district of the Inspector General of Fisheries for the north coast.

TRAKELL.

ROUNDING Kilcummin Head we find the sand has blown to a great extent over the western slope, and forms a line of downs, and a broad strand across the inlet between it and Rathlaken point; behind the sands is an extensive but shallow lagoon, dry at low water. The channel being in the middle of the strand, is too precarious to be of much use in navigation, or to admit of improvement although the grounds in the neighbourhood are fertile. This bay is called Trakell or the White Strand, and affords good supplies of flat fish; several yawls are kept on the western shore below Castle Lacken, but they must be hauled up a steep cliff into a paddock near Mr. Burke's house; nor does there seem any method of forming a harbour at any reasonable expense.

We are now in a district called the Laggan a country of limestone, alternating with sandstone, of considerable fertility, and a great population.

The roads terminate at Castle Lacken, which is also the last post office on this side, and from hence round to Dunfiny Bay, the coast is a precipitous cliff of horizontal beds of limestone, black plate and sandstone.

DUNFINY BAY.

ON the eastern point of this little bay is the remarkable insulated rock called Downpatrick Head. The perpendicular cliff affords five distinct sections of the horizontal strata aforesaid, having a remarkable resemblance to the general appearance of the coal measures, though I am not aware of any of that mineral having been discovered.

The western side of Dunfiny is lower, with fine quarries of sandstone fit for building along shore. The bottom of the bay is a broad strand, on which the swell beats heavily in northerly winds. No natural shelter exists here for boats, though a good many yawls of the Norway make are kept, and the valley is fertile and exceedingly populous.

The Board proposed in 1822 to lay out 300 *l.* for a boat pier in this bay. The most favourable situation appeared to be at Bunatragher, on the S. W. angle, near the mouth of the western stream; but though an excellent description of building-stone for such a purpose exists on the spot, I could not take it upon me to commence this work, as it could not be made useful without a considerable increase of funds.

The plan of the work, and estimate are as follows

THE PIER.

		£	s.	d.
320 purches dry masonry in quay-wall and end, at 8 *s.*	– –	128	–	–
1,800 yards .filling beach stones, at 1 *s.*	– – – – –	90	–	–
640 yards paving on back and end, with flagstone on edge, at 5 *s.*	–	160	–	–
200 feet of parapet wall over rough cordon, at 7 *s.* 6 *d.*	– –	75	–	–
600 yards paving on pier	– – – – – – –	30	–	–
Stairs and moorings	– – – – – – – –	20	–	–
		503	–	–
Contingencies, 20 per cent.	– – – –	100	12	–
	£.	603	12	–

For about three times this sum a work of much greater importance might be made by a pier from the edge of the limestone rock, fifty yards more to seaward, as this pier would have three or four feet at low water.

A country road exists to the bay of Dunfiny from Palmerstown Bridge on the river of Rathfran; another, from the village of Ballychashel, runs up the valley of Ballanglan, towards Killala and Crossmolina; but the country to the westward, as far as Broadhaven, had not even a footpath before the year 1822 ; since that a good carriage line of road has been made by Government, ten miles westward to the bounds of Erris, and to the village of Balderrig and is intended to be continued to Broadhaven.

From Dunfiny Bay the coast to the westward is a precipitous cliff, without any landing-place even for a boat, for several miles; the country behind it is for the most part an uncultivated bog, the property of John Knox, esq. of Castlerea; it is, however, capable of easy reclaim, having in general a subsoil of great fertility.

Some of the ravines exhibit sections of black mould, forty to sixty feet deep, mixed with limestone gravel. The same kind of soil continues over all the great moor forming the western part of Tyrawly, until we arrive at the foot of the primitive mountains which divide it from. Erris and Burrishoola. If once opened by roads, and provided with any kind of harbours, this country would soon rival the most fertile parts of Mayo.

BALDERIG.

To the westward the coast rises in lofty cliffs of quartz rock. There are two or three coves in which yawls are kept; but being open to the north, and so near the deep, the groundswell breaks heavily within them.

Flat fish are found in those coves, and they have the advantage of the neighbourhood of the cod ground of Bank Bwee, if craft fit for the offing could be kept in them.

At Balderig there appears a tolerable opportunity of having a small harbour, by running a short pier on the west side to cover a cut which might be made through the beach into the stream. behind, which is pretty level and deep for near half a mile up the valley, and would therefore be a tolerable harbour if the entrance could be kept open.

PORT TURLIN.

The cove of Port Turlin has also a few yawls; but lies too open to admit of any improvement, unless by cutting a canal or passage through the beach. Six or seven boats only are kept at Port Turlin.

PORT-A-CLOY.

THIS cove is deep, and affords good shelter in most winds in summer, and even from the north it is somewhat protected by the lofty rocks called the Staggs of Broadhaven, which are about a mile offshore: there is a depth of twenty-four fathoms at its mouth, and four fathoms close in shore.

There are about a score of row-boats in this creek, and the peninsula between it and Broadhaven is thickly inhabited; though the south side, near the village of Dunkeeghan, is mostly blowing sand, and the uplands moor.

FISHING GROUNDS ON THE NORTH COAST.

N.E. of the Staggs about six miles, in fifty fathoms water, is the Yellow Bank, or Bank Bwee, before mentioned; it extends from Glynck Hill, in Erris, about one fourth of the way across to Arran. It abounds with cod and skate, or maiden ray.

The whole north shore of Mayo is excellent fishing ground in harvest; but it cannot be pursued in sailing boats, as there are no ports to save them.

Off Erris Head, when you open Glenlard E. N. E. about five miles, is an excellent ling bank, in thirty-five to thirty-six fathoms. Small cod also abound in the entrance of Broadhaven.

BROADHAVEN

THE lofty coast continues to Benwee Head, and in the detached isle of Kid Island, between which and Erris Head we find a bay of six miles in breadth at the mouth, by four or five deep, being the entrance into Broadhaven. The coast on either side is a lofty precipice, but lowers as we pass up the bay, which has two principal arms.

That on the east side, named Sauffodacon, receives two considerable rivers; but filled with sand, and barred across the mouth.

A little on the outside of the bar is a point of land called Rinrua, near the village of Dunkeeghan, where an opportunity occurs for a small pier. It might be made similar to Tarmon, and would cost about 500 *l.* to 600 *l.*

The sand outside the bar, as usual, affords plenty of fiat fish.

Thirty or forty row-boats are kept about Dunkeeghan.

On the west side of the bay there is a small opening in the cliff, frequently mistaken for Broadhaven. The inlet, though land-locked, is dry at low water, and forms no harbour; it however possesses a few yawls.

Between these two is the proper entrance of the haven, less than half a mile in width, and four fathoms water; and within it the inlet continues in a curve form for near seven miles to the isthmus of Bellmullet.

There are two creeks which dry at low water, as does the upper part of the haven for several miles ; but the lower part is an excellent harbour, affording good shelter in three fathoms on soft sand, between the points of Rentagel and Inver. On either or both of these it would be desirable to have a small landing-pier.

Of these two places Inver appears to be the most useful, as there is to be a branch road from that point along the side of the lake of Carromore to join the central Erris road near the outlet of that lake, which will afford a ready communication to Castlebar and the interior.

The strands are also passable at low water from hence to Bellmullet, and serve as a road.

There is no other trade in Broadhaven but fishing, and that only pursued in row-boats for immediate consumption. Heretofore there were some sailing-boats which carried fish to Sligo, at present there are none.

The country would appear, from the description by Sir William Monson in his Naval Tracts, to have made very little progress in two hundred years. A better prospect is now opening from the measures taken by Government to open up this country by roads, in explaining which it will be necessary to give a brief description of Erris.

ERRIS.

THE western part of Mayo, between Achil island and the north coast is named the Half Barony of Erris, though in extent larger than the entire county of Dublin. it is for the most part a low plain of bog, on a bottom of mica slate, backed by an amphitheatere of lofty quartz mountains, which from Glynsk, on the north coast, pass southward as far as the isthmus of Corawn; Achil dividing this district from

Tyrawley and Burrishoole. The mountain range then turns westward through Corawn Peninsula and the island of Achil.

In the latter they are chiefly quartz, in the former conglomerate sandstone.

The two inlets of Broadhaven and Black Sod penetrate deeply into the plains of Erris, approaching from the north and south within 400 yards of each other at high water. The isthmus is called Bellmullet. The peninsular district beyond, about nine miles in length, is low, and where not overblown with sand exihibits considerable cultivation. The population is about 8,000, while the main of Erris contains 10,000.

Through the range of hills above mentioned the river Owenmore makes its way by a narrow valley from the upper plain of Tyrawley to the sea at the lower end of Black Sod Haven; its mouth is called Tullaghaan Bay, and is a haven for small vessels; but dries three fourths over from the east side at low water. Along the valley of this river is the only carriage road yet made into Erris.

One branch proceeds to it from Castlebar, another from Crossmolina: the former lately begun by the grand jury, has been completed by the funds for the employment of the poor: is a level line, and in general skilfully laid out; the other has all the defects of the ordinary country roads.

Through the valley above mentioned the original line has been improved, and after passing the mountains a new line goes off to the isthmus of Bellmullet, thence along the west side of Black Sod Haven to Elly Harbour. Other branches are also in progress from the same point to the west side of Broadhaven, and to the isthmus of Corawn Achil.

By the former of these branches a communication is opened to the north coast and Killala, and the latter unites with a new line, also in progress, from Newport to the extremity of Achil.

The change produced by the roads into Erris is already very perceptible. Previous to 1798 no way existed of entering that country but on foot or by sea; at least when a horseman attempted the journey it was necessary to have the assistance of six or eight guides on foot to cast the horse at every swamp, fasten his legs, and haul him over by ropes; an operation which is to me sufficiently familiar, being often resorted to from necessity in the course of our expeditions.

The journey could seldom be accomplished from Ballina in less than three days; and if the rivers happened to be in flood, the delay might be much greater. The expense, as might be readily conceived, in money, whiskey, and civil entreaty, much more than it is now to go with carriage and post-horses in one day from Castlebar to the extreme part of the Mullet.

Since the commencement of 1824 carriages have appeared within the Mullet; it is not uncommon to proceed there from Castlebar in one day, about fifty English miles.

Dennis Bingham, esq. a landed proprietor, resident within the Mullet, has built a considerable village on the north side of Saleen Bay, within the Mullet ; and here a small harbour has been begun by the Commissioners of the Fisheries. A corn store is also built at the harbour.

Since March last Mr. Carter, another proprietor, having come into possession of the farm of Bellmullet, has begun another village on the isthmus, which, being accessible by water on either side, seems likely to become a place of consequence. Already a good store has been built there by a Mr. Ivers, and filled with oats and bailey; two cargoes have been shipped this season, several small shops have been opened, and most of the building lots are taken.

The most important improvement here would be to cut a channel across the isthmus so as to permit boats to pass at high water from one haven to the other, as each of the inlets dry to a considerable distance at low water, and Broadhaven in particular is shallow at the head: it would be necessary to continue this cut for some way through the strand; but assuming only the depth of eleven feet, which is obtainable very near the shore in the Black Sod Bay, I find the cut would have to be about 120 perches in length, and might probably be executed for about 5,000 *l*. The bottom, for part of the way, is mica slate rock, which is, above high water, only bog and gravel.

This cut would permit the passage, in most tides, of such vessels as use the Forth and Clyde Canal, and would certainly be a great convenience in the time of the fishery; but I presume it is unnecessary in this place to enter much further into the detail of probable expense. A plan of the isthmus, and probable direction of the cut, is annexed to this Report.

As the tide in Black Sod rises earlier and somewhat higher than in Broadhaven, there would consequently be a current through the cut which might enable it to be kept clean without lockage; and this current, if not too strong so as to create bars, would probably improve the present channel through the upper part of Broadhaven. At all events there can be little difficulty in cutting that channel straight, and preserving it by jetties or gravel lining.

The sound of Achil, though it ebbs so far as to be fordable at low-water, permits vessels of the class of coasters to pass at high water, and with the proposed cut at Bellmullet we would have a kind of inland navigation from Westport to the bay of Donegal.

GENERAL REMARKS ON THE WEST COAST.

If the north coast of Connaught be scantily supplied with harbours, the same character by no means applies to the remainder of that province.

From Broadhaven to the county of Clare is one continued succession of havens, the finest indeed in Ireland, not excepting those in the county of Cork. Many of them are fit for vessels of war of the largest class, and they are now much frequented by the revenue cruizers.

The charts and nautical descriptions of these valuable harbours are still in an imperfect state; and the most beneficial purpose which this survey could be directed to, was the publication, by the Board of Fisheries, of proper charts of these fine harbours, as well for the purposes of fisheries as navigation generally.

The country in which these harbours are situated is poor, chiefly granite and primitive rock, with an immense proportion of bog. Between Broadhaven and Galway Bay there is little trade of any kind, not a single decked vessel is owned between Killala and Newport, or between Westport and Galway;. nay, even very few half-decked: there are, however, an immense number of open boats, as well for sailing as rowing, especially in the Connemara division. They are chiefly employed in the turf trade, excepting during the herring fishery.

The sea off this coast abounds in fish, being by far the most productive on the Irish coast; and it is regularly visited by decked wherries from Skerries, Rush, &c. from the east coast, and by hookers. From Kinsale and from the south.

These vessels being able to stand the heavy sea of the Atlantic, can run to the fishing banks in the offing, and of course succeed better than the country boats.

The Skerry men use long lines, also spillers in summer, and hand-lines in the winter fishery. One of the crew is left at some station on the shore to attend to drying the fish, and the wherry comes in with the fish caught at least once a month during the season, or until the cargo be up, when they make off for Dublin.

The Kinsale hookers use spiller lines of about 300 fathoms only; being unfit for running the long line, but succeed nearly as well as the Skerries men. They also bring mackerel nets, which they employ at night near the Bells Rocks with good success about Martinmas.

There are no other mackerel nets used about Clew Bay.

The superiority of the wherries and hookers over the country boats is equally conspicuous along all the west coast of Ireland and Scotland, and as far as the Isles of Shetland; and these craft are equal not only to stand the sea in the fishery, but also to be employed on the coasting trade, and to make considerable voyages, they seem to be deserving of encouragement in this maritime nation.

Exclusive of want of capital there may be one cause for the adoption of small boats on this coast, viz. that for the larger decked vessels it is necessary to have harbours with quays, to which they can be brought for the purpose of loading or discharging. Thus, on the coast south of Dublin, where there are no such harbours, the boats, as at Arklow, &c. are mostly open; whereas at the pier harbours, Rush, Skerries and Balbriggan, decked wherries and cutters are chiefly used.

The open boats are also better adapted to the trade in turf and sea-weed, which is the principal occupation for them when not employed in the fishery.

Within this range of coast there are two great inlets, Clew Bay and Galway Bay, which penetrate across the granite country to the interior limestone district, which is more fertile and better cultivated.

Here we have the seaports Westport and Galway, the latter an ancient maritime town, the other of more modern date: from each of these places is an export of above 10,000 tons of corn annually.

Kelp is also a staple of these towns, chiefly produced on the rocky skerries of the primitive districts, to the extent of about 7,000 tons; mostly sent from Galway, being carried thither by the country boats.

Since the construction of the new fishery piers this article has been begun to be exported from the spot where it is manufactured,— a circumstance which has produced an increased liberality in the dealings of the Galway merchants; and it will probably tend greatly to improve the quality of the article, which is now so much adulterated as to bring the least of any kelp in the kingdom.

Villages are also rising at some of these places; so that the inhabitants, acquiring a market at their homes, and saved the heavy expense of transport by sea or land they have been hitherto subjected to, may be expected soon to make considerable advances in every sort of industrious occupation. This has been particularly evinced, during these last two years, in the village of Clifden, as shall be detailed when we come to describe that place. Much benefit is also to be expected on this coast from the removal of the duty on salt, and the consequent vexatious regulations of the fishery laws. Herrings are often known to set in to some of the bays in vast abundance, yet from the want of salt, and the difficulty of transporting it across the country by the old bridle-roads, or round the headlands, in wintry weather, from Galway or Westport, as well as the enormous monopoly prices to which it is immediately raised, the herrings are left to rot on the shore in heaps of many thousands, and the unfortunate fisherman, whose little stock has been expended during summer in fitting out his boats and nets, sees himself ruined, with abundance apparently in his grasp.

Salt is now stored in Chifden, and on the isthmus of Bellmullet in Erris, and other places are in contemplation.

The white fishery commences on this coast with the season of Lent; spiller lines are used from Ash Wednesday; bait is procured in the strands, viz. slugs for cod and haddock; muscles or whelks are unknown, as also crabs and scollops. A piece of herring or other fish is commonly used, especially for turbot, which are found in Clew Bay outside the bars, Achil Sound, and on the other sandy bars; also on the great banks of the offing; though seldom sought there for sale, such as are got there by the wherries being for the most part consumed by the crew, or given away to the gentlemen's families on shore.

THE BOFFIN BANK.

THE principal white fish bank is situated between Boffin and Achil, about two and a half leagues north of the former, and three from Achil Head.

You are on it with the Crow Patrick Rock, north of Ennisturk and High Island, west of Ennishark, with about forty fathoms water; the water on either side, and towards the north-west end, deepens quickly to sixty fathoms, and a bottom of rock outside.

This great bank seems formed in the wake of Boffin Island, as there is a general northerly current from Sline Head to Erris Head, when three miles from the shore: it is abundantly supplied with all kinds of white fish, cod, ling, glassen, mackerel and gurnet, especially in spring, as also a few herrings. The haddock and whiting got there are only used as bait.

Whiting, pollock and other rock-fish are found near the Bells and other rocks; best bait for them are fresh-water eels; the line put lengthways through them, and the hook out at the navel.

On the 2d of May 1823 we found a few boats on the white fishing of Boffin: one sloop from Galway, one from Connemara, two or three wherries from Kinsale, and two from Westport, provided with long lines and nets.

ENNISKEA BANK.

NORTH of Boffin Bank, within the Bells, and along the shore of Achil, the ground is principally rocky. North of Achil Head we have a sand-track running up Black Sod Bay, affording as usual turbot and flat fish.

At the Black Rock rough ground; but beyond that lies the Enniskea ling bank, extending from five to eight leagues to the westward of these isles, abundantly stocked with fish, and much frequented by the wherries in summer, with spiller and long lines; hand-lines during winter.

The mark for it is the minister's house on Annack Head, in forty or sixty fathoms, red sand.

This bank is chiefly fished from May to August, and is too rough later in the season for the country boats.

South of Boffin Island the ground in the offing is rough to Sline Head, within which up to Galway Bay is clean ground, and an abundant white fishery.

SUNFISH BANK.

OFF this coast, about thirty miles, or just within sight of the high land of Achil Island, is the ground called the Sunfish Bank, we are on it with Sleavemore Achil, about E. S. E., per compass; seventy to ninety fathoms.

This bank is remarkable from the break of the tide on it, with ebb and flood, and is supposed to be a ridge of land extending from the Blaskets to

Erris Head, in about seventy fathoms. Half a mile further off we have fifteen fathoms more water, and the increase of depth is also considerable within it; the water outside deepens quickly to 100 fathoms and upwards; and the probability is, that this bank is near the edge of soundings.

This place is frequented in spring by the sunfish or basking shark, the best season being the last week in April or first in May: they come hither from the north, and are seen from Tory Island to beyond the Blaskets; in fine weather they come to the surface in the morning and evening, when they are pursued and struck with harpoons for the sake of their oil, of which one fish will yield to the extent of seven to ten barrels of liver, worth about 40 *l.* sterling.

The fish are in considerable numbers together, and easily got at in a fine day; but the weather at this season is very uncertain, and the swell so great, that frequently there are hardly any fish caught: in 1822, for example, there were only two killed.

Should a fine day or two occur during the season there may, however, be thirty to forty killed, but in such a case the fish make off speedily to the south, as the fishermen say, whenever they smell the blood.

Should any stragglers remain they get so lean as to be hardly worth killing.

The spear used for striking the fish in this business, is an iron rod of half inch bolt, about four and a half feet long, having a moveable barb at one end and a socket at the other, which receives a wooden handle about six feet long; between the socket and bar slides a thimble or ring, to which is spliced the end of a coil of two inch line, and which passes through two grummets or loops tied to the wooden handle. When prepared for striking the barb is turned into the line of the shank, and kept so by a turn of yarn round the tail, which, when the fish is struck, slips off, permitting the barb to turn at right angles, and thus keep the spear in hold; the handle being then pulled, slips out of the socket and slides along the line when the fish runs it out.

The animal lies quietly on the surface, and allows the boat to come along side, nor does it make off until pierced to the quick; the wooden handle serves as a buoy when all the line is let out: many fish however are struck without effect and the spears lost.

Indeed the whole business appears to be an unprofitable one, the outfit of so many boats as engage in this fishery cannot be estimated at much less than 2,000 *l.* in a season, and besides the loss of time of so many people at an important period of the year, the fish caught, even in a favourable year, has never, I believe, been above 1,200 *l.* or 1,500 *l.* and latterly not a tenth of that sum, and this chiefly got by the few decked vessels that can stand out waiting for good weather.

SKETCH OF THE SPEAR.

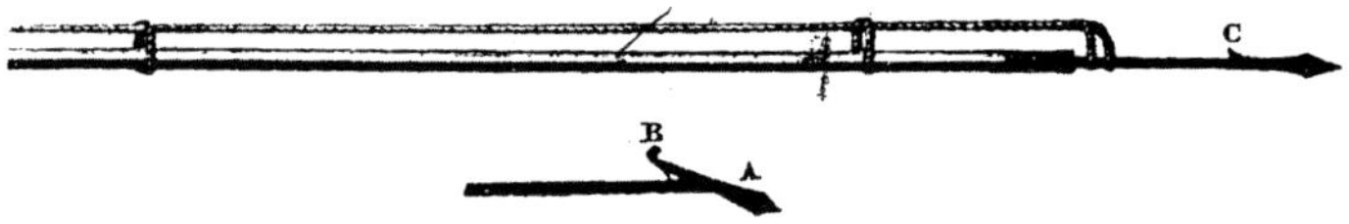

Spear from socket to point, 4½ feet long; shank of 1½ inch iron; outer part on a pivot, so that when in the fish, it opens as at A; part B rounded so as not to cut the fish when it pulls. The wood in handle comes out when the fish goes down, and slides on the rope. C bicket, which keeps the spear straight until the fish is struck.

Sir, Dublin,
November 9th, 1826.

HEREWITH I beg leave to transmit the further Report of the Fishing Stations, &c. of the West Coast as far as Black-head, in the county of Clare, comprehending all that has been actually surveyed. The charts of that survey are with the engraver in London. The plans of the several piers accompany the Report.

I observe by your letter of the 8th September that you were merely awaiting this Report to lay my accounts before the Board.

I have the honour to be, Sir, very respectfully, your's;

Alexander Nimmo.

H. Townsend, Esq.

REPORT on the Fishing Stations of the West Coast of Ireland, from Errishead southwards.

THE northern end of the peninsula of the Mullet is a precipitous and rugged cliff round by Errishead and Eagle Island, On which the N. W. Atlantic beats heavily into the rocky coves; the first inlet noticeable is

Portafranka.

This inlet is narrow and rocky at the entrance, and generally with a heavy breaker; it is only accessible with southerly and easterly winds; within the creek extends nearly three miles, but mostly dry at low water, and I do not perceive any set action at which a pier could be made of any great service; though once in, a small vessel is in good shelter.

The south side of this inlet is formed by the peninsula of Annagh, about three hundred acres, connected with the main by a very narrow isthmus of sand. This place appears to me capable of some improvement, and not ill adapted for prosecuting the fishery. At any rate it is the only position on the west side of the Mullet, the whole shore of which to the southward is a naked beach of sand.

To the south of Annagh-head about two miles, and nearly as much off land, are the small isles of Inish Glore and Inniskerach.

In the middle of the northern open is the heavy breaker called Monaster Ladiri, in the southern open is the half-tide rock Carriglaharan. The interior space, of about half a league square, is somewhat protected by all these; and a vessel may lie behind Inneskearach in tolerable shelter in four fathoms, good holding ground, though a heavy swell tumbles over the rocky points towards high water. This place is not sufficiently known, but may be of service to vessels embayed south of Eagle Island.

The bight of Annagh is further protected by two rocky ledges running off from the southern and western sides. Within the latter I should propose a small pier to be built, from, which vessels may pursue the Inneskea fishery in summer, and in winter they may lay up with great safety in Portafranka. The property belongs to Major Bingham.

A fishing village here would be within a couple of miles of the rising towns Binghamstown and Belmullet, and of course most conveniently situated for communication with the interior.

Estimate for Rough Pier at Annagh.

3000 yards at 2 s.	–	–	–	£. 300
Jetty pier of 20 fathoms	–	–	–	600
Contingencies	–	–	–	100
				£. 1,000

Breakwater on the ledge, of rough stone found on the spot, 100 yards long, 3 high, and 10 average wide:–

Should this place ever appear likely to sand up, it might be advisable to make a cut through the neck into Portafranka.

The range of cliff of the north-west of Erris, after being broken through at Annagh, seems to reappear successively in the islets of Innis Glore and Inneskereach, the Duffer Rock, the isles of Inniskea and the Black Rocks. The breaking up or debris of this chain seems to have formed or at least covered the peninsula of the Mullet, the greater part of it being a naked sand.

Isles of Inniskea.

These isles are interesting in this inquiry from their immediate vicinity to the great Lyng Bank; the two principal isles are together nearly four miles long, being separated only by a narrow sound not fifty fathoms wide.

The western side is a cliff with several wild coves; the eastern side, especially of the North Isle, is an extensive blowing sand; but as the isles have a tolerable anchorage on that side nearly opposite the sound, it would be extremely useful to have at least a landing-pier on one or other of them.

I have had the harbours on each surveyed for the purpose, but neither of them are very well suited to our purpose without a very considerable expense, which I suppose is not to be obtained.

The present boat-cove or landing-place is to the south of the little isle of Rusheena, and does not therefore communicate with the anchorage at low water, the strand between the isle and shore being dry at half tide; but it may be materially Improved by carrying out a pier beyond the present ledge of rock about the line on low water, say for about 300 *l.*

The North isle has a cove which might be materially improved; but as it is nearly a mile from the anchorage, I am of opinion that a small pier at the entrance of the sound will be found more eligible than any of them.

The Mullet.

The southern extremity of the peninsula rises in a considerable hill of red granite; several rocks and small isles of the same material extend two or three miles to the westward, on which the Atlantic beats heavily in most winds; within these, and somewhat sheltered by them, is a boat-cove at Nackel or Lurganian, frequented by the people of Enniskea, and where a few boats are kept hauled up: it seems too wild to admit of any material improvement.

Tarmon or Blacksod Pier.

Rounding Blacksod Point, or Tarmon, we have a small bight within a spit of stones at the entrance of a sandy bay, where a little landing-pier was begun in 1822. The neighbouring shore affords excellent blocks of granite, and thereby gave a temptation to the superintendent of this pier to form the face of the quay in hewn stone; and, as might have been expected, the sum allotted for the purpose was exhausted ere the pier was made of much use. I do not approve of what has been done here; and though this would be a useful place, it as yet only admits boats; and though the work is neatly executed, I am of opinion that the best thing to be done for making it a harbour, would be to take a part of it down, and re-build it according to a different plan at the same place, and make more use of the isle or spit; for the present pier being short, and with deep water beside it, it will be difficult to render it tolerably quiet without more room for the sea to spend away. There is a shallow bight in the spit well calculated for this, or for keeping boats in, but it has been shut out by the present construction; and the notion of deepening along the pier, which was entertained by Mr. Bald, must be given up on account of the expense. The present jetty, for one hundred and twenty-feet, being raised to high-water mark may stand, and the back of it, fourteen feet wide, walled up with rough stone. The portion between that and the isle, as yet only about six to eight feet high, to be taken down and relaid from the extremity of the isle, at an angle about sixty degrees from the present position, and down to low water, firmly paved on the back and end.

The inner part of the isle is to be cut away to the same line, and the materials laid across the hollow between the east end of the isle and the main as a rough mole, twenty-one feet broad at top, sloping three to one to seaward; a rough parapet wall, similar to that at Cleggan, is then to be carried round from the land to the pier-head, and the basin excavated as low as possible. It is chiefly bog.

This place will then have a tolerable inner harbour of one hundred and twenty feet square, and clean ground along the outer pier, which may be extended, if thought proper, at any future time. It will be a valuable station for the white fishery.

There is a considerable population about Tarmon, and the road is now extended to the pier; which will always be a convenient place for crossing to Achil or Ballycroy.

Farther up Blacksod Bay we find good shelter on the west, though the lesser bays are mostly filled with sand. The only one deep enough for vessels of burden is Elly Harbour, near Bingham Castle. Nothing has been done here, but a landing jetty would be of service.

Estimate:

	£	s.	d.
100 feet new quay wall, 16 feet at end, 10 feet at root			
14 × 5=70 – – – – – –	100	–	–
End wall, 24 feet, a' 1 *l.* 10 *s.* – – – –	36	–	–
Excavation along island,			
40 × 4 × 10=1,600			
50 × $\frac{4}{2}$ × 10=1,000			
Yds. 2,600 a' 1 *s.* – – – –	130	–	–
Filling pier, 36 × 7 × 3=324, a' 6 *d.* – – – –	8	2	–
Embankment or breakwater;			
50 × $\frac{3}{20}$ ∽ 14 × 2=1,960 yards a' 1 *s.* – –	98	–	–
Inner wall of jetty, 100 × 8–4, a' 10 *s.* – –	50	–	–
140 perches rough parapet – – – –	140	–	–
	562	2	–
Add 20 per Cent – – –	112	8	–
£.	674	10	–

Saleen

This place was one of those selected by the Mansion-house Committee for 1822, and a small pier was begun under the superintendence of Mr. William Bald, mainly for the advantage of the rising village of Binghamstown, which is built on a regular plan by Major Bingham, and already counts near one hundred houses.

The plan adopted for the harbour is not very favourable, but would answer suffi-ciently well for boats. A kind of rude beginning had already been made on the spot, and the intention of the plan was to enlarge and extend it.

When it is considered that the object in building piers within such a bay as Blacksod is not so much for shelter as to make landing-places convenient to the population, the value of Saleen Pier will be better understood. It has made considerable progress, though rather expensively formed, being faced on each side with hewn granite from Tarmon Point.

All the benefit to be expected from this pier will be obtained without a very great additional expense: it would be proper to deepen the inner dock, so as to receive a larger class of boats, and to make a return or kant to the pier-head. The channel along it should also be deepened; and as the strand runs a great way off, there would be no benefit of importance obtained by lengthening it.

I have already furnished a specification and plan for completing this work, and I consider it may be done for 378 *l.* 13 *s.*

Estimate for completing Saleen:

	£	s.	d.
100 feet new quay wall, 16 feet at end, 10 feet at root			
14 × 5=70 – – – – – –	100	–	–
End wall, 24 feet, a' 1 *l.* 10 *s.* – – – –	36	–	–
Excavation along island,			
40 × 4 × 10=1,600			
50 × $\frac{4}{2}$ × 10=1,000			
Yds. 2,600 a' 1 *s.* – – – –	130	–	–
Filling pier, 36 × 7 × 3=324, a' 6 *d.* – – – –	8	2	–
Embankment or breakwater;			
50 × $\frac{3}{20}$ ∽ 14 × 2=1,960 yards a' 1 *s.* – –	98	–	–
Inner wall of jetty, 100 × 8–4, a' 10 *s.* – –	50	–	–
140 perches rough parapet – – – –	140	–	–
	562	2	–
Add 20 per Cent – – –	112	8	–
£.	674	10	–

Above the point of Ardmore, near Binghamstown, Blacksod Bay is crossed by a reef of rocks which are mostly dry at half ebb. At the east end is a good anchorage behind the peninsula of Cleggan, with deep water and an extensive sandy bay within; but this place is so much out of the ordinary line of communication through the country, that it seems hardly to stand in need of any further improvement at present.

Further up the bay dries entirely at low water; but the narrow isthmus at Belmullet has appeared a convenient situation for a village, and in fact one is now rising there, as before observed, with comparative rapidity

About eleven feet water is obtainable here at high water; and Mr. Carter, the proprietor, and Mr. Ivers have begun a landing-pier, which does not appear however to be of a very durable construction.

This establishment appears deserving of encouragement; and an expenditure of about 300 *l.*, of which I doubt not the proprietor would contribute half, would answer every necessary purpose.

Ballycroy.

Descending along the eastern side of Blacksod Bay, we have little but low bare sandy shores and bog. There are two or three considerable inlets, but all dry at low water.

At Kenfennalty Point, opposite to Tarmon, is a sandy cove, protected by a point of rock, in which there is some shelter.

To the south of that, Tullaghaun Bay, which is the estuary of the Owenmore River, forms a natural harbour, up which vessels may pass for several miles and ride in safety. Were the road along the Ballycroy shore perfected, a ferry pier at Tullaghaun would be of service to travellers, and also to the fishing craft; but the southern side being an extensive strand at low water, its value must always be limited.

The remainder of Ballycroy affords us nothing of interest, though the extensive shallow bay between it and Coroan Achil is sometimes a seat of the herring-fishery.

Two little ferry piers at the narrow would be useful, they would connect that part of Ballycroy with the new Achil road, and be frequently of great service to the fishery. I have opened a branch line of road down to this place.

Achil.

The next district, called Achil, forms an interesting feature in our present inquiry. It is composed of two parts, viz, the island, which is the largest on the coast of Ireland, and the peninsula of Coroan Achil, or the Hook of Achil, connected with the main by the narrow isthmus of Mollyranny. Sir Neal O'Donnel is the chief proprietor, the Marquess of Sligo has likewise some proportion.

The island of Achil contains 22,500 Irish acres, eighty miles of coast, and a population of nearly four thousand souls. It is chiefly composed of lofty hills of quartz rock, especially on the north and west sides, with large flat bogs opposite to the main; little of it is yet cultivated, and there are no roads, though one is now in progress through the peninsula and island at the public expense.

Sea-weed and shell-sand abound on the shores of Achil, and are applied to the potatoe crop; and the isle possesses about one hundred boats of various kinds. The fishery however, though abundant in the neighbourhood, is little pursued by the Achil boatmen, their chief employment being the carriage of turf to the shores of Westport. The Clew Bay fishers consider the Achil men as quite ignorant of the business of the fishery; however by the rod and hand-line they contrive to catch enough for their own consumption.

The population is chiefly congregated in large irregular villages.

The western side of Achil is mostly a cliff, and the few coves too wild to admit of being made boat-harbours without works of greater magnitude than have yet been undertaken by the Fishery Board.

The eastern side,. on the other hand, at least within the sound, affords shelter almost every where.

A powerful tide runs in the sound at the narrows, called the Bull's Mouth, between Achil and Ennisbegil. The southern part of the sound is fordable at low water; and as this will necessarily be a constant thoroughfare, the roads on either side having been directed to it, it seems the natural place for establishing any thing like a village or sea-port town. The ferry quays will be of some service in that. way. Salt-works were at one time established at this point.

At the south end of Achil is a smaller island called Achil Beg, the channel between dries at the west end at low water, but the east end forms a natural anchorage called Achil Nole; safe in all weathers.

This being one of the stations selected by the Mansion-house Committee, a small pier was built there in 1822, on the main Achil side, under the superintendence of Mr. William Bald, and is completed to near low-water mark; it will become of more consequence when the roads in progress through Achil are brought up to it.

The harbour being frequented by coasters makes it an object of importance to have a proper landing place.

On the ocean side of Achil, though it would appear very desirable to have some harbour constructed for small craft, there are no situations which I can recommend as fitted for a winter harbour; the best appears at Keem, near Achil-head, and would be formed by a pier of two hundred or three hundred feet long, jutting from the southern end of the strand nearly in the direction of the meridian. It would cost at least 1,500 *l.*

Coraan Achil

This peninsula is chiefly composed of a mountain of conglomerate sandstone, with flat bogs on the north shore, is more barren and less peopled than the island; though about half its extent the population is not above a seventh, viz, six hundred. There are no harbours on the south side of Coraan Achil, which is a material disadvantage to the fishery of Clew Bay. The only place where one seems practicable is at Doobeg, behind the Spit of Boulders, at the mouth of two mountain streams; the passage should be cleared out; a breakwater of rough stones formed on the west side on a small landing jetty within. The expense would be about 600 *l.*

After this an opportunity occurs of forming a harbour at Poalnagorrin, where tile low reefs and inlets of limestone commence, at the foot of the sandstone mountains near the isthmus of Coraan.

A passage of fifty yards long, ten wide, and three deep—one thousand five hundred yards, at 2 *s.*—150 *l.*—being cut across the reef, will allow boats to pass into an extensive inlet or lagoon within, thereby making an excellent place of shelter, and the most to seaward of the north side of Clew Bay.

Clew Bay.

This fine inlet is from twelve to fifteen miles long, about seven broad, nearly all clear of danger; the mouth of it is covered for one thirds of its breadth by Clare Island, and the upper end abounds in a multitude of safe harbours and roadsteads; there are however none upon the northern or southern sides.

These shores are chiefly bold and rocky; the northern has just been described the southern is a shivery quartz rock, with serpentine, and towards the western end, granite. The district is named,

The Barony of Morisk.

The remarkable peak Croaghpatrick rises from the south shore to an elevation of two thousand six hundred feet; from its top we have a beautiful prospect of the neighbouring bay and islands, with the amphitheatre of mountains of Erris, Burrishoole and Connamara.

Behind Croaghpatrick the country is mostly bog, upon slate rock; but cultivation is advancing rapidly by the encouragement of the neighbouring market of Westport; limestone being brought on horseback by farmers from the beaches in the head of Clew Bay.

The southern part of Morrisk rises into steep mountains of slate and porphyry, of which Mulrea, two thousand eight hundred and fifty feet in height, is the highest mountain in Connaught.

Morrisk is divided from Connamara by the long narrow inlet called the Killery Bay or harbour.

A tolerable road exists along the north side of Morrisk to the village of Louisburg, and thence along the western shore nearly to the mouth of the Killery.

A carriage-road has been made from Westport to the village of Bundurragh, on the north side of the Killery, and a new line is now nearly completed from Westport to Leenane, at the head of the Killery, and into Connamara and Joyce's Country.

The interior of Morrisk is nevertheless still in great want of proper lines of communication, though the Marquess of Sligo, who is the chief proprietor, is making considerable exertions to remove that inconvenience, by employing the small tenantry in making narrow roads through their respective farms.

Oldhead Pier.

The west and north-west of Morrisk affords us no favourable opportunity of making any kind of harbour, excepting within the point of Oldhead, under the western head of Croaghpatrick. This place was adopted by the Mansion-house Committee for the construction of a pier, which was begun here in 1822, and made considerable progress; the front was made in sandstone, hewn ashlers, brought from the opposite shore, and in that mode it could not be completed for the estimated sum; but as this is the only anchorage in the outer part of Clew Bay, it would be an object of great utility to the trade of Westport, and especially to the fisheries, if completed.

For this purpose it is chiefly necessary to make a return-head to the present work, raise it a course, pave it, and give it a parapet.

It has been proposed to extend this pier one hundred feet farther; but I conceive such an extension would be of little service, as it could only give an additional foot of water, and there are now fourteen feet at high water along the pier, which is more than at Westport or Newport.

An opening into a small pond near the base of the pier might be of some service, either as a dock for laying up boats, or for penning up a breakwater to scour it clean of sand. There are some small points of rock which should be cut away in the course of the operation to low-water level.

The upper part of Clew Bay is occupied by a congeries of small isles nearly one hundred in number, among which and the adjoining creeks and inlets of the main are a variety of safe roadsteads and harbours, capable of admitting vessels of all classes.

Estimate for completing Oldhead Pier:

	£	s.	d.
200 cubic yards rock to be excavated – – – – – –	30	–	–
Raising present hammered face and coping the whole pier, 1,007 cubic feet, a' 1s. – – – – – – – – –	50	7	–
Rough masonry to back ditto, and in parapet wall, 90 per, a' 8s. –	36	–	–
300 yards paving at back, a' 2s. 6 d. – – – – – –	37	10	–
3 mooring posts, 1 l. 2s. 9d. each – – – –	3	8	3
Contingencies and superintendence, 20 per cent – –	31	9	–
£.	188	14	3

Estimate for adding a Kant of 40 feet long to the end of the present Pier

	£	s.	d.
40 feet lineal of hammered face and coping – – – –	60	–	–
130 perches dry hacking to ditto, a 7 s. – – – –	45	10	–
825 cubic yards of filling, a' 1 s. – – – – –	41	5	–
200 square yards of paving, a' 5 s. – – – – – –	50	–	–
70 feet lineal of parapet, with condor and moorings – – –	19	10	–
Cutting down rock for entrance to pier, 100 cubic yards, a' 2 s. 6 d. –	12	10	–
Contingencies, superintendence, &c. – – – – – –	49	7	–
	278	2	–
Total for Pier and Kant – – £.	466	16	3

It would be impossible by description alone to convey a proper idea of these valuable harbours; and as Mackenzie's chart of them is imperfect and erroneous, I have annexed a new and valuable delineation of them taken from Mr. Bald's survey of the county of Mayo, on which the soundings taken for the Board of Fisheries are carefully laid down, and by the help of which any person may navigate this curious labyrinth in safety.

These isles are composed of a deep loamy soil upon a basis of limestone, and are all fertile and well cultivated; a great many of them, especially on the north side, are accessible at low water. The towns of Newport and Westport are built at the upper end of two of the farthest inlets, and are provided with quays, to which vessels of nine or ten feet draft may come at high water, and of twelve feet at springs.

The latter town is the principal market, though distant about a mile from the quay. It is well built; has a population of about five thousand; exports about ten thousand tons of corn per annum, besides kelp, linen, &c.; and sends a good deal of cured fish to the interior.

Useful boat quays and wharfs may be made at many other points among the inlets, but they are not at all necessary for shelter; I shall only describe the most important stations.

Newport.

The entrance into Newport is spacious and direct, but several islets and rocks are to be observed as we pass upwards; and it seems the best adapted to large vessels, there being several good roadsteads on either side within the isle. A large vessel entering at low water must avoid the shoal ground off the entrance, on which there is only two fathoms.

Inishgowla Harbour, on the right hand, is within the island of that name, stretching eastward about a mile, with clean good ground in three to eight fathoms. The channel into it is along the north shore of Inishee and Inisgowla Isles.

The former having a remarkable clay cliff on the west side, of about ninety feet high, is readily known. Keep this cliff E. shutting in the isle of Inishfeash, you pass south of the shoal above mentioned; then a little clear of the two isles until you pass Inishgowla. From this harbour a passage exists at high water for vessels of twelve feet or under within the isles towards Westport, which may be traced on the chart.

Above this, and on the same side, is the harbour of Innisturk, between that isle and the peninsula of Ross, with two to eight fathoms good ground. Take either side of Inishlavkeil, but on the south keep within one third of it to avoid a rock one third from Inishcuttle, which dries at low water; keep mid-channel in the other sounds. In this harbour the deep water approaches closest to the main land near the village of Ross, and a landing-pier there might be advantageous, especially for sharp vessels.

On the opposite side of the Newport channel is another excellent roadstead, capable of a number of large ships, with good anchorage from two to six fathoms; you pass it along the isles forming the north side of the Newport channel, so as to avoid the rocks and breakers in the middle of the entrance.

Within this to the northward is the harbour of Rosstrunk, well sheltered and good ground over nearly a square mile, but with two to three fathoms only, and therefore best for small vessels, which may go farther up the bay to Newport and Burrishoole, &c. at half flood, the water becoming shallow, and though they take the ground in the way, they will receive no damage, the bottom being soft mud.

The inlets to the west of Rosstrunk are mostly dry at low water; but near the western extremity of the isles on that side is Roe Island

Harbour, within the isle of that name, which, though small, has depth of water for any ship; good ground. Roe Island is easily known from the steep clay cliff on the south side, but there are several rocks and shoals in the neighbourhood of this island to be attended to, which are accurately laid down on the chart.

North of Roe Island, one mile, is a little islet or ledge called Mynishbeg, about one hundred and fifty fathoms from Mynishinore Isle, which is of considerable importance, being much frequented during the fishing season by the boats from the north side of the bay. It is a semicircular beach of stones, the middle of which only is covered at high water, and within it is an excellent anchorage on blue mud, with four fathoms close to the shore, sheltered on all sides excepting when the swell tumbles over the middle at high water; this part ought to be raised to high-water level, which would cost not more than 100 *l.* a single wall on the inside, backed with stones, would answer also as a boat quay occasionally.

To the north of this, at the east end of the peninsula of Coraan, there is an inlet which a small expense would convert into a useful boat harbour. For this purpose it is necessary to cut across a bar of limestone and boulders for about one hundred feet, arid down to low water, to admit boats with the tide. This place would be two miles to the westward of any of the creeks among the isles.

Two miles farther west, on the shore of Coraan Achil, or Tuinard, a small stream from the hills falls into the sea at a place called Doobeg, forming at its mouth a shallow inlet or lagoon, covered by a shingle bank, as already observed. The east end of this may be deepened and quayed for about 600 *l.* and it would thus form an useful harbour for small craft on the lower part of the north shore at a moderate expense. Something of this kind is much wanted there, and no other place affords any opportunity for it.

Off Doobeg two miles is a shoal called Thanny-oar, with only feet water, the rest of the bay is about fifteen fathoms deep, chiefly clean sandy ground, affording turbot and other flat fish, also mackerel in plenty among the isles; the bottom is mud, and there are three reefs or ledges of rock about a mile or more west of the isles, on which rockfish are caught: these are Loarbaun and neighbouring rocks to the west of Roe Island on the north sides; Clogcormnick in the middle, a ledge stretching west from Inishee and Davilan, and Carrigadilisk near the south side, between Oldhead and the entrance to Westport.

But the most valuable fishing in this neighbourhood is outside of Clare Island, on the banks of Ennisturk and Boffen, before described. This fishery is, however, little pursued from Clew Bay, a few boats only being employed at Westport, mainly for the supply of the town.

Westport.

The islands and channels on the Westport side of the bay are protected from the sea by a very singular breakwater of shingle and boulder stones, which with little interruption runs from the entrance to Newport Harbour at Inishee Island, all the way to the southern shore under Croaghpatrick. There are in this line of beach six openings navigable; but by far the most important is that of Beulnascrona, nearly in the middle, which forms the ordinary channel to Westport and the anchorage below; and which is marked by a small lighthouse on the northern beach, built and supported at the expense of the Marquess of Sligo, since 1807.

The entrance is two hundred and forty fathoms wide, and six fathoms deep in the narrows; but there are shoals on each side, extending on the north side two hundred to three hundred fathoms, N. W. by W. of the light, and on the south side, or Doreinish, nearly half a mile in the same direction to seaward, which must be carefully avoided.

The intermediate channel is clear S. E. by E.

Anchorages.

When within the entrance a vessel may anchor any where behind the bar of stones on the south side, called Dore Inish, on mud, in two fathoms or less, and this is the ordinary place for vessels trading to Westport. Or by going on and turning round the east end of the isle, on the north side, you enter the harbour of Inish Gort, completely sheltered on all sides, in three to five fathoms. A part of the beach on the west side is covered at high water, but not so as to disturb the anchorage.

If we pass Inish Gort entrance, we may run round and anchor behind the isle on the left, called Inish Lyre, in two fathoms or less;

that depth continues all the way to the shore of Rossmunea, and up behind Cullenmore Island, in which direction there is a channel for small vessels to Inish Gowla Harbour, and Newport, as before described; but the deeper water is from the harbour of Inish Gort, between Cullenmore and Clynish Island, and between this last and Mynish-beg, and then avoiding the rocks of Carrigeennafranka, and that to the west of Irish-Cuttle.

The channel to the south side of Inish Lyre has a sand-bar, with three feet water, and those further in are rocky, and only fit for boats at high water; vessels running up to Westport, therefore, pass outside of Inish Lyre, through Dore Inish roadstead, between the islets of Inish Laughan and Inish Inemel; hence a clear channel, with five feet, to between Inish Goula and Inishraher, where an anchorage exists with two or three fathoms.

There is another passage to this anchorage, as will be perceived by the chart; viz, by the south side of Inishraher, through the narrow entrance between the point of Bertra and Inish Dagh, or by the north side of this last island; but this passage is more intricate for a stranger.

From the anchorage last mentioned a passage lies on either side of the shoals called Glashelon; that on the south is the most direct, and marked by perches; between it and Annagh-head is shallow, with only three feet water, but off the head small vessels may lie in eight or nine feet, whence the water shallows to the rocky entrances of Westport Harbour, between Rosslea and Rossbeg, where there are two distinct channels marked by perches, a bank of rock, dry at low water, lying between; after which the channel up to Westport Quay is clean, but nearly dry at low water.

This quay is extensive and well built, though still unfinished; it has not above ten feet at high water.

The extensive inlet below, though very safe, is not capable of much improvement; but a good deal would be gained by a breakwater out to Wallace's Rock, and by blasting out the channel at its head for, two or three feet deeper, and forming a scouring basin.

I have therefore made an estimate for these improvements; the cost would be several thousand pounds, and they do not seem to come under the province of the Board of Fisheries.

Having already described the pier of Oldhead, the only station on the south coast which seems to admit of improvement, I shall, ere we pass the Killery Bay, take a short view of the isles in the offing.

Clare Island

This island lies in the mouth of Clew Bay, of a triangular shape, about four miles long, and contains about one thousand four hundred inhabitants. It is the property of Sir Samuel O'Malley, baronet.

The rock is chiefly slate, and on the east end conglomerate sandstone; the north-west side rises to an elevation of several hundred feet, and is a precipitous cliff; the south side is lower, and the island is tolerably well cultivated.

A lighthouse has been built on the north-east point of the island, on a lofty cliff being considerably within Boffin and Achil Islands; it is doubtful if this be of much value as a sea-light; but it forms a good mark for Clew Bay, the deepest entrance to which is on that side.

The island affords good anchorage under its lee in S. W. winds; and in a little sandy bay, at the east end, a small pier was built by the father of the present proprietor, which admits fishing boats at high water. The late Mr. Munro Mackenzie, when on his nautical survey, of the coast of Ireland, having highly recommended the construction of a harbour in that place, for the benefit of the coasting and fishing trade, several applications have been made by the proprietor to have such a work undertaken.

The pier constructed by Sir Samuel O'Malley's father is only a narrow wall of about seven feet thick, and two hundred and twenty-eight feet long, covering a cove of one hundred yards by fifty, and the depth inside not more than nine or ten feet at high water at the pier-head; most of the cove shallow and rocky.

The simplest and most necessary improvement is to excavate the rock inside, and at the harbour-mouth, to low water, as it would be a pity to lose the present pier until something of much greater consequence were substituted for it, and in its present position it is incapable of being extended.

This excavation may, I think, be done for about 300 *l.;* a part of the materials may be built up as a jetty-head on the rock on the west side; the remainder of it rises in large masses, may be laid on and in front of the rock outside, as the commencement of a kind of breakwater, which would tend, to quiet the present entrance, and give a landing-place at low water at the back of the pier.

If funds were obtained for a more extensive work, this breakwater may be continued as a rough mole across the mouth of the bay, which is three hundred yards wide; and after the first two hundred feet, would be in two fathoms at low water.

The slate rock on the present site is not well calculated for such a structure, but good blocks of conglomerate sandstone may be had on the opposite side of the bay, and could be brought by water.

This mole or breakwater would hardly be effectual until two hundred feet at least beyond the rock; it would cost about 20 *l.* per foot for the rough mole only, or 4,000 *l.*; and with quays, pavement, parapet, &c. at least 5,000 *l.*

		£	*s.*	*d.*
Or thus, if quayed from low water only:				
250 feet of rough filling to low water, only from point of rock, 4 yards deep, and 18 average width, a' 5 *l.* or 6 *l.* – –		1,500	–	–
A quay wall on ditto, and on rock -at base, 300 feet, a' 2 *l.* coping included – – – – – –		600	–	–
309 feet back wall for ditto, with parapet, 21 feet in height, and 8 feet thick, a' 2 *l.* 10 s. – – – – –		750	–	–
Filling 8 X 100 X 5 = 4,000 yards, a' 1 s. - – – –		200	–	–
		3,050	–	–
Mooring posts and contingencies, ⅕ – –		610	–	–
	£.	3,660	–	–

There would be nine or ten feet water behind this pier; and as vessels will not lay along side, the wall need not be hammered. The old pier and jetty inside should be allowed to remain, thus covering a kind of inner dock or bason, which should be deepened at low water, and an entrance opened next the rock.

If this pier in deep water be still thought too expensive, a pier may be built in three feet water only, to shelter the present cove; by cutting off one hundred and twenty feet of the present wall, and running out one hundred and fifty feet parallel to the shore from the rock on which the present wall is built, with a small jetty pier on the opposite point, so as to leave eighty feet opening, and excavating the cove as already described, the space within will be nearly one hundred and an half acres, and may be entered by fishing craft at all times. This would be very effectually done for about 1,500 *l.*

There is no other place on this isle capable of being made into a harbour for sailing craft; but a very useful creek for yawls could be formed at Port-a-Chonille, near the ruins of the abbey, in the middle of the south side, of which I have given a design; for this purpose the hollow parts of the Skerry Rock should be filled up in the way which I have specified, and the knobs of rock within cut away. Expense about 250 *l.*

Boats land at a little creek near the west end of this isle, called Lackuy, but it seems incapable of much improvement.

Between Clare Island and Ennisturk is the islet Caher, inhabited, but having no harbour.

Ennisturk.

The isle of Ennisturk is situated about six miles from the west coast of Mayo, about half way from Boffin to Clare Island, two miles long by a mile broad; chiefly of slate rock; the western shore is a steep cliff, the eastern side lower, with several narrower coves. The only landing-place or harbour is at the east end of the isle in a small bay, affording tolerable anchorage in westerly winds in summer, in four fathoms in strong clay. This little bay is about one hundred fathoms broad, and fifty deep; the upper part ending in two coves with ragged slate shores, each of which is dry at low water; the landing-place is in one of these on the north side nearly land-locked. Some small rocks, which show at low water, occupy half the mouth of this cove, the other half of the cove is nearly dry.

Although this cove appears to be land-locked, yet the run within it in easterly winds, and at all times during winter storms was so great, that no boat was ever allowed to be there in bad weather; the few yawls they possessed were universally hauled up on the land; and in a gale previous to my visit, most of them had been broke by the sea washing up near the houses.

As it was impracticable to keep any large sailing boat in Ennisturk, though in the centre of the most valuable fishery on the coast, the inhabitants caught little more than for their own consumption, by means of two or three yawls they had; and they were frequently for weeks together deprived of communication with the main land.

This state of things, and the great benefit of having a pier in Ennisturk, was represented to me, as engineer of the western district, on the part of the inhabitants, by Mr. S. Knight, an intelligent country surveyor, who had made a survey of the island, I presume on account of Lord Lucan; and accordingly I proposed this among others to the Commissioners of Fisheries, when the additional piers for the western coast were determined on.

Not then prepared to say what ought to be done in Ennisturk, I afterwards suggested to the Commissioners to have the sum intended for this island laid out on the island of Boffin; but as the resolution had been passed for Ennisturk, it did not appear to them advisable to rescind it.

I visited this island in 1823 with Lieutenant Dundas, R. N. in the surveying cutter; and on examining the cove, and hearing the report of the inhabitants, it appeared to me that the run was sent into the cove principally by the point A. projecting beyond the point of rock B, and thus forming a kind of funnel to collect the swell.

The most obvious method of preventing this would be to run a pier from the point B. overshooting A., so as to direct the swell into the southern cove; but to this there were two objections, viz. the pier, being in deep water and exposed to the immediate stroke of the sea, would cost more than any funds at the disposal of the Fishery Board could defray; and secondly, from the point B. the shoal-water extended to a considerable distance, so that after forming this expensive pier, the entrance to this harbour would after all be dry at low water.

Seeing this to be impracticable, the next object was to shut out a part of the swell by narrowing its entrance into the creek, for which a favourable opportunity was offered by the rock in mid-channel, The space on the south side of it being unfit for navigation; and by running a breakwater over these, the portion of swell thus shut out would be caught, and spend itself on the rough coves C. and D. immediately outside.

When the water was thus made comparatively still within the cove, I proposed to have a small jetty from the point of the upper bight E., to send the remaining swell into the north bight F.; and by cutting down the rock near the houses, a quay for the use of small craft could be formed, at which boats at least might be moored with safety behind the jetty; the water is too shallow for large craft at this place.

With this view, therefore, and seeing that the jetty could be of no use until the breakwater aforesaid was executed, I directed the space between the rock and the north shore to be filled up with a substantial work of rough stone, being all that could be done against the first winter; this work could have been effected for half the estimate at most, and the remainder next summer to be laid out on the jetty. I left Mr. William Bald, of the county of Mayo, to begin this work, and direct its subsequent progress; I have not seen it since that time; but on finding, from the representation of the Board of Fisheries, that the funds were rapidly exhausting and yet this breakwater not nearly completed, and that a number of masons and stone-cutters, who were wholly unnecessary, had been sent into the isle for this work, I directed an experienced assistant,. Mr. Alexander M'Gill from Dunmore Harbour, to make his way into the island, check this improper expenditure, and cause the work to be prosecuted agreeable to the original intention. Since that time I have also sent another of my assistants, Mr. O'Hara from Killough Harbour to visit this work; and with some exertion the breakwater has been completed, though not until the grant from the Fishery Board was exhausted. It appears, by the change of construction, the work has been enabled to withstand the sea; and I have heard of no injury done to any boat within the harbour since, but find, on the contrary, that additional new ones are building.

I do not doubt but the run appears at the mouth or entrance of the harbour to be greater to the islanders than before, the very object of the work being to prevent the water from rising and falling so fast within the harbour as without; and I dare say the run into the bight E. now appeals as great as into the bight F., as this last, from the position of the pier, must necessarily be much diminished but if the jetty I propose for E. be executed, so as to shut off the interior wave, I am satisfied we must have as quiet a small harbour here as the place can possibly admit, (unless the expensive work just proposed by me be undertaken), whatever opinion may be. given to the contrary.

Ennis Boffin.

This island, the most important with respect to the fisheries of any upon the Irish coast, is about four miles in length, by two average broad; population, one thousand one hundred; (it was the property of the Marquess of Clanricard, until lately, I believe, exchanged to the Marquess of Sligo); distant three miles from Cleggan, in Connamara, but forms part of the county of Mayo. The island of Ennishark nearly half its extent, is a dependent upon it; also the lesser isles of Davilin, Lgin, &c.

The north coast of Boffin is a precipitous cliff having two bays or inlets, which are too wild to be used as harbours; but the western one has at its upper end an interior lagoon, or lake, divided from the sea by a beach of stones, through which if an opening was made to tide-level, a harbour would be obtained for sheltering craft.

The southern side of Boffin has an excellent natural harbour, with nine or ten feet water, and well suited, therefore, for such vessels as pursue the fishery; and the finest cod bank on the Irish coast lies about two or three leagues to the northward, nearly midway from Boffin to Achel. head, as already described.

This harbour appeared so important in the time of the Commonwealth as to have been fortified by a block-house, still existing, on a detached isle at its mouth. It faces the south-west, but in that direction is protected by some low rocks, and also by the isle of Ennishark. The upper end is land-locked, and the craft lie on soft mud.

This harbour has, however, some defects which it would be difficult to remedy altogether; with southerly winds it is entirely shut, and in consequence the isle is frequently cut off for many days from communication with the main, or proceeding to the fishing grounds.

It is very desirable, therefore, that another harbour should be formed on the isle, and a situation for that object appears to the east end facing the main. I have had this place examined, and consider that a pier-harbour may be constructed there, at an expense of about 1,500 *l*.., which would be of great benefit in the fishing season.

Boffin is every way deserving of attention from the Commissioners of Fisheries and of the various grants for improvements on the west coast, none has ever been applied to it.

Ennishark, though equally near the above mentioned valuable fishing-bank, has no harbour or proper landing-place, and is therefore of little consequence in the present inquiry. Population near two hundred.

The other small isles near Boffim are also cliff, and too insignificant to attract our attention.

Killery Bay.

The northern side of Morrisk, outside of Oldhead, is too much exposed to afford any good site for a harbour, and the western side is chiefly beaches of blowing sand, which continue as far as the mouth of the Killery;

This inlet divides the counties of Mayo and Galway, runs ten miles into the interior between steep and lofty mountains, and is from a quarter to half a mile in width; it is throughout an excellent harbour, capable of holding large ships, having for a great part of its length ten to twelve fathoms water. There are squalls from the hills at times, which render it unsafe for sailing boats. An excellent fishery exists in the bay for cod, haddock, &c. and a great fall of herrings comes into it in the early part of the season, from 1st November to February, at which time it is much frequented by jotters for the supply of the interior.

This place, until lately, was most imperfectly provided with roads, having a dangerous bridleroad only on the Galway side; but this inconvenience is now nearly done away by the works carried on by Government for the improvement of this district, a carriage-road being now opened, with bridges over the dangerous mountain torrents from Westport to the head of the bay, where it joins the old Galway country road. which is also under improvement; and from Leenane near the head of the bay another carriage-road is opened through the hills to the head of Lough Corrib, where one branch runs along the northern shore of that lake to Cong and the plains; and another branch is in progress across the centre of Connamara, where it falls on the Galway road, and is also to be continued to the southern coast at Costello Bay.

Bundurra.

Previously to this Lord Sligo had obtained from the county, by means of a loan from the consolidated fund, a road to the middle of the north shore of the Killery Bay, near which he has a romantic fishing lodge; and at the termination of this road a pier has been built by the assistance of the Fishery Board, and a village is in tolerable progress.

The first application for the construction of this pier was in May 1822, and I had a survey and design made for it. Lord Sligo, in July 1823, expressed his willingness to bear half the expense. From various causes of delay it was only undertaken, by contract, by Mr. O'Hara, one of my assistants, in 1824, after having been publicly advertised. Mr. O'Hara has since executed this work in a very substantial manner. It consists of a good quay along shore from the inlet and jetty-pier, of seventy-five feet, projecting from the shore. The stair at the end is accessible by boats at low water, and the quay itself admits any coasting vessel to load or discharge. The harbour is covered outside from the bay by a point of land, and within there is a natural dock in the mouth of the river well calculated for laying up craft in winter.

As there is a valuable salmon fishery on this river, Lord Sligo has desired to have a landing-place or slip constructed for hauling the nets. It should be so managed as not to interfere with the layer for vessels.

As yet there is no road along this shore; that from Bundurra passes through the mountains, and along a remarkably rocky precipice towards Westport, but is sufficiently level and since the bridge on the Bundurra River was finished by the government funds, carriages have no difficulty in passing from this pier to the town of Westport; it is consequently much frequented in the time of the fishery.

Leenane Joyce's Country.

Near the head of the inlet an opening exists through the mountains on the Galway side, by which a road has been carried to the banks of Lough Corrib, as above mentioned, through the district called Joyce's Country, where this road departs from the Old Killery bridle-road; I had occasion to correct the latter, and made it along the shore with a rough jetty-pier at the head. This work, although rough, answers as an excellent boat quay, and is of considerable value in the time of the fishery. It is upon the property of the provost of Trinity College, and the site seems well adapted to become a thriving village, if the objections to improvement on that kind of property, could be obviated. The neighbouring glen is populous and pretty well cultivated, and the interior, since the new road has been opened, is rapidly improving.

Bunowen.

Further down that side of the Killery is a small village called Bunowen, at the mouth of a little river coming from Conamara, which forms a kind of natural harbour.

Salruck.

A rocky bridleroad leads from hence along the south shore and across a mountain pass to the inlet called Salruck, or the Little Killery, which is a very beautiful and safe natural harbour of half a mile long, five or six fathoms water, covered with a small isle at its entrance.

There is a considerable population, and several sail-boats here, and the slate-rock ledges afford natural landing-quays. Hard by is a ferry over the mouth of the Great Killery into Morisk.

The bridleroad continues along shore, and gradually improves as far as the house of Mr. Blake, near the point of Renvile. No opportunity for constructing any harbour occurs along this part of the coast, though a frequent seat of the fishery, except at a small point called the White Strand, where a pier could be formed at the expense of 600 *l.* or 700 *l.*

The isles of Innistegel, on the opposite shore, possess a good anchorage for small craft.

Conamara.

We are now entered on the district of Conamara, one of the greatest importance with respect to the fisheries, possessing the finest harbours, and in the vicinity of the most valuable fishing grounds on the coast of Ireland; and as the works undertaken by the Board, in conjunction with those directed by the Government, for the. improvement of the interior and employment of the poor, have already produced remarkable results, a brief description of Conamara generally may not be here out of place.

Having been employed by the Commissioners for the Inquiry into the nature and extent of the Bogs in Ireland to survey and report upon this district, the result of which has been printed in the Fourth Report of these commissioners, I was well aware, when appointed by his Excellency the Lord Lieutenant to lay out and superintend the public works of the western district, of what operations were most necessary for the improvement of this country, and in conducting them I have been equally anxious to keep in view the benefit of the fisheries as the extension of agriculture

The whole tract between Lough Corrib and the Atlantic Ocean is frequently named Conamara: it is, however, subdivided into three separate districts, viz. Conamara, Jar Connaught, and Joice's Country, which are nearly if not exactly counterminous with the baronial divisions, Ballynahinch, Moycullen, and half barony of Ross. The isles of Arran also form a half barony. Of these Conamara Proper lies along the Atlantic, having on the east a line drawn from the head of Kilkerran Bay by Lough Oured and Shanonafold Mountain along the ridge to the Killery. The name signifies the 'bays of the ocean,' and in a loose sense is understood to comprehend as far as Costello Bay. The triangular district to the eastward, between Kilkerran Bay and Lough Corrib, is named Jar Connaught; and the northern portion, between the heads of Lough Corrib, Lough Mask, and the Killery, is named Joyce's Country, from the prevailing family name of the natives.

This district appears not undeservedly to be considered as among the most uncultivated parts of Ireland. On a general view it seems a continued tract of bog and mountain, the arable land not a tenth, if even a twentieth, of the whole surface. The entire population is about fifty thousand, of which Conamara Proper has twenty thousand chiefly resident on the coast, those of the interior not exceeding three or four hundred families, and mostly on the bridleroads which have been made through the district: that step in improvement having been carried much farther than in the similar district of Erris. On the shore of Jar Connaught are about eight thousand. The remaining population being chiefly along Lough Corrib. The interior is almost desert.

Though the general improvement of Conamara would seem an undertaking of the most arduous description, it is not without facilities which might upon a candid consideration make it appear a subject more worthy of attention than many other of the waste lands of the kingdom.

The climate is mild, snow being little known during the winter; the cattle are never housed; the mountains on the north, and general variety of surface, afford considerable shelter. The summers, however, are wet, and it is exposed to heavy westerly winds.

Although Conamara be mountainous it is by no means an upland country like Wicklow; at least three-fourths of Conamara Proper is lower than one hundred feet over the sea. Great part of Jar Connaught rises from the shore of Galway Bay in a gently sloping plain to about three hundred feet, at the upper edge of which there are some hills of about seven hundred feet, and beyond them a low limestone country extends to the edge of Lough Corrib, and but little elevated above its level, which is only fourteen feet higher than the sea. But Joyce's Country, on the other hand, is an elevated tract, with flat topped hills of one thousand three hundred feet to two thousand, interspersed with deep sand narrow vallies.

The district is nearly surrounded by the sea on the south and west, and the great lakes Mask and Corrib on the east, the latter navigable into the town of Galway, and could be easily made so to the sea. Various great inlets penetrate the district, so that no part of it is distant four miles from existing navigation. There are upwards of twenty safe and capacious harbours, fit for vessels of any burthen; about twenty-five navigable lakes in the interior, of a mile or more in length, besides hundreds smaller; the sea coast and all these lakes abound with fish. The district, with its islands, possess no less than four hundred miles of sea shore. On Lough Corrib it has about fifty miles of shore, so that with Lough Mask, &c. there are perhaps as many miles at shore of the sea or navigable lakes as there are square miles of surface.

This extent of shore is particularly important from its produce in sea-weed, either for manure or the manufacture of kelp; the value of this last article a few years ago amounted to about 50,000 *l.*; at present, from the low prices, it does not exceed 16,000 *l.*

There are extensive banks of calcareous sand round the coast in almost every bay, and in the interior are numerous beds of limestone, nearly all the navigable lakes having some on their banks. The supply of fuel is evidently inexhaustible.

Conamara is very destitute of wood, a few scrubby patches only being scattered thinly through it.

The country, however, possesses an extensive stool of timber, for in almost every dry knole or cliff the oak beach and hazel appear shooting in abundance, and require only a little care to rise into valuable forests; several bloomeries, which were erected about a century ago, consumed much of the timber, and copsing was afterwards neglected. The sheltered vales, navigations and abundant water-power, would form great advantages in the cultivation of timber.

If we draw a line from Galway to Oughterard, and thence to the bay of Ardbear the southern part is a continued granite moor covered with bog of various depth. It contains no limestone; considerable quantities are brought from Arran and the county of Clare, as ballast by turf boats, and thrown out on the shore; this supplies what is necessary for building, it has not yet been applied to agriculture, but could be procured for that purpose, and burnt on the shore for somewhat less than 1 *s.* per barrel. There are banks of shell and coral sand on all the coast, but especially in the bays of Kilkerran, Birterbuy, Bunown, Mannin, &c. This sand is raised by dredging, and by beaching the boat on it at low water; that of Kilkerran, Birterbuy and Mannin is pure corraline. There are also dry banks of calcareous sand on the coast, especially at the western extremity, which are accessible by land nevertheless calcareous sand is not much applied to agriculture as yet, though its value is generally acknowledged.

The great suppiy of manure on the coast is the red sea-weed, which is cast ashore in considerable abundance, and frequently cut in the deep water by people in boats; two or three boat loads, of about six tons each, are usually applied as manure over an acre of potato ground.

The original population of this district seems to have been entirely confined to the coast. This is in a great measure yet the case. The old churches and chapels are all on the shore, and the only occupation was fishing. Even now there are very few people who can be considered as farmers alone. Farming and fishing, it is well known, do not assort well together, and however active the natives appear in the latter occupation, they are little inclined to exertion in the former. The pursuit of the sunfish or basking shark, in the months of April and May, employs a good many hands at a season particularly inconvenient. The usual size of boats is about nine tons, cost 40 *l.*.exclusive of tackle; and the number being considerable (for scarce a farm but has one or two of these, besides smaller) shows that a considerable capital is applied to navigation, though very little in agriculture; for the only implement of husbandry is the spade.

The manufacture of kelp from the black weed is now very general, and though tolerably productive, abstracts a great supply of manure from the purposes of agriculture, in so much that it is very questionable, the great advantage of reclaiming the waste being considered, whether it would not be better for the proprietors to apply the whole of the kelp used to the land: The quantity manufactured in the whole of Conamara is about four thousand tons, which may require about fifty thousand tons of the sea-weed, and it is probable might suffice for manuring four thousand acres of land; and this, after one course of cropping, would be let in permanent pasture, worth 5 *s.* to 10 *s.* an acre at least; while during the cropping the produce cannot be reckoned at less than 10 *l.* to 12 *l*, and the rent to the landlord 2 *l.* to 3 *l* per annum.

The present price of kelp is under 4 *l.* per ton, from the inferiority of manufacture it does not bring so much as the Scotch kelp; the expense of cutting, burning, &c. is reckoned at: 30 *s.* or 2 *l.* per ton. but the truth cannot be ascertained being combined with the rent of the lands for kelp; burning is not a particular profession, but is always done by the tenants and cottagers on the spot; they begin cutting in May, after finishing the tillage, and employ on it all the time that can be spared from turf and potatoes until Michaelmas; an able kelper may make three tons, the average is two; and three men in one house may make from seven to eight in a season: the number employed, would therefore appear to be two thousand.

About twenty days work are required. to cut and land the weed for one ton; the quantity of weed which makes one ton and a half is amply sufficient to manure an acre, which would be done at the rate of 45 *s.* seeing that the expense of carrying out the weed to the field is as great as that of drying and burning the kelp.

In Galway sea-weed for manure is usually sold at half a guinea per ton.

Though the red weed is equally fit for making kelp, yet as it comes mostly in bad weather, where it cannot be dried, it is seldom made use of for that purpose.

The kelp in 1808 sold in Galway at 13 *l.* per ton, freight thither from the bays 5 *l.* per ton; at present the price is so low as 3 *l.* 10 *s.* to 4 *l.;* so that many of the farmers find it more for their interest to employ the weed in agriculture.

This disposition is likely to become general in the present state of the markets, and seems deserving of encouragement. The benefit that would accrue to Conamara from the transfer of the manure and labour to the improvement of the land is perhaps not rated too high, when we say it would be annually as much as the present rental.

A good deal of turf is cut on these shores and carried to Galway, or sold on the spot to boats from Arran and Clare; turf in Galway is worth about a guinea per four ton boat-load, price in the bays 1 *s.* per ton of the boat. A man cuts in one day six ton boat-load or slaue, two spread it, one foots; one stacks, and twelve take it to the shore, producing 1 *s.* 1 *d.* or 1 *s.* 2 *d.* per day for the wages.

The district between Ballynakill Bay and Ardbear is a mica slate and quartz country interspersed with veins of primitive limestone, and the same formation runs in a belt through the hills all the way to Lough Corrib near Oughterard; it contains also very beautiful serpentine and verd antique, some of which has been lately quarried for ornamental purposes, and promises to introduce a new source of employment into Conamara; the principal quarry belongs to Mr. Martin, and is situated in the middle of the Twelve Pins Mountains, about five miles from the sea.

The northern part on the banks of the Killery is chiefly hard red sandstone and brown porphyry, with clay slate, some of which seems very fit for roofing, but is not wrought for that purpose.

From Galway there exists a tolerable carriage-road to Oughterard along the limestone plain; but from thence into the centre of Conamara it was led over rocks and bogs in such an unskilful manner, and generally in such bad order, as to be scarcely passable for any sort of carriage. From this central line several narrow. bridleroads branched in different directions across the mountains, but these, at least in winter, very hardly passable even for footmen.

Since my survey in 1813 a good deal of exertion has been made by the proprietors of Conamara to obtain a better carriage-road into the district, and a loan had been obtained for executing a considerable part of the central line, and some other portions have since been completed by the government funds; but along the coast, where the principal part of the population were seated, there was only a narrow road from Galway to the shore of Costello Bay, but beyond that hardly a footpath.

Leaving, therefore, the completion of the central road to the exertions of the country gentlemen, the first object of Government has been to open a proper line of carriage-road along the coast, by improving this last mentioned, and prolonging it by the heads of the several bays, with bridges over the great rivers, as far as the village of Clifden, at the head of the bay of Ardbear; and thence northwards, in like manner, to the head of Ballynakill, where some tolerable country roads are met, having been lately improved under the care of Mr. Blake. It is also intended to improve and alter these, as far as the new line already described, which comes to the head of Killery Bay: thus giving a complete line of carriage-road round the whole of Conamara

From this line of coast-road branch lines have been laid out to the different fishery piers on the coast; and in several other places, where the main line approaches to the sea, I have made it an object of attention to convert the road itself into a boat quay. The operations on this road are chiefly as yet confined to the western part of the country, where a very good road is now opened from the north coast, round Ballynakill Bay by Streamstown to Clifden, thence by Ardbear to Ballynaboy, across the moor to the head of Besterbuy Bay, and the prolongation to Costello is in hands. A branch runs from this to the quay at Derrinver. Another branch is completed, of several miles extent, to the fishery pier at Cleggan, which forms the nearest point of communication with Boffin. The old road, along Mannin Bay to Bunowen, is improved. A branch is also in great forwardness to Roundstone Pier; another into the isle of Garumna by causewaying across the tide fords; one to Costello Pier, and a short one to Speddle Pier.

A cross road has also been completed from Speddle Pier over the moor to Moycullen, and thence to the ferry of Knock upon Lough Corrib, where it communicates with the limestone plains.

Another road is in progress across the narrow part of the district, from the coast-road at the head of Kilkerran Bay to the head of Lough Corrib, where it joins the road from the Killerys to Cong, thereby opening a carriage-way to Westport in one direction, and in the other to the plains.

The effect of the various public works, as well roads as piers, and of paying the labourers in hard cash, upon the general prosperity of this country, since 1822, is quite surprising to every one; and though I may have been supposed somewhat sanguine in my expectations, the result has far outrun what I had anticipated. Trade and commerce have been introduced into the extremity of Conamara, capital has accumulated in a surprising degree, and with it enterprise and exertion in the fisheries, in agriculture, and in foreign trade; cargoes of country goods, salt, timber, &c. are now imported direct into Conamara: the poor village of Clifden has become a considerable market town, exports herrings, marble, &c. The peasantry, who were almost entirely unacquainted with money, now pay their rents in cash, a thing heretofore unknown; they are well clad, evidently in new stuff; they bring to the markets of Clifden a variety of articles for sale, and never depart without making some purchases. A few bad years, previous to 1822, had left them in rags and nakedness; and being forty miles from the nearest market town, over bogs and bad roads, without money to buy, they were in want of most of the conveniences of civilized life.

Ballynakill Bay.

Rounding the point of Renvyle, we come to the inlet of Ballynakill, an excellent harbour, capable of great ships, and covered at its mouth by the island of Truchelaun or Heath Isle, along the north side of which is the channel, with five to six fathoms water. Smaller vessels may round the point of Ballynakill on the south side, and be in a safe bight with two fathoms on mud completely land-locked. The beach there is clean gravel, but a small landing jetty would be a useful improvement, and might be made for about 100 *l*.

Further up, this bay divides into two or three inlets; that on the south is clean and pretty deep, the northern side more shallow.

Derrinver Pier.

On that side, at the end of the new roads, a small pier has been constructed by the government funds, for the purposes of the fishery and convenience of that peninsula.

It is partly formed by the rock, and consists of a basin of about one hundred feet square; the extremity of the pier is accessible at low water; it forms a convenient point of access with the anchorage of Ballynakill, in the direction of Westport, and by the new road through Joyce Country towards Galway and Cong.

This pier is on the estate of Mr. Blake of Renvyle, and would be a good site for a fishery village. The coast-road communicates with the shore of Ballynakill Bay at several other places.

Cleggan Bay and Pier.

The bay of Cleggan is next to the south of Ballynakill, about two miles in length, more open than this last, though it derives some protection from the isle of Ennesbofin; it has five to ten fathoms water, and is capable of great ships; the bottom is all clean sand and shells, good ground; the country from Cleggan to Ballynakill tolerably cultivated.

This bay was one of those selected by the Mansion-house Committee for the construction of a small fishery pier in 1822; such a thing was much wanted, as the bay was too open for small craft, though otherwise a good outlet, and the nearest harbour to Boffin.

After examining, the shore of the whole bay, I adopted a small bight on the south side, where a cut might be made into a little bog hollow, so as to form an inner dock out of the swell, the entrance to which might he covered by a short pier.

This work, and several of the neighbouring piers, I placed under the management of Mr. Alexander Hay, a young architect of talent and good character, though not of great experience in such works, which indeed it was difficult then to obtain. This excavation and pier proceeded with tolerable success, until a heavy gale in October, with an unusual high tide, filled part of the entrance with gravel, and damaged part of the pier. I then sent one or two workmen of more experience, to make up the pier in the most substantial way the case admitted; but the grant for that pier was now exhausted, and the work had accordingly to be suspended until additional funds could he obtained. This was only agreed to in January 1824, when I found the storms of that winter were likely to destroy all that had been done unless immediate steps were taken to put it out of danger; Cleggan pier being particularly exposed. I therefore deemed it absolutely necessary to employ a person of skill and experience on this service, and sent for Mr. Alexander M'Gill from Dunmore, who, having been for twenty years constantly and successively employed in that kind of work, was one on whose judgment I could rely, and whom I had already in 1822 employed to examine the stations on the southern coast. I instructed him particularly to put the piers at Cleggan and Roundstone in a state of safety, and subsequently to do the same at Ennisturk, but to undertake nothing more until the determination of the Board was known. To his skill and exertion it is owing that these works have been preserved.

Cleggan is now a stout pier of two hundred feet long, and fourteen feet high at head, with an inner basin not quayed, of one hundred feet by sixty, the entrance to which is not yet deepened so far as would be desirable, but it is only a bank of gravel. The pier has a parapet wall and sloping pavement to seaward, rough but substantial work, and has stood the gales of two winters without injury; it should however have a small jetty, or return at the head, and a course of coping along the quay wall. It affords a very convenient landing-place to the boats of Boffin and the western fishery, and having an excellent road from thence to Clifden, is very much frequented. The adjoining farm being in lease, buildings have not been yet undertaken at this place, but there is little doubt of its becoming a thriving village. The proprietor is Thomas Martin, esquire, of Ballynahinch.

From the south point of Cleggan Bay we have a low granite shore, with rocks in the offing as far as Aghris Point. Outside are the small isles of Crua, High Island and Friar Isle, lofty cliffs, and exhibiting nothing remarkable but some old monastic remains.

Omey Island is a low isle of granite, connected with the main at low water by a strand on the west side; the channel forms a natural harbour at each end: population, two hundred and fifty.

Streamstown Bay or River.

This is a long narrow inlet, navigable for nearly five miles, but almost dry at low water. There is a copper mine and a marble quarry near the upper end.

Kingstown Harbour.

This is a safe creek, a little south of Streamstown, little known or frequented, except by smugglers, on account of rocks at its entrance. An accurate survey has been made of these, and laid on the chart of Ardbear.

Isles of Tarbet and Innisturk.

These also are tolerably cultivated, the former having one hundred, the latter seventy inhabitants. From Tarbert towards Kingstown a small bar extends, only passable by small vessels. The sound of Innisturk is deeper, and forms a harbour.

Ardbear, or Clifden Harbour.

South of Tarbert Isle are the rocks of Carrigaroan, on which a heavy breaker commonly exists, marking the entrance to Ardbear. These rocks are steep on the south side, which is the ordinary passage; but in the progress of the survey, a good though narrow channel was discovered by us on the north side, and used in southerly winds; it is marked on the chart, as are all the numerous rocks in the entrance to this valuable harbour, as also of Mannin Bay, and to Slyne Head.

Passing the rocks of Carrigaroan, you bring Mr. D'Arcey's house, called Clifden Castle, just clear of Renard Point, which keeps you off all the rocks towards Mannin Bay; a little past the narrows there is a spit of sand from the north side, with three fathoms in the channel south of it; this place breaks sometimes in heavy westerly winds; within there is excellent anchorage, in six to eight fathoms opposite the castle.

Farther up the inlet divides into two parts, of which that on the southern side, or Ardbear, has a bar of one fathom at its entrance, but within is. deep, with good anchorage; the upper part is rocky; there is a good salmon fishery at its head. The northern arm runs up to the village of Clifden, and is dry at low water; but the tide flows twelve feet to Clifden Quay.

Doobeg Wall.

At the north of the Clifden arm a small bank or islet of gravel lies off the north shore, and dries into it at low water; this bank being a mile below Clifden, and accessible at its head by boats at low water, appeared a desirable situation for sheltering fishing boats, and in 1822 it was connected with the main, by running a wall across the low part and backing it up with gravel; this has stood well, and affords a convenient landing-place for vessels in the lower harbour, and shelter inside for boats. Cargoes have been landed there, and Mr. D'Arcey has a small storehouse on the spot.

Cljfden Quay.

Farther up this inlet narrows with a deep passage between rocks, and then opens in a basin of twenty acres, completely land-locked with the village of Clifden at its head, on the north side. The channel of the river running along the north shore of this basin makes that side the most suitable for the construction of a quay.

Mr. D'Arcey, the proprietor of Clifden, had applied to the Board for aid to construct a quay or pier for this place in 1821, and a contract was agreedon with him for this purpose; nothing however was done until 1822, when I was authorized by Government to employ the poor of Clifden in commencing this work, and a part of Doobeg wall, and of the upper quay, was built by them.

Mr. D'Arcey next season prosecuted this quay pursuant to his contract, and although still unfinished, a considerable extent is already in use The upper end, by being placed in as deep water as the lower end, has been more expensive than the boat quay at first intended; but this will ultimately be an advantage, as the great and rapid progress which Clifden has since made could not then have been anticipated.

In 1822 the village of Clifden exhibited nothing but scenes of misery and despair. It had made no progress for ten years, having only one slated or two-story houses, and a few thatched cabins. It now contains above one hundred tolerable houses, mostly two stories, and roofed with Bangor slates; about thirty country shops, several of the owners worth 500 *l.* to 1,000 *l.*, who were not worth 50 *l.* three years ago.

I had the curiosity to procure a list of the various things sold in 1824, and find that there is at least 3,000 *l.* a year of direct revenue paid to Government.

A brewery and distillery, and milling concerns, have been built since 1823, and are now (1826) in action; this has afforded a market for the grain, which could only be before disposed of by private distillation. Oats, also hay and straw, are regularly brought for sale to Clifden market; until last season, 1825, we were obliged to bring the oats wanted for our horses from Galway in barrels. Agriculture is now rapidly extending.

Two new stores have been built on the quay; they are occupied by herrings from this season fishery; two cargoes have been exported.

The removal of the salt duty has permitted a direct import trade from Liverpool, and three cargoes of that and country goods have arrivedduring the winters of 1824 and 1825. Seven cargoes were imported direct into Clifden in 1825, for the use of the country; two more are expected, one of them timber, direct from America

The harbour of Ardbear or Clifden is much frequented by the cruizers; there are many rocks near the entrance, which are badly expressed on Mackenzie's chart, but as they are all carefully laid down on our survey, I have made one of the charts to contain this harbour, and the whole from Aghris Point to Slyne Head, which I have no doubt will be an acceptable present to navigators.

Mannin Bay.

This bay is rocky and more exposed than Ardbear; it has only one good anchorage, and is but little frequented; and extensive kelp shore, and valuable fishery for herrings, bring many boats to it occasionally.

Bunowen and Slyne Head.

The peninsula of Bunowen is among the best cultivated and most populous parts of Conamara; there is on either side a considerable tract of blowing sand, and a number of low rocks and isles, which are chiefly valuable for their kelp, and very abundant lobster fishery. The isles which form Slyne Head, and multitude of adjoining rocks, are now delineated for the first time with any accuracy; and it will be seen by our chart that there are navigable sounds among them, which if known might be the means of frequently preserving life and property, numerous shipwrecks occurring here almost every year.

A light upon Slyne Head seems also an object most desirable; the nearest sea lights, viz. Arran and Clare Island, being too far embayed to be of service to vessels which fall in with this dangerous headland, and before reaching them the fine harbours of this coast are all passed.

The south shore is a blowing sand until we round Ross Point, when there is a good anchorage in the bight of Bunowen in S. W. winds, with two and a half fathoms, and mud. The rocks in the offing also shelter it materially, but they are not to be passed by a stranger unless by attending to the directions given in the chart above mentioned. This place was recommended by me for a small quay in 1822, but was not one of those adopted by the Mansion-house Committee. One sufficient for the purpose of the place may be built at the storehouse for about 600 *l*; proprietor, Mr. Gahagan O'Neill.

Roundstone Bay and Pier.

This harbour is formed between the main of Errisbeg and the isle of Innisnee; the sound at the upper end of which is fordable at low water, but rocky. The entrance to the bay is covered against the S. W. by the island of Innis Lacken, which is best passed on the east side; within it the harbour runs up three miles clear good ground, with five to two fathoms water; and about two thirds up, on the west side, is the new fishery quay or pier; it is a wharf along the side of a bight of one hundred and fifty feet, with a jetty of sixty feet long, at the head of which there is five feet water, and the quay would admit a large ship; the Plumper, gun-brig, has lain at it to careen.

This being one of the stations adopted by the Mansion-house Committee, the pier was begun by day labour in 1822, under the management of Mr. Hay above mentioned, and was in that season raised to spring-tide level in rough granite brought by boat, so as to give more extensive occupation, which probably added to its expense. A good deal of business was however done at it in shipping kelp, curing herrings, &c. Being within Slyne-head it is a convenient place for those of the western part of Conamara, to communicate with Galway. A branch of the coast-road to it is nearly completed.

In 1824 my assistant, Mr. M'Gill, was employed, as before mentioned, to make up the defective part of this pier, which was beginning to receive some injury, which he did to the level of coping; and subsequently, as the tenant of the farm on which this pier is situated was very clamorous for damages alleged to be sustained by him during the progress of the work, I ventured at my own expense to purchase up his interest in the lease, as the most likely way to settle his claim.

I now hold it by lease under Mr. Thomas Martin, and expect soon to have a tolerable fishing village; several people are already settled there, and I am building a store for the purposes of the fishery.

Ballynahinch River.

On the east side of Innishee is an inlet navigable with three fathoms to near the bridge, where the new coast-road crosses the Ballynahinch River; hard by is an abundant salmon fishery; the road afterwards coming close to the shore at some steep rocky ground, an opportunity has been taken of forming a small quay of a part of it, near Cloonisle, which may yet be of importance to the district, especially for the shipment of the beautiful green marble from the Twelve Pins Mountains.

The peninsula of Rossma, mostly bog, intervenes between this and the next inlet, or

Birterbuy Bay.

This beautiful harbour properly begins at Innisnee-isle, the last-mentioned inlet being only an arm of it; it runs up for five miles, half a mile wide at the entrance, then from one to two broad, with deep water and fine ground. There is a shoal of coral sand in the upper part, near the small isles, which cuts off the deep water of the upper arms, but at the part below we have five to ten fathoms, so that it is capable of the greatest ships, and it may be completely fortified; it is, however, very little frequented; there is no road from it; the coast-road now in progress touches the upper end.

The peninsula of Orresanagh which intervenes between Birterbuy and Kilkerran Bays, is also without any road, though it has a considerable population, especially on the coast. There are two good harbours, Ardboy and another, on the east side of Mynish Island, for small vessels, much frequented during the herring. Fishery and many safe boat creeks, but no pier or quay. The isles of Mason, Mynish and Tynish, at the south end of this peninsula, are all inhabited and engaged in fishery and kelping: Mynish, five hundred inhabitants; Mason, seventy-one; Tynish, one hundred and thirty-one. In Elan Macdara there are some curious monastic remains; Cruanakeely isle is high, without any proper landing; it is uninhabited, and used as a deer park by Mr. Martin. Within these isles is a roadstead or exterior harbour for great ships, outside of Birterbuy.

The Skird Rocks in the offing are lofty, and much frequented by sea-fowl; they form an useful beacon mark for Roundstone and Birterbuy.

Kilkerran Bay.

This bay has not been nautically surveyed for the fisheries, but I had a good outline of it from the bog survey; this is an extensive inlet, and fit for great ships; the upper arm, on the east side, called Commas

Bay, is shallow and rocky; the whole possesses a most productive kelp shore of nearly one hundred miles, isles included. A series of shallow fords, passable at low water, connect the isles on the east side, so that one may walk from Bealandangan into Annaghvane Isle, Lettermore, Garumna, Lettermullin, Knappough, Furnish, &c.; these fords are passable by boats at high water, so as to communicate with Keigal and Greatman's Bay; and this is the common route pursued to Galway. by the small craft with turf, &c. from Kilkerran Bay.

At present no carriage-road approaches within several miles of any part of Kilkerran Bay or Greatman's Bay; but the new coast-road, which is in progress, passes along the head of Kilkerran Bay to the eastern shore of Costello Bay, and to Galway. A branch or cross road is also forming from the north-east arm, or Commas Bay, across the central road of Conamara to the head of Lough Corrib, where it joins the road made from the Killery-head to Cong, and this will open a communication with Westport, and also with the plains of Connaught.

The islands of Garumna, Lettermullin, Lettermore, &c. in Kilkerran Bay, contain about two thousand five hundred, inhabitants, who, with those on the main, are principally employed in making kelp and preparing turf for sale in Galway and the county of Clare, where on the limestone land this article is scarce and dear; some limestone is returned as ballast, and the trade thus formed occupies a great many boats, which on a favourable appearance of herrings are all employed in the fishery.

Except an imperfect and nearly impassable bridleway, which penetrates the penninsula of Rossmuck, there is no road of any kind to the shore of Kilkerran Bay.

The transport of goods or people is therefore entirely done by water. I have already said that this bay contains within it a hundred miles of shore; the situations where landing-places may be found are of course innumerable, yet such is the backward situation of the district that there was no attempt at any thing of the kind, nor was there a place about Greatman's Bays where boats could be safely laid up, on my visit to this place, as engineer to the western district, during the distress of 1822; therefore it appeared that the construction of a harbour of that kind, which was prayed for in a memorial from the neighbourhood, was the most useful work on which the population could be employed, and I caused a creek in Garumna Island to be cleared and quayed, with a wall to the south side, formed by dry stone transported by boats. This harbour had made considerable progress when the necessity for this kind of expenditure ceased; it would be very desirable to have it completed, but even in its present state it forms a valuable place of shelter for the boats of the bay in their voyages to and from Galway, or to lay them up in.

The pier is nearly five hundred feet in length, with a return-head of one hundred feet; one half is wa lled, but not entirely filled behind, the rest is only a breakwater. A stone perch marks the extremity, at which there is four feet at low water, so that boats can enter at all times.

This work was entirely executed by the funds provided by the Government, and exclusive of any aid from the Board of Fisheries.

Greatman's. Bay is a safe harbour for vessels of moderate draft, but there are rocks in the upper part which must be attended to, and the entrance is frequently broken with southerly winds. It has been carefully surveyed, and the soundings marked in the neighbourhood of the new pier, engraved on the same sheet with Costello, and a sketch of the neighbouring isles is inserted from my bog surveys.

Casheen Bay is much frequented by the cruizers, being easy of access, good shelter and water, fit for great ships. It has not been surveyed. It lies on the west side of Garumna Island, between that and the isles of Lettermullin and Furnish, which are accessible from each other at low water.

Kiegal Bay is a harbour on the south side of that ford, but being more open to southerly winds is little frequented.

Costello Bay is divided by the peninsula of Killun from Greatman's Bay, and is the most eastern of the harbours of Conamara, the coast from thence to Galway being a straight shore with low skares of granite, among which are some sandy covers. This harbour admits large. ships, and, excepting the upper part, is clean and with good holding ground. A very minute and accurate survey has been made of it, and is now engraved, in which the position of the new fishery pier, and all the dangers, with marks for avoiding them, are carefully laid down.

This bay has been thought worthy of being fortified, a strong martello tower being built on the east side. Off this tower is a rock not noticed in Mackenzie's chart, nor is the shoal culled Ruallachd, near the entrance, and several others in the upper end.

This being one of the stations chosen by the Mansion-house Committee in 1822, I laid out a pier on the eastern side, so as to communicate more readily with Galway, within the peninsula called the

Lion, where there is excellent shelter and deep water. The work was at first conducted by Lieutenant Dundas, R. N. who resided in that neighbourhood, and was secretary to the local committee. In consequence of some difference between this gentleman and Mr. Blake, the guardian of the proprietor of the ground, who was a minor, I found it necessary in September to give up the system of day labour here, and make a contract with a builder from Galway for finishing the pier. It was also found necessary to increase its dimensions, partly to cover some rock near the head, and on account of the extraordinary rise of tide in this bay, which is greater than any other part of the west coast.

The pier is nearly three hundred feet long, twenty-one feet broad, coped with hewn limestone from Arran; has four feet water at the head at low water, so that it is accessible by sailing boats at all times, and at high water will admit of any merchant vessel: it is thus found particularly convenient for the boats trading to Galway, or those of the fishery; nearly one hundred boats belong to this bay. The chief defect as yet is, the want of a road; but a branch from thence to the coast-road is in progress, the expense of which, and of the additional work of the pier, has been defrayed out of the government funds to the amount of 375 *l.*

From the south of Costello Bay there is a road along the coast to Galway, which has been greatly improved and rendered fit for carriages by the government grants. Two points on this coast attracted attention in 1822 as a half-way station for the boats passing to Galway; of these Invern Creek has been improved by the proprietor, Mr. Blake, and now affords shelter to small boats. The other station adopted by the Fishery Commissioners is,

Spiddle.

This place is situated half way between Costello Bay and Galway, and a rude pier was once attempted here, which, when I saw it, was a ruin. It is sandy cove among the low granite rocks, near the mouth of a small river, on which there a good salmon fishery.

The deputation of the Commissioners of Fisheries in 1822, having appropriated 150 *l.* towards the building a pier here, and having the promise of a contribution from some neighbouring proprietors, I had the work begun by day, labour, under the management of Mr. Morris of that place, and workmen from Dunmore Harhour.

When the fishery grant was expended a further aid was obtained from the government funds, of 26 *l.*, and a subscription of 100 *l.* from Sir Robert Staples; the other subscriptions were not made good. Mr Morris, however, contributed his time and attention.

This harbour is dry at low water, and has about ten feet at high water; the entrance is about one hundred and fifty feet wide, between rocks, dry at low water; then rounding the pier-head we have a clean sandy cove of about one acre and a half.The quay is of rough granite, and is well protected to seaward by some rocks covered at spring tide. This harbour is much frequented by the coasting and fishing craft, but on account of the minority of the proprietors, Messrs. Cumming, there has not yet been any increase in the neighbouring village. A new road hasbeen made by the Government from this harbour across the great moor of Jar Connaught to the ferry on the middle of Lough Corrib; and whenever leases can be obtained, I have little doubt but there will be speedily a thriving village built. Some of the best turbot banks of the Bay of Galway lie just in the offing.

Furbagh.

Three miles nearer Galway Colonel Blake built, in 1822; a little boat harbour on this shore; it is not in the most favourable situation, but is occasionally useful for small craft; I accommodated him with tools and implements for the work.

Barna.

Halfway from hence to Galway, the late Mr. Lynch of Barna had built a boat pier and fishing village, called Freeport. This pier was ruined by a storm many years ago, and in 1822 the poor inhabitants were in great distress. To afford employment to the poor of this place, General Elly was furnished with 150 *l.* of the government fund to commence the repair of this pier; he was residing there for the benefit of sea bathing, and until his departure very kindly undertook the management of these funds; subsequently the work was prosecuted by day labour. The deputation of Commissioners of Fisheries appropriated 300 *l.* towards it, the returns of expenditure for which were made to the Board.

The greater part of the pier has thus been rebuilt in a far more substantial than before, viz, three hundred feet in length. The extremity is still in a ruined state, and should be repaired, though it seems unnecessary to carry it to the original extent. I have made out a plan and specification for completing it in the most economical form.

This pier is dry at low water.

Nearer Galway there is a creek or lagoon of about sixty acres, covered at high water, but at low, mostly dry; as this place is about to be embanked in the progress of the new coast-road, it is unnecessary further to notice it than to observe that the bridge or dyke which encloses it forms a boat quay: the place at present is little frequented by any craft.

A small quay fit for yawls only, has been built near Galway, on the east of Mutton Island, at the private expense of Mr. O'Hara, recorder of that town.

Galway.

We are now arrived at Galway, a town of great note in the Irish fishery, and where that business forms more exclusively the occupation of a great part of the population than in any other town on the west coast. The fishermen of Galway inhabit a particular quarter of the town, the Claddagh, a beach on the west side of the river, and are near a thousand families. They have sundry peculiar customs, and are in the habit of promulgating laws and regulations for the conduct of the fishery, and compelling their observance by the other boatmen of the bay.

The boats used in Galway are stout open or half-decked hookers, from five to fifteen tons burthen, carvel-built, generally good sailors, but seldom well found, or with sufficient fishing gear. In accommodating these, the harbour of Galway is very defective, and it were much to be wished that something were done to remedy it. In strong freshes from Lough Corrib the boats can hardly enter at any time, and especially at low water, and they then lay in the roads, or in the pool below, where till of late they were much exposed to S. and S. E. winds, until the tide flowed to enable them to get up the Claddagh.

Several years ago two narrow jetties were built across the beach at the Claddagh, which permitted more easy access to the boats when they were fleeted.

In 1822 the poor of Claddagh were employed to deepen this beach, raise the jetties, and form a rough quay along the shore; and I drew up a plan and specification for improving these jetties, so as to protect these boats more effectually, but as the Commissioners of Fisheries, from some insubordinate conduct of the boatmen, declined giving any further assistance at that period, nothing further was done.

In the mean time being commanded by his Excellency the Lord Lieutenant to report upon an application from the merchants and fishermen of Galway for the improvement of the port, it appeared to me that the construction of a pier or breakwater on the slate rock, which forms the western side of the entrance, was the most desirable measure of general benefit that could be attempted at a moderate expense, and I was ordered to proceed with that work, which is now nearly completed.

The slate pier extends five hundred feet from the point, with a return jetty of forty feet; the inside is faced with hewn limestone, with a sloping pavement to seaward: according to the accounts of the fishermen, the water on the bar of Galway has been deepened by this pier, as boats now get up to the jetty, and into Shelton, at low water; it is therefore of material importance to the fishing craft; but in fact the character of Galway Harbour has been entirely changed by this pier, as the pool between it and the old docks forms now an outer harbour, which admits a vessel to lay afloat at low water in shelter, and the old dock quays are themselves completely covered against southerly and S. W. gales, by which vessels laying there have formerly often been wrecked. The pier-head also forms a distinct object to steer upon in entering the harbour, preventing the risk of striking on the ledge of the slate, which was a common occurrence, and it affords a good layer for outward bound vessels.

The further improvement of the old quays of Galway, and the construction of a wet dock for which there is a good opportunity, though highly desirable for the mercantile interest, is less connected with our present inquiries; none of the fishermen living on that side; besides, as the property on these quays is private, it does not seem a proper subject for public expenditure: but on the Claddagh side a further operation would be of the greatest benefit to the fishermen, and which I should beg earnestly to recommend to the notice of the Board.

The point of the slate is separated from the Claddagh or fishing town by a marshy creek of twenty acres, which is overflowed at high water. I propose to run a road or dyke across the mouth of this creek, three hundred yards long, to be a short cut to the slate pier by near half a mile at high water, and also an additional quay for the fishing craft. In this to be an opening for admitting boats crossed by a stone arch or small swivel bridge, and thus a dock will be formed of about four or five acres, in which all the fishing boats of Galway could be laid up in complete security.

The marsh is separated from the sea by a long narrow spit of shingle, which is very rarely overflowed, but which could be easily raised over all tides, if thought worth the while.

Estimate of the Expense of the proposed Dyke at the Claddagh of Galway.

	£	*s.*	*d.*
20 perches embanked roadway at each end, at 2*l.* p' – – – –	80	–	–
30 perches retaining wall on each side, average 8 feet high, 4 thick, at 5*s.* p' foot – – – – – – –	315	–	–
Stone arch of 15 feet span, or swivel bridge – – – –	75	–	–
4,000 yards filling, shingle, at 10 *d.* p' yard – – – –	166	13	4
Coping wall, limestone, at 3*s.* p' – – – – – – –	189	–	–
Mooring-posts – – – – – – – – – –	20	–	–
	845	13	4
Contingencies 20 p' cent – –	169	2	8
£.	1,014	16	–

In 1771 Parliament granted 500 *l.* for repairing the Market Quay, and in 1783 1,000 *l.* for the further improvement of the harbour of Galway.

In addition to the fishermen residing in Galway, it is common for the boatmen of Lough Corrib, especially those residing about Menlough, and who are chiefly employed in the turf trade, upon the setting in of the herrings, to haul their boats across the town and proceed to the sea-fishery. A canal between the lake and the town has sometimes been talked of, and could certainly be accomplished at a moderate expense. It would much facilitate the transfer above mentioned, as well as the export of the produce of the lake, and the return of kelp for manure, and of fish; but the discussion of this matter seems foreign to our present object.

The roadstead of Galway is protected by the small and low isle of granite called Mutton Island, upon which the Ballast Office maintain a lighthouse, half a mile off the shore, and into which the strand dries at three quarters ebb; between it and Hare Island, on the west side, there are from three to four fathoms at low water, good holding ground. Hare Isle, which is a small knob of limestone gravel, is also connected with the main by a spit of gravel, dry at low water, so that this roadstead is already covered, as it were, by natural half-tide piers.

A singular improvement would be to raise the strand of Mutton Isle to high-water level, by a breakwater of rough stone, which would not cost above 7 or 8,000 *l.*, and which would enable vessels in the roadstead to communicate more readily at any time with the shore.

But although this appears too extensive an undertaking to be attempted by the Board of Fisheries, yet a landing jetty on the inside of Mutton Island is well worthy of attention; boats are frequently obliged to stop in the roads for the flood-tide, as the freshes from Lough Corrib run very strong in the ebb; and by means of this jetty they would be enabled. to communicate with the shore at low water. The stone may be got on the spot by quarrying the rock into the pool on the east of the lighthouse, and thus a good boat harbour formed at the same time. A work of this kind would be a singular convenience to the fishermen. Hare Island has also a creek which could be made a boat harbour, but is too inconvenient for the town.

Indeed I should consider the improvement of this roadstead as a national object; for after a careful survey I have made of the whole of Galway Bay, I find no other place fit for large vessels, except, indeed, the inside of Blackhead, which has ten and eleven fathoms close on shore, and is the only place for men of war.

The other inlets of Galway Bay, above Costello Bay, though some of them afford excellent shelter, as will be immediately explained, are all either shallow or barred at the entrance.

In Mackenzie's chart of Galway Bay, no notice is taken of the Santa Margaretta Rock, unless the shoal he marks be intended for it, and if so, it is in a very wrong position, and fully a mile too far inland; this danger is about halfway between Kilcorgan Point and Barna Cliff, and is the more important, as the ordinary sailing direction, viz. to keep Doochnish Castle halfway between Mutton Island and Hare Isle leads right over it, with Elem Edy Castle on south side Kilcorgan Point, Barna House or Barna Cliff and Mutton Isle, on Joice's House Merview, you are on it; least water two fathoms; it has proved fatal to several ships; but with all the other dangers of the bay, is truly represented in the survey now engraving for the fisheries, and where the line of two fathoms shows distinctly the extent of safe navigation.

Oranmore

Above Galway the sea penetrates four miles to the head of the creek or bay of Oranmore, where there is a village of five hundred inhabitants; this inlet is shallow and rocky, unfit to admit any thing but small vessels, which get up to a kind of quay at the old castle; this quay is private property, and anchorage dues are levied by the owner, Mr. Blake; no improvement of course was attempted a good deal of business is done there occasionally in landing turf and seaweed for manure, of which this bay affords an abundant supply; and herrings occasionally frequent the bay.

The north shore is remarkable for several powerful springs, which appear part of the waters of Lough Corrib, escaping through the limestone; on one of these is a salmon fishery.

Renvyle, or New harbour.

To the south-west of Oranmore two miles is the islet called New Harbour, or Renvyle. It is a beautiful and safe harbour for small ships, and frequently resorted to in westerly winds, and in winter by vessels from Galway Roads, about a mile long and a quarter of a mile broad. The bottom is clean, the inner part mud skirted with fine limestone gravel.

The deputation of Commissioners selected this harbour for one of their quays, which has accordingly been built on a bight of the north shore, where it is quite land-locked; the quay is one hundred and fifty feet long, with a return of sixty feet for boats, with a jetty and stair at the head fifteen feet at high water, and faced with hewn limestone. There is not as yet any public road to this quay, otherwise probably a great deal of business would be done there, but it is very much used by vessels, especially for careening or wintering.

Ardfry.

A long narrow peninsula, on which is the demesne and house of Lord Wallscourt divides this last from the next inlet, which is about two miles long, but narrow, and the upper part shallow and rocky; small craft only frequent this inlet, in which no pier has been built.

Between Renvyle and Ardfry there is a long spit or reef, terminating at the cockle rocks opposite to Hare Island. This reef should be taken care of in navigating Galway Roads; it is cleared with Oram Castle open of Saben Point.

Kilcorgan Point.

In the middle of the bottom of the bay the land jets out to a considerable extent in Kilcorgan Point, until opposite Mutton Island, and thereby covers Galway Roads against S. and S. E. winds. The greater part of this point of land is occupied by a remarkable lagoon, where the limestone rock sinks beneath the sea level, and long narrow spits of beach have been washed across the inlet. This lagoon is too shallow to be navigated by any thing but boats, and these only get through the entrance at Blackwater; small boats can go quite across the point; the figure of this lagoon will be best judged from the chart; it affords a plentiful supply of seaweed for manure.

Ballinacourty.

The spit of beach on the south side shoots away up the bay, so as to form a natural pier or breakwater, covering the pool or harbour of Ballinacourty. This is one of the finest and safest stations in the bay; it has fourteen feet at low water, with mud and grassy banks; but the entrance is narrow and somewhat intricate, so that it is considered as a bar harbour. Here is one of the piers, adopted by the deputation above mentioned, at the village of Ballinacourty on the north side.

It is formed by a cut across a narrow bank of gravel into a pool behind; the cut is quayed on the east side in hewn limestone, and will admit coasting vessels; the pier is prolonged at half-tide level down to low-water neaps; if this place: becomes of more consequence, this part may be advantageously raised to high water, and a return-head made to it.

A groin or pavement should also be made on the opposite side of the cut. The pool behind is a good place for laying up small craft, of which a considerable number belong to this place. A better road is much wanted up to Claren-bridge.

This inlet runs up four miles above Ballinacourty, dividing into two arms, one of which terminates at Claren-bridge, the other at Kilcolgan, on the road from Galway to Gort. These inlets are shallow, and little frequented, except by boats with sea-weed and turf, great quantities of which are brought from Conamara. The country round is a naked shallow limestone soil, which when cleared and cultivated is fertile, and grows the finest wheat in Ireland; much of it is in sheep-walk. The population is chiefly confined to the shores, where the abundance of sea-weed encourages agriculture, and where fuel can be had by water, the interior having little or no bog.

Sradbally.

A mile below Claren-bridge, on the south side, is a village called Stradbally, with a small pier, and a few boats, which are occasionally employed in the herring fishery.

St. Kitts.

Two miles further, beyond the house of Tyrone, is a village, named Killeen, or New Arran, of three hundred inhabitants, having also several yawls. This place appeared a favourable situation for a boat pier; and was adopted by the deputation in 1822. The pier named St. Kitts was begun by day labour, and afterwards completed by contract. It is a neat quay of hewn limestone, one hundred and twenty feet long. With a return-head of twenty feet; at the upper end is a small bridge, through which boats pass into a pool behind the quay, covered from the sea by a gravel bank, where they may be laid up in complete shelter. This place is much frequented by the Galway fishermen during the herring season, and with N. W. winds. The pier is covered in front by the small isle of Mynish, three hundred yards distant, and some strong reefs, and has two or three entrances at high water; at low water it is dry. It is nearly opposite Ballinacourty.

The road is not completed into it; but when it is, this, as well as Ballinacourty, may be expected to have a considerable trade; as this inlet of Kilcolgan, being covered on the S. W. by island Eady; forms one of the safest harbours in the bay of Galway, and it penetrates the farthest into the interior.

Elan Edy.

This consists properly of three islands, connected by spits of shingle. On the eastern is an old castle, which is an useful sea mark; the spits form safe boat harbours for each side of the isle, which has seventy-two inhabitants, and several sailing craft, occasionally employed in the fishery; outside of it is an extensive turbot bank.

Kinvara.

South of island Edy is the bay of Kinvara; the exterior part faces the W. but is somewhat covered by the shoal between Deer island and the main.

It then turns to the S. E. over a bar of eight feet, within which is a pool of two and a half fathoms, land-locked; the upper part is shallow and rocky, only used by small craft, which pass to the quay of Kinvara at the head. This quay was built originally by county presentment, and is rather in bad order; a good deal of corn is shipped there in small vessels, and turf landed from Conamara. It is rather far up to be of much importance to the fisheries. Gort is a considerable inland town; three thousand inhabitants; is within seven miles.

Crushna, about a mile lower, is a village, with some fishing yawls; a road was made into it by the government funds in 1822, and fish are frequently landed there for the supply of the interior.

Duras.

Beyond this is the peninsula of Duras; and connected by a narrow spit of beach, at its extremity, is the peninsula of Aghnish, a detached part of the county of Clare. At Duras, on the north side, the proprietor, Mr. Lynch, proposed to the Board of Fisheries to be at half the expense of a small pier, which was agreed to; the work was contracted for by Mr. Lynch, and completed in 1823. The pier is neatly built in hewn limestone, one hundred and twenty feet long, with a kant or return-head of forty feet, and substantial parapet. The shore being a flat sand, we have only ten feet water at this pier, which nevertheless is found to be a great convenience to the fishing craft. A good deal of produce is shipped here for Galway, by the Messrs. Lynch and Martin.

From the spit of Aghnish a shoal of foul ground runs off to a small rock called Deer Island, and round the point of Aghnish. no situation occurs for any landing-pier, though a martello tower has been built on the shore.

COUNTY OF CLARE.

Burrin Quay.

Within Aghnish an inlet runs up four miles to Curranrue-bridge and tide-mill; the mouth is narrow, about two hundred yards wide; the interior lagoon has some good oyster beds; the tide ebbs and flows at the entrance of this inlet with considerable force, forming a channel of two and a half fathoms deep, but without is a bar of nine feet only; the upper part of the inlet is shallow. This a favourite bathing place, and there is a considerable population, especially in summer. At the entrance of the inlet on the south side two attempts have been made to build a quay. The village was also named New Quay, but both of them are in ruins. Indeed I doubt it they were ever of any use.

This place was selected by the deputation in 1822 for a small quay, and 250 *l.* appropriated to it. I laid out the work in the vicinity of the innermost of the old jetties, and left a workman there who had been employed at Killough Harbour, who began it very substantially in hewn limestone, and proceeded some time by day labour; the funds, however, were inadequate, though some aid was obtained from the public; the work was suspended for a season or two, and ultimately was finished by contract.

The quay is now neatly built in hewn limestone, one hundred and five feet along shore, with a return of seventy-two feet for boats, and jetty pier-head of seventy feet, extending to low water, with a good parapet. This pier is fifteen feet high, and admits coasting vessels. A great deal of business is done here in shipping corn, landing turf, and also in the fisheries; a packet sails regularly in summer to Galway, and some good houses have been built at the quay; the wharf is commonly filled with craft.

Pooldoody and Ballyvaughan.

From the bar of Burrin Quay the coast is a limestone cliff round Fenvard Point and into Ballyvaughan Bay; the great part of this bay is shallow, and only fit for boats, and a small quay for their use was built there many years ago; it is hardly worthy of much improvement.

The eastern part of the bay runs up by a narrow channel to the lagoon of Pooldoody, in which boats are completely land-locked; this lagoon or pool contains an extensive or celebrated oyster bed.

Beyond Ballyvaughan the shore becomes steep, rising at length into the lofty limestone cliffs of Blackhead. As deep water is nearer the shore here than at any part of Galway Bay, and that great ships may ride close to the land, a small landing jetty somewhere about Glanning Castle would appear a very useful work, and could be made for about 200 *l.*

Beyond Blackhead the coast of Clare affords no opportunity of constructing a place of shelter for several miles. The interior consists chiefly of hills of limestone, much of it naked, with singular crags and ravines. The inhabitants are few, the whole barony of Burren containing less than eight thousand persons.

The Isles of Arran.

The isles of Arran seem to be a continuation of the limestone country of Burren, from which they are about four miles distant. They are three principal islands, Arranmore, St. Gregory's Isle, and Innishschir, also the smaller branch isles at the western extremity.

The range is about twelve miles in length, and the broadest part of Arranmore is nearly two miles; the whole population about three thousand. The southern side of Arran is a precipitous cliff of limestone, and the isle declines gradually northward to the waters edge, on which in the main isle there are two inlets, but only one of them capable of forming a harbour. On this inlet are two or three villages, of which that called Killaney contains about one thousand inhabitants, who are chiefly employed in the fishery, especially for cod and lyng, which they dry on the benches of limestone rock along the shore.

In 1822 there was a good deal of distress in Arran, and some relief was afforded by employing the people to quarry ashlers of limestone for the public works on the main.

The deputation of Commissioners having resolved on building one of the quays at Killaney in Arran, I laid it out, and had it begun in the usual way by day labour, but as there was considerable difficulty in communicating with the isle, the plan I gave was not perfectly understood, at least not exactly followed, and the fund exhausted ere it was completed. It was, however, found of great service during the ensuing winter, and has since been completed by contract.

As the sea ebbs to a considerable distance at Killaney, I restricted the pier to covering the upper part only, to which it gives complete shelter, but the road to it from the old castle is also made so as to serve as a quay in off-shore winds.

It is built in hewn limestone, of which the isles afford excellent blocks, similar in execution to the pier of Duras, (*illegible*) feet long, with a return head or jetty, and a summer quay along shore as far as the old port. The exterior part of the bay is a good roadstead in all but north east winds, and in them good shelter will be found behind the pier, where a hundred boats at least may be laid in safety.

The entrance to Killaney is wrong described in Mackenzie's chart and Pelham's map of Clare, being on the west, not on the south of Straw Island.

References

Preface

[1] Seán de Courcy, *Alexander Nimmo: Engineer extraordinary*, in *Connemara: Journal of the Clifden & Connemara Heritage Group*, Vol.2 No.1, (Clifden 1995); Ted Ruddock, Alexander Nimmo, FRSC, MRIA (1783-1832), in *A Biographical Dictionary of Civil Engineers in Great Britain and Ireland, Vol 1: 1500 – 1830*, Ed Prof Sir Alex. W. Skempton, (London 2002), pp483-9

Chapter 1

[1] Robert G. Preece, Archivist, Royal Inverness Academy, History of Inverness Royal Academy, Unpublished Manuscript

[2] *Annual Register*, 1832, pp181-2

[3] Ruddock, 'Nimmo', *op.cit.*; *Oxford Dictionary of National Biography*, Vol. 40, Ed by H C G Matthew and Brian Harrison, (Oxford 2004), pp914-5; Ted Ruddock, *Arch Bridges and their builders*, 1735 - 1835 (1979); Preece, *op.cit.*

[4] *The Dictionary of National Biography*, Volume XIX, (Oxford 1973) pp489-494; Samuel Smiles, *The Life of Thomas Telford*, (London 1867)

[5] Alexander Nimmo, Journal of Survey of Inverness-shire, 1806, NLS, Manuscript Collection, Adv. Ms. 34 4 20

[6] *Ibid.*

[7] Ruddock, 'Nimmo', *op.cit.*

[8] *Annual Register*, 1832, pp181 - 2

[9] George O'Brien, *The Economic History of Ireland from the Union to the Famine*, (London 1921), p147

[10] *Fourth Report of the Commissioners Appointed to enquire into the Nature and Extent of the several Bogs in Ireland, and the practicability of draining and cultivating them*, 1813 –14 (131) vi part ii, 167(Hereafter cited as *Fourth Report of the Bog Commission*), pp14-15

[11] J.H. Andrews, *A Paper Landscape: The Ordnance Survey in Nineteenth-Century Ireland*, (Dublin 2002), p5

[12] *First Report of the Commissioners Appointed to enquire into the Nature and Extent of the several Bogs in Ireland, and the practicability of draining and cultivating them*, 1810 (365) x, 389, (Hereafter cited as *First Report of the Bog Commission*); *Second Report of the Commissioners appointed to enquire into the nature and extent of the several Bogs in Ireland, and the practicability of draining and cultivating them*, 1810-1811 (96) vi. 579

[13] *First Report of the Bog Commission*

[14] Skempton, *op.cit.*, pp274-5

[15] *Ibid.*, pp34-5

[16] *Fourth Report of the Bog Commission*, Appendix No 1

[17] *Fourth Report of the Bog Commission*

[18] *First Report of the Bog Commission*

[19] *Fourth Report of the Bog Commission*, Appendix No 1

[20] *Minutes of evidence taken before the select committee of the House of Lords appointed to examine into the nature and extent of the disturbances in Ireland*, 1825 vii (Hereafter cited as *House of Lords inquiry into disturbances in Ireland* (1825)), evidence of Alexander Nimmo, p161

[21] Dr Arnold Horner, *Iveragh Co. Kerry in 1811: Alexander Nimmo's map for the Bogs Commissioners of 1809-1814*, (Dublin 2002), p4

[22] *Fourth Report of the Bog Commission*

[23] *Ibid.*

[24] *Ibid.*

[25] Patrick Knight, C.E., *Erris In the Irish Highlands and the Atlantic Railway*, (Dublin 1836), pp1-5

[26] Edward Wakefield, *An Account of Ireland Statistical and Political*, Vol. 1, (London 1812), pp33-4

[27] Daniel Augustus Beaufort L.L.D., *Memoir of a Map of Ireland*, pp73-83; *House of Lords inquiry into disturbances in Ireland* (1825), evidence of Alexander Nimmo

[28] James Hardiman, *The History of the Town and County of the Town of Galway*, (Galway 1958), pvi

[29] Robert Fraser, *Report on the County Surveys, and the Best Means for the Further Encouragement of the Fisheries of Ireland*, (Dublin 1818)

[30] *Second Report of the Select Committee on the State of Disease and Condition of the Labouring Poor in Ireland* (Session 1819 – Reprint), 1829 (347) iv (Hereafter cited as *Report of the Select Committee on the State of Disease in Ireland* (1819)), p97

[31] O'Brien, *op.cit.*, p123

[32] James McParlan, M.D., *Statistical Survey of the County of Mayo*, (Dublin 1802), pp85 - 6

[33] Hardiman, *op.cit.*, pp302-7

[34] Wakefield, *op.cit.*, Vol. 11, p97

[35] Hardiman, *op.cit.*, pp302-7

[36] *Ibid.*, p302

[37] Arthur Young, *A Tour in Ireland 1776 - 1779* Vol. 1 & 2, Ed by A. W. Hutton. (Shannon 1970), Vol.1 p276

[38] O'Brien, *op.cit.*, p20

[39] Hely Dutton, *A Statistical and Agricultural Survey of the County of Galway* (Dublin 1824), p343

[40] McParlan, *op.cit.*, pp85-8

[41] McParlan, *op.cit.*, p86

[42] Dutton, *op.cit.*, p354

[43] Wakefield, *op.cit.*, Vol. 11, p754

[44] *Report from the select committee on the employment of the poor in Ireland*, 1823 (561) Vol. vi, 331, p7 (Hereafter cited as *Report from the select committee on the employment of the poor in Ireland*, 1823)

[45] *Report of the Select Committee on the State of Disease in Ireland* (1819)

[46] Wakefield, *op.cit.*, Vol. 1, p32

[47] *Ibid.*, pp259-60

[48] McParlan, op.cit., p37; pp158-9

[49] *Report from the select committee appointed to consider the best mode of apportioning more equally the local burthens collected in Ireland, and to provide for a general survey and valuation of that part of the United Kingdom*, 1824 (445), viii (Hereafter cited as *Report on survey and valuation of Ireland* (1824)), evidence of Alexander Nimmo

[50] Alexander Nimmo, Report on the Bogs in that Part of the County of Galway to the West of Lough Corrib, Appendix No 12, *Fourth Report of the Commissioners on the Bogs in Ireland* (Hereafter cited as Alexander Nimmo's Bog Report on Co Galway) p188

[51] *Dictionary of Land Surveyors and Local Mapmakers of Great Britain and Ireland 1530 - 1850* (London 1997), Vol. 11

[52] Alexander Nimmo's Bog Report on Co Galway

[53] *Ibid.*, p199

[54] Alexander Nimmo's Bog Report on Co Galway

[55] Chevalier De Latocnaye, *A Frenchman's walk through Ireland*, (1796-7), translated by John Stevenson, (Dublin 1917), p170

[56] John D'Arcy to Marquess of Wellesley, Lord Lieutenant, CSORP 1833 6269

[57] Alexander Nimmo's Bog Report on Co Galway, p204

[58] Inhabitants of Ballynahinch Barony Memorial to Lord Lieutenant, May 1825, CSORP 1825 12059

[59] Alexander Nimmo's Bog Report on Co Galway, p187

[60] Kathleen Villiers-Tuthill, *Patient Endurance: The Great Famine in Connemara*, (Dublin 1997), pp3-4; Kathleen Villiers-Tuthill, *Beyond The Twelve Bens: A History of Clifden and District 1860-1923*, (Galway 1986), pp123-5

[61] Shevawn Lynam, *Humanity Dick Martin 'King of Connemara' 1754 - 1834*, (Dublin 1989), pp172-3

[62] Maria Edgeworth, *Tour in Connemara and the Martins of Ballinahinch*, Ed by Harold Edgeworth Butler, (London 1950), p52; p58

[63] Alexander Nimmo's Bog Report on Co Galway, p187

[64] *Ibid.*

[65] Wakefield, *op.cit.*, pp753-4

[66] Alexander Nimmo's Bog Report on Co Galway, p187

[67] *Ibid.*, p190

[68] *Ibid.*, p167

[69] Alexander Nimmo's Bog Report on Co Galway, p192

[70] Alexander Nimmo's Journal 1806

[71] *Ibid.*, p189

[72] Wakefield, *op.cit.*, Vol. 1, pp753-4

[73] Alexander Nimmo's Bog Report on Co Galway p190

[74] Wakefield, *op.cit.*, Vol. 1, pp753-4

[75] Alexander Nimmo's Bog Report on Co Galway p190

[76] *Ibid.*, p191

[77] *Ibid.*, p191

[78] *Ibid.*, p188

[79] *Ibid.*, p209

[80] *Ibid.*, p201

[81] *Ibid.*, p202

[82] *Ibid.*

[83] *Ibid.*, p203

[84] *Ibid.*, p199

[85] Maurice Semple, *Reflections on Lough Corrib*, Galway 1974, pp7-15; for more on Samuel Usher Roberts see Kathleen Villiers-Tuthill, *History of Kylemore Castle & Abbey*, (Galway 2002)

[86] Alexander Nimmo's Bog Report on Co Galway, p210

[87] *Ibid.*, p189

[88] *Ibid.*

[89] Report of Mr Richard Griffith, on Bogs in the Baronies of Tyrawly, Erris, and Burrishhole, in the County of Mayo; and of Tyreragh, in the County of Sligo, Appendix No 11, *Fourth Report of the Bog Commission* (Hereafter cited as Report of Richard Griffith on Bogs in Mayo and Sligo)

[90] *Ibid.*

[91] *Ibid.*, pp185-6

[92] McParlan, *op.cit.*, p93

[93] Report of Richard Griffith on Bogs in Mayo and Sligo, p183

[94] *Ibid.*, pp182-3

[95] *Report of the Select Committee on the State of Disease in Ireland* (1819), p97

[96] Report of Richard Griffith on Bogs in Mayo and Sligo, pp 182-3

[97] *Second Report of the Bog Commissioners, Appendix No 2*; *Report on survey and valuation of Ireland*, evidence of William Bald

[98] *Report on survey and valuation of Ireland*, evidence of William Bald. Efforts to locate the models have failed to date

[99] Report of William Bald, on the bogs in Mayo, *Third Report of the Commissioners Appointed to enquire into the Nature and Extent of the several Bogs in Ireland, and the practicability of draining and cultivating them*. 1813-14 (130) vi pt 1, Appendix No 7; No 8 & No 9

[100] Knight, *op.cit.*; *Fourth Report of the Bog Commissioners*; R.B. McDowell, *The Irish Administration 1801–1914*, (Connecticut 1964), pp200-201

[101] Alexander Nimmo, *Report and estimates on draining and cultivating the bog on the Cashan or the Northern division of Kerry*, (Dublin 1818)

Chapter 2

[1] *Report of the Select Committee on the State of Disease in Ireland* (1819), evidence of Alexander Nimmo, p101

[2] *Ibid.*, p103

[3] Alexander Nimmo, *Report on the means of improving the river and harbour of Cork*, (Cork 1815)

[4] *Report on survey and valuation of Ireland*, evidence of Alexander Nimmo

[5] *House of Lords inquiry into disturbances in Ireland* (1825), evidence of Alexander Nimmo, p172

[6] *Accounts of the Commissioners for advancing money from the Consolidated Fund for Public Works*, 1824 (278) xxi 61

[7] Michael Barry, *Across Deep Waters: Bridges of Ireland*, (Dublin 1985), p151

[8] *Report from Commissioners*: Dunmore Harbour, 1821 (492) xi

[9] Seventh and Final Account of Expenditure by Alexander Nimmo on Public Works in the Western District, drawn up by John & George Nimmo, 1832, OP 663/9, (Hereafter cited as Seventh and Final Account of Alexander Nimmo, 1832)

[10] *House of Lords inquiry into disturbances in Ireland* (1825), evidence of Alexander Nimmo, p156

[11] *Ibid.*, p152; pp166-7

[12] *Report from the select committee on the employment of the poor in Ireland*, 1823, p8; *House of Lords inquiry into disturbances in Ireland* (1825), evidence of Alexander Nimmo, p160

[13] *House of Lords inquiry into disturbances in Ireland* (1825), evidence of Alexander Nimmo, pp130-2

[14] *Report of the select committee on the State of Disease in Ireland* (1819), p96; evidence of T. Telford, p111

[15] *Ibid.*, pp95-99

[16] *Ibid.*, evidence of Alexander Nimmo, p102

[17] *Report of the Select Committee on the state of the poor in Ireland*, 1830 (667) vii

[18] *Report of the select committee on the State of Disease in Ireland* (1819), evidence of Alexander Nimmo, p101

[19] *Ibid.*, p104

[20] *Ibid.*, evidence of R. Fraser, p116; Fraser, *op.cit.*

[21] Fraser, *op.cit.*

[22] Alexander Nimmo to William Gregory, 17 August 1820, CSORP 1820 19

[23] *Third Report of the Commissioners of the Irish Fisheries, of their Proceedings, for the year 1821*, 1822 (428) xiv (Hereafter cited as *Third Report of the Commissioners of the Irish Fisheries*)

[24] *Report on survey and valuation of Ireland*, evidence of Alexander Nimmo

[25] Appendix No 16, *Third Report of the Commissioners of the Irish Fisheries*; Appendix No 15, *Fourth Report of the Commissioners of the Irish Fisheries, of proceedings for the year 1822*, 1823 (383) x; Appendix No 13, *Fifth Report of the Commissioners of the Irish Fisheries, of their proceedings for the year 1823*, 1824 (283) xii; Appendix No 13, *Sixth Report of the Commissioners of the Irish Fisheries, of their proceedings, for the year 1824*, 1825 (385) xv; Appendix No 9, *Seventh Report of the Commissioners of the Irish Fisheries for the Year Ending 5 April 1826*, 1825 - 26 1826 (395) xi; Appendix No 7, *Eight Report of the Commissioners of Irish Fisheries, for the year ending 5 April 1827*, 1826 – 27 (487) xi

[26] *Report on survey and valuation of Ireland*, evidence of Alexander Nimmo

[27] *Ibid.*

[28] Dutton, *op.cit.*, p509n

[29] Knight, *op.cit.*, p5

[30] Alexander Nimmo, Report on the Fishery Stations of the Coast of Connaught, *Sixth Report of the Commissioners of the Irish Fisheries, of their proceedings for the year 1824*, 1825 (385) xv; (Hereafter cited as Alexander Nimmo's Report on the Fishery Stations of the Coast of Connaught)

[31] Seventh and Final Account of Alexander Nimmo, 1832; *Dictionary of Land Surveyors and Local Mapmakers*, Vol. 11, *op.cit.*

[32] Alexander Nimmo's Report on the Fishery Stations of the Coast of Connaught (p181of appendix to this book)

[33] *Sixth Report of the Commissioners of the Irish Fisheries*, p34
[34] Roderick O'Flaherty, *A Chorographical Description of West or H-Iar Connaught*, written in 1684, Ed by James Hardiman, (Dublin 1846), p72-3
[35] Alexander Nimmo's Report on the Fishery Stations of the Coast of Connaught, (p200 of appendix to this book)
[36] William H. King, *Account of The Fisheries in 1822 on the Western Coast of Ireland*, CSORP 1826 13598
[37] *Sixth Report of the Commissioners of the Irish Fisheries*
[38] *Ibid.*
[39] Alexander Nimmo's Report on the Fishery Stations of the Coast of Connaught (p200 of appendix to this book)
[40] *Ibid.*
[41] Hardiman, *op.cit.*, pp302-7
[42] *First report of the commissioners of inquiry into the state of the Irish Fisheries with the Minutes of Evidence and Appendix* 1837 [77] xxii.i, p75-77
[43] William H. King, Account of the Fisheries on the Western Entrance of the River Shannon, extending from Killala Bay to Kerry Head, in *An Account of the Present State of the Fisheries on the different parts of the Coast of Ireland*, (Dublin 1820), pp3-4
[44] *Ibid.*
[45] Alexander Nimmo, Report on the Fishing Stations on the North Coast of Ireland, *Fourth Report of the Commissioners of the Irish Fisheries, for the year 1822*, 1823 (383) x (Hereafter cited as Alexander Nimmo's Report on the Fishing Stations on the North Coast of Ireland) p24
[46] *Ibid.*, p23
[47] *Ibid.*, p26
[48] *Ibid.*, p23
[49] *Ibid.*, p34
[50] *Report on survey and valuation of Ireland*, evidence of Alexander Nimmo
[51] Alexander Nimmo's Report on the Fishing Stations on the North Coast of Ireland, p48
[52] *Ibid.*, p34
[53] *Third Report of the Commissioners of the Irish Fisheries*
[54] *Seventh Report of the Commissioners of the Irish Fisheries; Report on survey and valuation of Ireland*, evidence of William Bald, p65
[55] *Fourth Report of the Commissioners of the Irish Fisheries*

Chapter 3

[1] *Fourth Report of the Commissioners of the Irish Fisheries*
[2] Census of Ireland 1851, Vol. V, pp190-210
[3] *Ibid.*
[4] *Report from the select committee on the employment of the poor in Ireland*, 1823, p5
[5] *Ibid.*
[6] *Correspondence and Accounts relating to the different occasions on which measures were taken for the relief of the people suffering from scarcity in Ireland, between 1822 and 1839*, 1846 (734) xxxvii (Hereafter cited as *Correspondence and Accounts relating to relief measures taken between 1822 and 1839*), pp5-6
[7] *Report from the select committee on the employment of the poor in Ireland*, 1823, p4; *Correspondence and Accounts relating to relief measures taken between 1822 and 1839*, p5
[8] *Fourth Report of the Commissioners of the Irish Fisheries*
[9] Alexander Nimmo to Alexander Mangin, 10 June 1822, CSORP 1822 858
[10] *Fourth Report of the Commissioners of the Irish Fisheries*
[11] Alexander Nimmo's Report on the Fishing Stations on the North Coast of Ireland
[12] *Fourth Report of the Commissioners of the Irish Fisheries*
[13] Alexander Nimmo's Report on the Fishery Stations of the Coast of Connaught; Alexander Nimmo's Report on the Fishery Stations of the West Coast of Ireland, from Errishead southwards, Appendix 10, *Eight Report of the Commissioners of Irish Fisheries, for the year ending 5 April 1827*, 1826 – 27 (487) xi (Hereafter cited as Alexander Nimmo's Report on the Fishery Stations of the West Coast of Ireland)
[14] Alexander Nimmo's Report on the Fishing Stations on the North Coast of Ireland
[15] Census of Ireland 1851, Vol. V, pp190-210; R.B. McDowell, *Public Opinion and Government Policy in Ireland, 1801–1846*, (London 1952), pp71-2
[16] Henry Goulburn to Alexander Nimmo, 6 June 1822, CSORP 1822 970; Memorial of the Administrators of the Estate of Alexander Nimmo, 10 December 1832, CSORP 1832 6376
[17] Alexander Nimmo's Report on the Fishery Stations of the Coast of Connaught
[18] Mayor of Galway to Henry Goulburn, 20 Sept 1822, CSORP 1822 1157
[19] Nimmo to Mangin, 13 June 1822, CSORP 1822 859
[20] Nimmo to Mangin, 10 June 1822, CSORP 1822 857
[21] Brian Jenkins, *Henry Goulburn 1784-1856, A Political Biography*, (Liverpool 1996), p135
[22] Henry Goulburn to Nimmo, 6 June 1822, CSORP 1822 970
[23] *Ibid.*
[24] Nimmo to Lord Ffrench, 23 June 1822, CSORP 1822 865
[25] Nimmo to Mangin, 13 June 1822, CSORP 1822 859
[26] Nimmo to William Gregory, 19 June 1822, CSORP 1822 861
[27] Nimmo to Mangin, 10 June 1822, CSORP 1822 857
[28] Nimmo to Gregory, 19 June 1822, CSORP 1822 861
[29] Inhabitants of Ballinahinch Barony Memorial to Lord Lieutenant, 3 May 1825, CSORP 1825 12059
[30] Nimmo to Gregory, 22 June 1822, CSORP 1822 864; Nimmo to Gregory, 19 June 1822, CSORP 1822 861; Nimmo to Marquess of Anglesey, Lord Lieutenant, 20 April 1831, OP 973/82
[31] Nimmo to Mangin, 13 June 1822, CSORP 1822 859
[32] Nimmo to Gregory, 22 June 1822, CSORP 1822 864
[33] Nimmo to Mangin, 27 June 1822, CSORP 1822 868
[34] *Return, showing the Gross Expenditure in the Counties of Mayo and Galway, under the Directions of Messrs. Nimmo and Killaly, Civil Engineers, from the year 1821 to 1833*, 1833 (285) xxxv 575, p3 (Hereafter cited as *Return, showing the Gross Expenditure in the Counties of Mayo and Galway from*
35 *1821 to 1833*)
36 Nimmo to Wellesley, 30th June 1822, CSORP 1822 868
An Account of all sums of money advanced out of the Consolidated Fund, on the recommendation or with
37 *the sanction of the Irish Loan Commissioners*, 1831 (184) xvii
38 Nimmo to Wellesley, 30 June 1822, CSORP 1822 868
39 Nimmo to Gregory, 22 June 1822, CSORP 1822 864
House of Lords inquiry into disturbances in Ireland (1825), evidence of Alexander Nimmo, p166
[40] Gregory to Nimmo, 25 June 1822, CSORP 1822 864
[41] Wellesley to Mangin, CSORP 1822 868
[42] *House of Lords inquiry into disturbances in Ireland* (1825), evidence of Alexander Nimmo, p171
[43] *Ibid.*
[44] *Ibid.*, p168
[45] Alexander Nimmo's Report on the Fishery Stations of the Coast of Connaught, {p187 of appendix to this book)
[46] *First Report of select committee appointed to inquire into the amount of advances made by the Commissioners of Public Works in Ireland*, 1835 (329) xx 145, Evidence of Richard Griffith, p223; Seventh and Final Account of Alexander Nimmo, 1832
[47] *Ibid.*
[48] *House of Lords inquiry into disturbances in Ireland* (1825), evidence of Alexander Nimmo, p161

[49] Memorial of the Administrators of the Estate of Alexander Nimmo, 10 December 1832, CSORP 1832 6376

[50] Clifden Union Death Registers; Prerogative Grant Book 1839, Folio 16a; Register of Wills

[51] Ruddock, Nimmo, *op.cit.*

[52] *Monies expended under the direction of Alexander Nimmo, on Public Works in the Western District of Ireland*, 1830 (624) xxvii (Hereafter cited as *Return, showing expenditure under A. Nimmo in Western District* (1830))

[53] *Return, showing the Gross Expenditure in the Counties of Mayo and Galway from* 1821 - 1833, p2

[54] Seventh and Final Account of Alexander Nimmo, 1832

[55] *Report on survey and valuation of Ireland*, p77; p83

[56] Nimmo to Edward Smith Stanley, 27 April 1831, OP 973/80; Appendix No 1, *Fourth Report of the Bog Commissioners; Dictionary of Land Surveyors and Local Mapmakers, op.cit.*

[57] Daniel Mackin to chief secretary, 1832, CSORP 1832 4140

[58] Nimmo to Stanley, 27 April 1831, OP 973/80; *Dictionary of Land Surveyors and Local Mapmakers, op.cit.*

[59] Nimmo to Stanley, 27 April 1831, OP 973/80

[60] Seventh and Final Account of Alexander Nimmo, 1832

[61] Alexander Nimmo's Report on the Fishing Stations of the Coast of Connaught; Seventh and Final Account of Alexander Nimmo, 1832; *Dictionary of Land Surveyors and Local Mapmakers, op.cit.*

[62] Seventh and Final Account of Alexander Nimmo, 1832; *Dictionary of Land Surveyors and Local Mapmakers, op.cit.*

[63] Alexander Nimmo, *New Piloting Directions for St George's Channel and the Coast of Ireland*: Written to accompany the chart of St George's Channel and the coasts of Ireland, Drawn for the Corporation for Improving the Port of Dublin, (Dublin 1832)

[64] Seventh and Final Account of Alexander Nimmo, 1832; *Dictionary of Land Surveyors and Local Mapmakers, op.cit.*

[65] James White & Kevin Bright, *Treasures of The Royal Dublin Society: A Summary Catalogue of the works of art in the collection of The Royal Dublin Society*, (Dublin 1998)

[66] Nimmo to Stanley, 27 April 1831, OP 973/80; Alexander Nimmo's Report on the Fishing Stations of the Coast of Connaught

[67] Nimmo to Stanley, 27 April 1831, OP 973/80; Seventh and Final Account of Alexander Nimmo, 1832; Nimmo to Sir William Gossit, 3 October 1831, OP 973/87

[68] Seventh and Final Account of Alexander Nimmo, 1832

[69] *Dictionary of Land Surveyors and Local Mapmakers, op.cit.*

[70] Alexander Nimmo's Report on the Fishery Stations of the Coast of Connaught; Seventh and Final Account of Alexander Nimmo, 1832

[71] *Dictionary of Land Surveyors and Local Mapmakers, op.cit.*

[72] *Second Report from the Select Committee appointed to inquire into the amount of advances made by the Commissioners of Public Works in Ireland, with the minutes of evidence, appendix and index*, 1835 (573) xx 169, p191 (Hereafter cited as *Second Report from the Select Committee on the amount of advances made by the Commissioners of Public Works in Ireland* (1835)); Seventh and Final Account of Alexander Nimmo, 1832

[73] Edgeworth, *op.cit.*, p55

[74] *Second Reports from the Select Committee on the amount of advances made by the Commissioners of Public Works in Ireland*, (1835), p191; *Dictionary of Land Surveyors and Local Mapmakers, op.cit; Report from the Select Committee on Connaught Lakes; with the minutes of evidence*, 1835 (354) xx; *First Report of select committee appointed to inquire into the amount of advances made by the Commissioners of Public Works in Ireland*, 1835; Grant of Administration for the Late Alexander Nimmo Jnr. Prerogative Grant Book 1839 Folio 16a

[75] *House of Lords inquiry into disturbances in Ireland* (1825), evidence of Alexander Nimmo, p179

[76] *Ibid.*, p160

[77] *Ibid.*, pp130-2

[78] *Ibid.*, p165

[79] *Ibid.*, pp130-2

[80] *Ibid.*, pp162-3

[81] *Ibid.*,pp130-2

[82] *Ibid.*, pp130-2; p148

[83] *Ibid.*, p160, p166

[84] *Ibid.*, p161

[85] *Ibid.*, pp130-2

[86] *Ibid.*, pp178-9

[87] Nimmo to Mangin, 27 June 1822, CSORP 1822 868

[88] Alexander Nimmo's Report on the Fishery Stations of the West Coast of Ireland

[89] Blake Family of Renvyle House, *Letters from the Irish Highlands of Connemara*, (Clifden 1995)

[90] Nimmo to Gregory, 19 June 1822, CSORP 1822 861

[91] Alexander Nimmo's Report on the Fishery Stations of the West Coast of Ireland

[92] Nimmo to Gregory, 19 June 1822, CSORP 1822 861

[93] Nimmo to Gregory, 22 June 1822, CSORP 1822 864

[94] *Ibid.*

[95] Nimmo to Mangin, 27 June 1822, CSORP 1822 867

[96] Nimmo to Gregory, 19 June 1822, CSORP 1822 861

[97] Nimmo to Marquess of Anglesey, Lord Lieutenant, April 20 1831, OP 973/82

[98] Alexander Nimmo's Report on the Fishery Stations of the West Coast of Ireland

[99] Nimmo to Mangin, 27 June 1822, CSORP 1822 868

[100] *Account of the different Public Works in the Western District of Ireland, commenced under the direction of Alexander Nimmo, Civil Engineer, in the summer of* 1822, 1830 (624) xxvii (Hereafter cited as *Account of the different Public Works in the Western District commenced under Alexander Nimmo in* 1822)

[101] Alexander Nimmo's Report on the Fishery Stations of the Coast of Connaught; Knight, *op.cit.*, pp125-6

[102] Nimmo to Wellesley, 30 June 1822, CSORP 1822 868

[103] Knight, *op.cit.*, p94

[104] *Account of the different Public Works in the Western District commenced under Alexander Nimmo in* 1822

[105] Alexander Nimmo's Report on the Fishery Stations of the Coast of Connaught; McParlan, *op.cit.*

[106] Nimmo to Wellesley, 30 June 1822, CSORP 1822 868; *Account of the different Public Works in the Western District commenced under Alexander Nimmo in* 1822

[107] Alexander Nimmo's Report on Progress of Works in 1830, OP 973/73 (Hereafter cited as Nimmo's Progress Report 1830)

[108] Alexander Nimmo's Report on the Fishery Stations of the Coast of Connaught

[109] Samuel Lewis, *A Topographical Dictionary of Ireland*, (London 1837), p104

[110] Nimmo's Progress Report 1830

[111] *Account of the different Public Works in the Western District commenced under Alexander Nimmo in* 1822, p3

[112] Nimmo to Dublin Castle, 13 August 1822, CSORP 1822 890

[113] *Account of the different Public Works in the Western District commenced under Alexander Nimmo in* 1822, p3

[114] *Ibid.*

[115] *House of Lords inquiry into disturbances in Ireland* (1825), evidence of Alexander Nimmo, p169; Alexander Nimmo's Report on the fishing stations of the Coast of Connaught

[116] Nimmo to Goulburn, 19 July 1826, CSORP 1826 14676; Alexander Nimmo's Report on the Fishery Stations of the Coast of Connaught

[117] *Account of the different Public Works in the Western District commenced under Alexander Nimmo in* 1822, p3

[118] Nimmo to Mangin, 28 August 1822, CSORP 1822 1296

[119] Daniel Mackim to chief secretary, CSORP 1832 4140
[120] Nimmo to Mangin, 28 August 1822, CSORP 1822 1296
[121] Nimmo to Goulburn, 4 September 1822, CSORP 1822 1092
[122] Nimmo to Mangin, 12 September 1822, CSORP 1822 1295; Nimmo to Goulburn, 26 September 1822, CSORP 1822 2715
[123] J. Mahony M.D., to T. L. Whistler M.D., 7 September 1822, CSORP 1822 1047
[124] Dr Whistler to Goulburn, 14 Sept 1822, CSORP 1822 1047
[125] Goulburn to the Commissioners for the Relief of Distress Dublin Castle, 14 Sept 1822, CSORP 1822 1047
[126] Francis Barker to Goulburn, 26 September 1822, CSORP 1822 1157
[127] Nimmo to Goulburn, 5 October 1822, CSORP 1822 2717
[128] Seventh and Final Account of Alexander Nimmo, 1832
[129] Nimmo to Goulburn, 5 October 1822, CSORP 1822 2717
[130] Nimmo to Goulburn, 26 October 1822, CSORP 1822 2588
[131] King, *Account of The Fisheries in 1822*
[132] *Fifth Report of the Commissioners of the Irish Fisheries*
[133] King, *Account of The Fisheries in 1822*
[134] Alexander Nimmo's Report on the Fishing Stations of the Coast of Connaught
[135] King, *Account of The Fisheries in 1822*
[136] *Return, showing expenditure under A. Nimmo in Western District* (1830)
[137] Nimmo to Goulburn, 5 December 1822, CSORP 1822 3745
[138] *Ibid.*
[139] Nimmo to Goulburn, 28 December 1822, CSORP 1822 3857
[140] Nimmo to Goulburn, 19 April 1823, CSORP 1823 7454; *Return, showing expenditure under A. Nimmo in Western District* (1830)
[141] Goulburn to Marquess Wellesley, 23 May 1823, Add 37301, Wellesley Papers, Vols. xxv-xxxiv, British Library
[142] *Ibid.*
[143] *Return, showing expenditure under A. Nimmo in Western District* (1830)
[144] *Connaught Journal*, 20 March 1823, p2; 24 March 1823 p3; 3 April 1823, p3
[145] *House of Lords inquiry into disturbances in Ireland* (1825), evidence of Alexander Nimmo, pp148 - 51
[146] Ibid., p152
[147] Ibid., pp147-51; p154
[148] *Report on survey and valuation of Ireland*, evidence of Alexander Nimmo

Chapter 4

[1] Alexander Nimmo's Report on the Fishing Stations of the Coast of Connaught
[2] James Donnell's Report, Appendix No 10 *Tenth Report of the Commissioners of the Irish Fisheries, for the year ended the 5 April 1829*. 1829 (329) xiii (Hereafter cited as Donnell's Report, Appendix 10, *Tenth Report of the Commissioners of the Irish Fisheries*)
[3] Alexander Nimmo's Report on the Fishing Stations of the Coast of Connaught
[4] James Donnell's Report, Appendix 11, *Eighth Report of the Commissioners of the Irish Fisheries*
[5] *Eighth Report of the Commissioners of Irish Fisheries*
[6] Alexander Nimmo's Report on the Fishing Stations of the West Coast of Ireland
[7] *Ibid.*; James Donnell's Report, Appendix 11, *Eighth Report of the Commissioners of Irish Fisheries*
[8] Alexander Nimmo's Report on the Fishing Stations of the West Coast of Ireland
[9] *First Report of the Commissioners of Inquiry into the State of the Irish Fisheries, with minutes of evidence*, 1837 [77] xxii.i, Appendix xix (Hereafter cited, *First Report of Inquiry into the State of the Irish Fisheries*, 1837), p125; Villiers-Tuthill, *Patient Endurance*, p54
[10] *Ibid.*
[11] Registry of Deeds, Martin to Evans, 865.193.576193. Thomas Evans had three sons, Thomas, James and Patrick
[12] Alexander Nimmo's Report on the Fishing Stations of the West Coast of Ireland; James Donnell's Report, Appendix 11, *Eighth Report of the Commissioners of Irish Fisheries*
[13] *First Report of Inquiry into the State of the Irish Fisheries*, 1837, evidence of Lieut. Warren, R.N., Coastguard Officer, p97
[14] Alexander Nimmo's Report on the Fishing Stations of the West Coast of Ireland
[15] Dutton, *op.cit.*
[16] *House of Lords inquiry into disturbances in Ireland* (1825), evidence of Alexander Nimmo, pp168-9
[17] Alexander Nimmo's Report on the Fishing Stations of the West Coast of Ireland; Villiers-Tuthill, *History of Clifden*, pp16-17
[18] Nimmo to Gregory, 18 March 1826, CSORP 1826 13598
[19] Alexander Nimmo's Report on the Fishing Stations of the West Coast of Ireland
[20] Nimmo to Gregory, 18 March 1826, CSORP 1826 13598; *First Report of Inquiry into the State of the Irish Fisheries*, 1837, Appendix xix
[21] Alexander Nimmo's Report on the Fishing Stations of the West Coast of Ireland
[22] James Donnell's Report, Appendix 11, *Eighth Report of the Commissioners of the Irish Fisheries*
[23] Alexander Nimmo's Report on the Fishing Stations of the West Coast of Ireland; Hardiman, *op.cit.*, pp302-7
[24] Hardiman, *op.cit.*, p304
[25] *First Report of the Commissioners of the Irish Fisheries, of their Proceeding for the year* 1819, 1821 (32) xi; *Second Report of the Commissioners of the Irish Fisheries, of their Proceedings, for the year* 1820. 1821 (646) xi (Hereafter cited as *Second Report of the Commissioners of the Irish Fisheries*)
[26] *Second Report of the Commissioners of the Irish Fisheries*
[27] Alexander Nimmo's Report on the Fishing Stations of the West Coast of Ireland
[28] Nimmo to Goulburn, 27 June 1822, CSORP 1822 868; Nimmo to Gregory, 29 July 1822, CSORP 1822 879
[29] Alexander Nimmo's Report on the Fishing Stations of the West Coast of Ireland
[30] *Ibid.*
[31] James Donnell's Report, Appendix 10, *Tenth Report of the Commissioners of the Irish Fisheries*
[32] Alexander Nimmo's Report on the Fishing Stations of the West Coast of Ireland
[33] *House of Lords inquiry into disturbances in Ireland* (1825), p172
[34] *Connaught Journal*, 11 January 1830 p2; 14 January 1830; 11 March 1830; *Report of Alexander Nimmo on the Progress of the Public Works in the Western District of Ireland in the year* 1829, 1830 (199), xxvii (Hereafter cited as Nimmo's Progress Report, 1829); Maurice Semple, *Some Galway Memories*, (Galway 1973)
[35] Maurice Semple, *Reflections on Lough Corrib*, (Galway 1974), pp7-15
[36] Alexander Nimmo's Report on the Fishing Stations of the West Coast of Ireland; James Donnell's Report, Appendix 11, *Eighth Report of the Commissioners of Irish Fisheries*
[37] *Ibid.*
[38] Alexander Nimmo's Report on the Fishing Stations of the West Coast of Ireland
[39] James Donnell's Report, Appendix 11, *Eighth Report of the Commissioners of Irish Fisheries*
[40] Alexander Nimmo's Report on the Fishing Stations of the West Coast of Ireland; James Donnell's Report, Appendix 11, *Eighth Report of the Commissioners of Irish Fisheries*
[41] *Ibid.*
[42] *Ibid.*; James Donnell's Report, Appendix 10, *Tenth Report of the Commissioners of the Irish Fisheries*
[43] Alexander Nimmo's Report on the Fishing Stations of the Coast of Connaught, (p193 of appendix to this book)
[44] Alexander Nimmo's Report on the Fishing Stations of the West Coast of Ireland
[45] *Ibid.*, {p217 of appendix to this book)
[46] *Ibid.*
[47] *Ibid.*
[48] Dutton, *op.cit.*, pp374 - 5
[49] *Seventh Report of the Commissioners of the Irish Fisheries*

[50] Nimmo, *Piloting Directions op.cit.*
[51] *Ninth Report of the Commissioners of the Irish Fisheries, for the year ended 5th April 1828.* 1828 (465) xii
[52] *Tenth Report of the Commissioners of the Irish Fisheries*

Chapter 5

[1] *House of Lords inquiry into disturbances in Ireland* (1825), evidence of Alexander Nimmo, p161
[2] Nimmo's Progress Report 1830
[3] Seventh and Final Account of Alexander Nimmo, 1832
[4] Nimmo's Progress Report, 1830
[5] Edgeworth, *op.cit.*, pp27-8
[6] Sir W. Wilde, *Lough Corrib, Its Shores and Islands*, (Galway 2002), p275-6
[7] *House of Lords inquiry into disturbances in Ireland* (1825), p160
[8] *Ibid.*, pp148 – 51, p153
[9] James Donnell's Report, Appendix 11, *Eighth Report of the Commissioners of Irish Fisheries*
[10] *Ibid.*
[11] *A Return of the Coastguard (formerly called the Preventive Water Guard) stationed in Ireland, containing the number and distribution thereof in the several Maritime Districts, in the year ending 5 January 1834*, 1834 xlii, 117
[12] Alexander Nimmo's Report on the Fishing Stations of the Coast of Connaught
[13] Knight, *op.cit.*, p5
[14] Alexander Nimmo's Report on the Fishing Stations West Coast of Ireland, (p205 of appendix to this book)
[15] *Ibid.*; Knight, *op.cit.*, p175; Rita Nolan, *Within The Mullet*, (Mayo 1998)
[16] Alexander Nimmo's Report on the Fishing Stations of the Coast of Connaught
[17] *Tenth Report of the Commissioners of the Irish Fisheries*
[18] Nimmo's Progress Report, 1830
[19] King, Account of the Fisheries in 1822
[20] Alexander Nimmo's Report on the Fishing Stations of the Coast of Connaught, (p200 of appendix to this book)
[21] Anthony Marmion, *The Ancient and Modern History of the Maritime Ports of Ireland* (London 1858), p440; Nolan, *op.cit.*, p78
[22] Alexander Nimmo's Report on the Fishing Station of the West Coast of Ireland
[23] Villiers-Tuthill, *History of Clifden*, pp 15-25
[24] Dutton, *op.cit.*, pp385 - 392
[25] Memorial of Clifden tradesmen to lord lieutenant, 15 July 1823, CSORP 1823 6088
[26] Dutton, *op.cit.*, pp385 - 392
[27] *Report of the National Fishing Company*, (Dublin 1825), pp24-5
[28] *First report of the inquiry into the state of the Irish Fisheries*, 1837
[29] Alexander Nimmo's Report on the Fishing Station of the West Coast of Ireland, (p220 of appendix to this book)
[30] *Ibid.*
[31] *Ibid.*, (p217 of appendix to this book)
[32] James Donnell's Report, Appendix 10, *Tenth Report of the Commissioners of the Irish Fisheries*
[33] Alexander Nimmo's Report on the Fishing Stations West Coast of Ireland
[34] Encumbered Estate Court Rentals, Vol. 17 35
[35] Seventh and Final Account of Alexander Nimmo, 1832

[36] Alexander Nimmo's Report on the Fishing Stations West Coast of Ireland
[37] Clifden Union Death Register; Register of Wills, 9996 11
[38] James Donnell's Report, Appendix 10, *Tenth Report of the Commissioners of the Irish Fisheries*
[39] Alexander Nimmo's Report on the Fishing Stations West Coast of Ireland
[40] Lewis, *op.cit.*, p406
[41] Ruddock, 'Nimmo'
[42] Barry, Michael, *op.cit.*, p103
[43] Ruddock, 'Nimmo'; *Second Annual Report of Commissioners for Public Works in Ireland*, 1834 (240) xl, 233
[44] Alexander Nimmo, On the Application of the Science of Geology to the purposes of Practical Navigation, in *Transactions of Royal Irish Academy*, XIV, Part 1, Pl 1, 1823
[45] Ruddock, 'Nimmo'
[46] Nimmo to Richard Stevenson, 6 April 1829, NLS MS 19988

Chapter 6

[1] Nimmo to Gregory, 30 June 1827, CSORP 1827 153; *Return, showing expenditure under A. Nimmo in Western District* (1830)
[2] *Return, showing expenditure under A. Nimmo in Western District* (1830)
[3] Nimmo to William Lamb, chief secretary, 14 July 1827, OP 972/48
[4] Nimmo to Lamb, 5 September 1827, OP 972/48
[5] The Account of all Sums of Money received from His Majesty's Treasury by the late Alexander Nimmo and his Representatives, and expenditure thereof on Public Works, for Relief of the Poor in the Western District, pursuant to 3 geo. 4, c. 34, from the 6th day of January 1831 to the 11th day of December 1832, *Twenty-first Report of Commissioners for Auditing Public Accounts in Ireland*, 1833, (102) xvii, 1 (Hereafter cited as The Account of all Sums of Money received from His Majesty's Treasury by the late Alexander Nimmo and his Representatives, *Twenty-first Report of Commissioners for Auditing Public Accounts in Ireland*, (1833))
[6] Seventh and Final Account of Alexander Nimmo, 1832
[7] *First Report of the Proceedings of the Commissioners appointed for providing for the repairing, maintaining and keeping in repair, certain roads and bridges in Ireland.* 1826 (102) xi (Hereafter cited as *First Report of the Commissioners appointed for keeping in repair, certain roads and bridges in Ireland*, 1826)
[8] *ibid.*
[9] Ibid.; *Third Report of the Commissioners Appointed under the Act 6 Geo. IV.c.101. to provide for the repairing, maintaining and keeping of repair, certain roads and bridges in Ireland*, 1828 (126) xii
[10] *First Report of the Commissioners appointed for providing for keeping in repair, certain roads and bridges in Ireland*, 1826
[11] *Second Report for the Commissioners appointed under the Act 6 Geo. IV.c.101. to provide for the repairing, maintaining and keeping in repair, certain roads and bridges in Ireland*, 1826 –27 (171) xi
[12] Ibid.; Goulburn to Directors General of Inland Navigation, 15 September 1826, CSORP 1826 14676
[13] Major Denis Bingham to Lamb, 29 September 1827, OP 972/49
[14] Nimmo to Lamb, 20 October 1827, OP 972/49
[15] Nimmo to Gregory, 3 June 1828, CSORP 1828 865
[16] Daniel Mackin to Nimmo, June 1828, CSORP 1828 865
[17] Nimmo to Marquess of Anglesey, Lord Lieutenant, 20 April 1831, OP 973/82
[18] Anglesey to Richard Martin M.P., 25 April 1826, CSORP 1826 14337
[19] Nimmo to Anglesey, 20 April 1831, OP 973/82
[20] *Ibid.*
[21] This site is presently occupied by the Clifden Community School
[22] Edgeworth, *op.cit.*, pp 52 – 56; Kathleen Villiers-Tuthill, Rags to Riches in Clifden 'Society', in *Connemara, Journal of the Clifden and Connemara Heritage Group* Vol.1, No.1, (Clifden 1993)
[23] *Third Report of the Commissioners Appointed to provide for keeping in Repair, certain Roads and Bridges in Ireland*

24 Ibid.

25 *Fourth Report of the Commissioners appointed under Act 6 Geo. 4. c. 101. to provide for the repairing, maintaining and keeping in Repair, certain Roads and Bridges*. 1829 (132), xiii

26 *Return, showing expenditure under A. Nimmo in Western District* (1830)

27 Nimmo's Progress Report 1829

28 Ibid.

29 Ibid.

30 Nimmo's Progress Report, 1830

31 Nimmo's Progress Report, 1829

32 Edgeworth, op.cit., pp74-5

33 Nimmo's Progress Report, 1829

34 Ibid.

35 Ibid.

36 Ibid.

37 Ibid.

38 *Connaught Journal*, 11 February 1830

39 Census of Ireland 1851, Vol. V, pp190-210

40 Nimmo's Progress Report, 1830

41 Ibid., Sir Robert Kane, M.D., *The Industrial Resources of Ireland*, (Dublin 1845), p134

42 Nimmo's Progress Report, 1830

43 Ibid.

44 Ibid.

45 Ibid.

46 Ibid.

47 Ibid.

48 Ibid.

49 *Return, showing expenditure under A. Nimmo in Western District* (1830)

50 Nimmo to Stanley, March 1831, OP 973/73

51 *Ninth Report of the Commissioners of the Irish Fisheries*

52 *Tenth Report of the Commissioners of the Irish Fisheries, for the year ended the 5th April 1829*. 1829 (329) xiii

53 John D'Arcy to Chief Secretary, 1 October 1830, CSORP 1830 6088

54 *Report of the Select Committee on the state of the poor in Ireland* (1830), p40

55 Ibid., p39

56 Ibid.

57 *First Report of select committee appointed to inquire into the amount of advances made by the Commissioners of Public Works in Ireland*, 1835

Chapter 7

1 Oxford Dictionary of National Biography, Vol. 42, (Oxford 2004), pp355 - 359

2 *Correspondence and Accounts relating to relief measures taken between 1822 and 1839*

3 *Galway Advertiser*, 9 April 1831, p2

4 Ibid.; *Connaught Journal*, 11 April 1831

5 *Connaught Journal*, 11 April 1831

6 *Connaught Journal*, 11 January 1830, p2; 14 January 1830; 11 March 1830

7 *Connaught Journal*, 11 April 1831

8 Ibid.

9 Edgeworth, op.cit., pp28-36

10 *Connaught Journal*, 18 April 1831, p2

11 Ibid.

12 *Galway Advertiser*, 16 April 1831

13 Marquess of Sligo to Stanley, 16 April 1831, OP 973/81

14 Ibid.

15 Nimmo to Stanley, 20 April 1831, OP 973/82

16 Ibid.

17 Ibid.

18 Nimmo to Stanley, 23 March 1831, OP 973/76

19 Marquess of Sligo to Stanley, 16 April 1831, OP 973/81

20 Nimmo to Stanley, 27 April 1831, OP 973/80

21 Ruddock, 'Nimmo'

22 Alexander Nimmo Report on works at Courtown Harbour, NLS MS 19971; Samuel Lewis, *A Topographical Dictionary of the Parishes, Towns and Villages of Cork City and County*, (Cork 1998), p437; Alexander Nimmo, Report and Estimates on the intended railway line between Dublin & Kingstown, OP 974/137; *Second Annual Report of Commissioners for Public Works in Ireland*, 1834

23 *Second Reports from the Select Committee on the amount of advances made by the Commissioners of Public Works in Ireland* (1835), evidence of Alexander Nimmo, (nephew), pp190 - 93

24 Stanley to John Killaly, 28 May 1831, OP 1830 973/77

25 The Account of the Receipts and Expenditure of Monies issued from the Treasury in Ireland to John Killaly, on Account of Public Works in the West of Ireland, from 1st April 1831 to 14 April 1832, *Twenty-first Report of Commissioners for Auditing Public Accounts in Ireland*, 1833, (102) xvii

26 Stanley to John Killaly, 28 May 1831, OP 1830 973/77

27 Ibid.

28 *Connaught Journal*, 2 June 1831 p2

29 *Correspondence and Accounts relating to relief measures taken between 1822 and 1839*, p7

30 Nimmo to Sir William Gosset, 3 October 1831, OP 1830 973/87

31 Nimmo to Stanley, 27 July 1831, OP 973/82

32 J. Killaly to Thomas Taylor, Dublin Castle, 24 September 1831, OP 973/83

33 Gossitt to Stanley, 28 September 1831, OP 973/84

34 Nimmo to Stanley, 18 November 1831, CSORP 1831 3197

35 *Correspondence and Accounts relating to relief measures taken between 1822 and 1839*

36 J. Killaly to Gossit, 30 September 1831, OP 973/86

37 Ibid.

38 Ibid.

39 Wilde, op.cit., pp275-6

40 J. Killaly to Gossit, 30 September 1831, OP 973/86

41 *Report of Hamilton Killaly on the Roads in the Western District*, 30 April 1833, 1833 (285) xxxv 575

42 CSORP 1834 3207

43 *Galway Weekly Advertiser*, 28 January 1832, p 3; *The Times*, 26 January 1832, p 4

44 The Account of all Sums of Money received from His Majesty's Treasury by the late Alexander Nimmo and his Representatives, *Twenty-first Report of Commissioners for Auditing Public Accounts in Ireland*, (1833)

45 Seventh and Final Account of Alexander Nimmo, 1832

46 The Account of all Sums of Money received from His Majesty's Treasury by the late Alexander Nimmo and his Representatives, *Twenty-first Report of Commissioners for Auditing Public Accounts in Ireland*, (1833)

47 Seventh and Final Account of Alexander Nimmo, 1832; George & John Nimmo to Anglesey, 10 December 1832, CSORP 1832 6376

48 The Account of all Sums of Money received from His Majesty's Treasury by the late Alexander Nimmo and his Representatives, *Twenty-first Report of Commissioners for Auditing Public Accounts in Ireland*, (1833)

49 *Dictionary of Land Surveyors and Local Mapmakers of Great Britain and Ireland 1530-1850*, Vol. 11, op.cit.

50 *Return, showing the Gross Expenditure in the Counties of Mayo and Galway from 1821 to 1833*

[51] Thomas Martin eventually succeeded in acquiring sufficient funds from the lord lieutenant, to combine with his own contribution, to complete Nimmo's south coast road in 1842, CSORP 1842 W2080

[52] *Report of Hamilton Killaly on the Roads in the Western District*, 30 April 1833

[53] *Second Annual Report of Commissioners for Public Works in Ireland*, 1834, pp6-7

[54] *Ibid.*

[55] *Third Annual Report of Commissioners for Public Works in Ireland*, 1835 (76) xxxvi 339

[56] *Ibid.*

[57] *Second Annual Report of Commissioners for Public Works in Ireland*, 1834, pp6-7

Select Bibliography

Primary Sources

Unpublished Manuscripts
Preece, Robert G., History of Inverness Royal Academy

National Archives of Ireland
Chief Secretary's Office, Registered Papers, 1811–1832
Official Papers 1811–1832
Landed Estates Court Rentals, Vol. 17 35
Prerogative Grant Book 1839
Register of Wills 1832–39

Galway County Library
Map of the County of Galway, by William Larkin, printed 1819
Ordnance Survey Maps of County Galway, First Edition, 1840–41

Mayo County Library
Ordnance Survey Maps of County Mayo

Trinity College Dublin
Map Library
Ordnance Survey Maps of Counties Galway and Mayo

British Library
Wellesley Papers, Vols. xxv–xxxiv

National Library of Scotland
Manuscript Collection
Alexander Nimmo, Journal of Survey of Inverness-shire, 1806
Letters and Reports of Alexander Nimmo

Institute of Engineers of Ireland
Chart of St George's Channel and the coast of Ireland, drawn for the Corporation of Improving the Port of Dublin, Alexander Nimmo, Dublin 1832

Church of Ireland Parish Records
Clifden Union Death Registers

Registry of Deeds
Deeds relating to Connemara 1822–1832

Valuation Office
Cancelled Books

Printed Contemporary Records

National Library of Ireland
Annual Register, 1832
Census of Ireland 1851, Vol. V
Grierson, George and John, *New and Correct Irish Atlas*, Dublin undated. A study of place names and roads would suggest a pre-1825 date.
Digest of Evidence taken before Her Majesty's Commissioners of Inquiry into the State of the Law and Practice in respect to the occupation of Land in Ireland (The Devon Commission) Dublin 1847
National Fishing Company: *Report of a deputation from the directors of the National Fishing Co, on the subject of the fisheries on the south and western coast of Ireland* Dublin 1825

British Parliamentary Papers

First Report of the Commissioners Appointed to enquire into the Nature and Extent of the several Bogs in Ireland, and the practicability of draining and cultivating them, 1810 (365), x, 389
Second Report of the Commissioners appointed to enquire into the nature and extent of the several Bogs in Ireland, and the practicability of draining and cultivating them, 1810–1811 (96), vi, 579
Third Report of the Commissioners Appointed to enquire into the Nature and Extent of the several Bogs in Ireland, and the practicability of draining and cultivating them, 1813–14 (130), vi, pt 1
Fourth Report of the Commissioners Appointed to enquire into the Nature and Extent of the several Bogs in Ireland, and the practicability of draining and cultivating them, 1813–14 (131), vi, part 11, 167
Second Report of the Select Committee on the State of Disease and Condition of the Labouring Poor in Ireland (Session 1819 – Reprint), 1829 (347), iv
Report from Commissioners: Dunmore Harbour, 1821 (492), xi
A Bill to authorize certain temporary Advances of Money, for the Relief of the Distress existing in Ireland, 1822 (541), iii

Report from the select committee on the employment of the poor in Ireland, 1823 (561), vi, 331

Report from the select committee appointed to consider the best mode of apportioning more equally the local burthens collected in Ireland, and to provide for a general survey and valuation of that part of the United Kingdom, 1824 (445), viii

Accounts of the Commissioners for advancing money from the Consolidated Fund for Public Works, 1824 (278), xxi, 61

Minutes of evidence taken before the select committee of the House of Lords appointed to examine into the nature and extent of the disturbances in Ireland, 1825 vii

First Report of the Commissioners of the Irish Fisheries, of their Proceeding for the year 1819, 1821 (32), xi

Second Report of the Commissioners of the Irish Fisheries, of their Proceedings for the year 1820, 1821 (646), xi

Third Report of the Commissioners of the Irish Fisheries, of their Proceedings for the year 1821, 1822 (428), xiv

Fourth Report of the Commissioners of the Irish Fisheries, of proceedings for the year 1822, 1823 (383), x

Fifth Report of the Commissioners of the Irish Fisheries, of their proceedings for the year 1823, 1824 (283), xii

Sixth Report of the Commissioners of the Irish Fisheries, of their proceedings for the year 1824, 1825 (385), xv

Seventh Report of the Commissioners of the Irish Fisheries for the year Ending 5 April 1826, 1825–26, 1826 (395), xi

Eight Report of the Commissioners of Irish Fisheries, for the year ending 5 April 1827, 1826–27 (487), xi

Ninth Report of the Commissioners of the Irish Fisheries, for the year ended 5 April 1828, 1828 (465), xii

Tenth Report of the Commissioners of the Irish Fisheries, for the year ended the 5 April 1829, 1829 (329), xiii

Eleventh Report of the Commissioners of the Irish Fisheries, for the year ended the 5 April 1830, 1830 (491), xv

Report from the Select Committee on Proceedings for Regulating Grand Jury Presentments, 1826–27 (555), iii, 745

First Report of the Proceedings of the Commissioners appointed under the Act 6 Geo. IV.c.101. to provide for the repairing, maintaining and keeping in repair, certain roads and bridges in Ireland, 1826 (102), xi

Second Report of the Commissioners appointed under the Act 6 Geo. IV.c.101. to provide for the repairing, maintaining and keeping in repair, certain roads and bridges in Ireland, 1826–27 (171), xi

Third Report of the Commissioners Appointed under the Act 6 Geo. IV.c.101., to provide for the repairing, maintaining and keeping of repair, certain roads and bridges in Ireland, 1828 (126), xii

Fourth Report of the Commissioners appointed under Act 6 Geo. IV. c. 101. to provide for the repairing, maintaining and keeping in repair, certain roads and bridges in Ireland, 1829 (132), xiii

Sixth Report of the Commissioners appointed under Act 6 Geo. IV. c.101. to provide for the repairing, maintaining, certain roads and bridges in Ireland, 1830–31 (295), iv

Report of the Select Committee on the state of the poor in Ireland, 1830 (667), vii

Account of the different Public Works in the Western District of Ireland, commenced under the direction of Alexander Nimmo, Civil Engineer, in the summer of 1822, 1830 (624), xxvii

Report of Alexander Nimmo on the Progress of the Public Works in the Western District of Ireland in the year 1829, 1830 (199), xxvii

An Account of all sums of money advanced out of the Consolidated Fund, on the recommendation or with the sanction of the Irish Loan Commissioners, 1831 (184), xvii

Sums expended, 1824–1830, on the several Roads in Ireland, under the direction and control of Supervisors, and levied by Grand Jury Presentments, 1831 (211), xvii, 543

Return, showing the Gross Expenditure in the Counties of Mayo and Galway, under the Directions of Messrs. Nimmo and Killaly, Civil Engineers, from the year 1821 to 1833, 1833 (285), xxxv, 575

Twenty-first Report of Commissioners for Auditing Public Accounts in Ireland, 1833 (102), xvii

First Report of the Commissioners on Public Works, Ireland, of their Proceedings, and an Abstract of their Expenditure, for the year 1832, 1833 (75), xvii, 373

Second Annual Report of Commissioners for Public Works in Ireland, 1834 (240), xl, 233

Third Annual Report of Commissioners for Public Works in Ireland, 1835 (76), xxxvi

First Report of select committee appointed to inquire into the amount of advances made by the Commissioners of Public Works in Ireland, 1835 (329), xx, 145

Second Reports from the Select Committee appointed to inquire into the amount of advances made by the Commissioners of Public Works in Ireland, with the minutes of evidence, appendix and index, 1835 (573), xx, 169

A Return of the Coastguard (formerly called the Preventive Water Guard) stationed in Ireland, containing the number and distribution thereof in the several Maritime Districts, in the year ending 5 January 1834, 1834 xlii, 117

Report from the Select Committee on Connaught Lakes; with the minutes of evidence, 1835 (354), xx

First Report of the Commissioners of Inquiry into the State of the Irish Fisheries, with minutes of evidence, 1837 (77), xxii.i

Correspondence and Accounts relating to the different occasions on which measures were taken for the relief of the people suffering from scarcity in Ireland, between 1822 and 1839, 1846 (734), xxxvii

Secondary Sources

Aalen, F. H. A., Kevin Whelan and Matthew Stout (eds), *Atlas of the Irish Rural Landscape*, Cork 1997

Andrews, J.H. *A Paper Landscape: The Ordnance Survey in Nineteenth-Century Ireland*, Dublin 2002

Anglesey, The Marquess of, F.S.C., *One-Leg, The Life and Letters of Henry William Paget First Marquess of Anglesey K.G. 1768–1854*, London, 1961

Bald, William, *An Account of a Trigonometrical Survey of Mayo*, Dublin 1824

Beaufort, Daniel Augustus, L.L.D., *Memoir of a Map of Ireland; Illustrating the topography of that Kingdom, and containing a short account of its present state, Civil and Ecclesiastical*, Dublin 1792

Bentley, Richard, *The Angler in Ireland: or An Englishman's Ramble through Connaught and Munster, during the summer of 1833, Vol 1 & 11*, London 1834

Blake Family of Renvyle House, *Letters from the Irish Highlands of Connemara*, Clifden 1995

Boylan, Thomas A. and Timothy P. Foley, *Political Economy and Colonial Ireland*, London 1992

Brabazon, Wallop, *The Deep Sea and Coast Fisheries of Ireland*, Dublin 1848

Collins, Timothy, *Transatlantic, Triumph & Heroic Failure: The Story of the Galway Line*, Cork 2002

Coulter, Henry, *The West of Ireland: Its existing condition and prospects*, Dublin 1862

Cox, R.C. & M.H. Gould, *Civil Engineering Heritage Ireland*, London 1998

Cunningham, John, *A town tormented by the sea: Galway, 1790–1914*, Dublin 2004

De Coursy, Seán, Alexander Nimmo; Engineer extraordinary, in *Connemara: Journal of the Clifden & Connemara Heritage Group, Vol.2 No.1*, Clifden 1995

De Latocnaye, Chevalier, *A Frenchman's Walk through Ireland 1796–7*, translated from the French by John Stevenson, Belfast 1917

Dictionary of National Biography Volume XIV, Oxford 1973

Dictionary of Land Surveyors and Local Mapmakers of Great Britain and Ireland 1530–1850, Vol. 1 & 11, London 1997

Dutton, Hely, *Statistical Survey of the County of Clare with observations on the means of improvement*, Dublin 1808

Dutton Hely, *A Statistical and Agricultural Survey of the County of Galway with observations on the means of improvement; drawn up for the consideration, and by the direction of The Royal Dublin Society*, Dublin 1824

Edgeworth, Maria, *Tour in Connemara and the Martins of Ballinahinch*, ed. Harold Edgeworth Butler, London 1950

Fraser, Robert, *Report on the County Surveys, and the Best Means for the Further Encouragement of the Fisheries of Ireland*, Dublin 1818

Freeman, T.W., *Pre-Famine Ireland: A Study in Historical Geography*, Manchester 1957

Greenhill, Basil, *The Merchant Schooners*, London 1978

Hall, Mr & Mrs S.C., *Alexander Ireland: Its Scenery, Character, &c*, London 1842

Hardiman, James, *The History of the Town and County of the Town of Galway*, Galway 1820, reprinted Galway 1958

Horner, Dr Arnold, *Iveragh Co. Kerry in 1811: Alexander Nimmo's map for the Bogs Commissioners of 1809–1814*, Dublin 2002

Horner, Dr Arnold, *Kenmare River Co. Kerry in 1812: Alexander Nimmo's Map for the Bogs Commissioners*, Dublin 2003

Horner, Dr Arnold, *Wicklow & Dublin Mountains in 1812: Richard Griffith's Map for the Bogs Commissioners*, Dublin 2004

Inglis, Henry D., *A Journey Throughout Ireland, during the Spring, Summer and Autumn of 1834*, Vol 11, London 1834

Jenkins, Brian, *Henry Goulburn 1784–1856: A Political Biography*, Liverpool 1996

Jordan, Thomas E., *An Imaginative Empiricist – Thomas Aiskew Larcom (1801–1879) and Victorian Ireland*, New York 2002

Kane, Sir Robert, M.D., *The Industrial Resources of Ireland*, Dublin 1845

King, William H., Account of the Fisheries on the Western Entrance of the River Shannon, extending from Killala Bay to Kerry Head, in *An Account of the Present State of the Fisheries on the different parts of the Coast of Ireland*, Dublin 1820

Knight, Patrick, *Erris in the Irish Highlands*, Dublin 1836

Leerssen, Joep, *Remembrance and Imagination: Patterns in the Historical and Literary Representation of Ireland in the Nineteenth Century*, Cork 1996

Lewis, Samuel, *A Topographical Dictionary of Ireland*, London 1837

Lewis, Samuel, *A Topographical Dictionary of the Parishes, Towns and Villages of Cork City and County*, Cork 1998

Lynam, Shevawn, *Humanity Dick Martin 'King of Connemara' 1754–1834*, Dublin 1989

McDowell, R.B. *Public Opinion and Government Policy in Ireland, 1801–1846*, London 1952

McDowell, R.B., *The Irish Administration 1801–1914*, Connecticut 1964

McParlan, James, M.D., *Statistical Survey of the County of Mayo*, Dublin 1802

Marmion, Anthony, *The Ancient and Modern History of the Maritime Ports of Ireland*, London 1858

Martineau, Harriet, *Letters from Ireland*, London 1852

Matthew, H. C. G., and Brian Harrison (eds.), *Oxford Dictionary of National Biography*, Vol. 40, Oxford 2004

Maxwell, W.H., *Wild Sports in the West*, London 1832

Michs W.L., *An Account of the constitution, administration and dissolution of the C.D.B., of Ireland*, Dublin 1925

Nimmo, Alexander, *Report on the means of improving the river and harbour of Cork*, Cork 1815

Nimmo, Alexander, *Report and estimates on draining and cultivating the bog on the Cashan or the Northern division of Kerry*, Dublin 1818

Nimmo, Alexander, The Application of the Science of Geology to the purposes of Practical Navigation, in *Transactions of Royal Irish*

Academy, XIV, Part 1, Pl 1, 1823
Nimmo, Alexander, *New piloting directions for St George's Channel and the coast of Ireland*. Accompany the chart of St George's Channel and the coast of Ireland, drawn for the Corporation of Improving the Port of Dublin, Dublin 1832
Nimmo, Alexander (nephew), *Report to the directors of the Great Central Irish Railway, Dublin*, Dublin 1836
Nolan, Rita, *Within The Mullet*, Mayo 1998
O'Brien, George, *The Economic History of Ireland from the Union to the Famine*, London 1921
O'Flaherty, Roderick, *A Chorographical Description of West or H-Iar Connaught*, written in 1684, ed. James Hardiman, Dublin 1846
O'Keeffe, Peter J., 'Richard Griffith: planner and builder of roads', in G.L. Herries Davies and R. C. Mollan (eds.), *Richard Griffith 1784–1878*, Dublin 1980
O Luing, S, Richard Griffith and the roads of Kerry. in *Journal of the Kerry Archaeological & Historical Soc.* No 8 (1975); No 9 (1976)
Ó Tuathaigh, Gearóid, Aspects of Nineteenth-Century Galway, in Diarmuid Ó Cearbhaill, (ed), *Galway Town and Gown 1484–1984*, Galway 1984
Oxford Dictionary of National Biography, Vol. 42, Oxford 2004
Ruddock, Ted, 'Alexander Nimmo FRSC, MRIA (1783–1832)' in Professor Sir Alex. W. Skempton (ed), *A Biographical Dictionary of Civil Engineers in Great Britain and Ireland*, Vol. 1: 1500–1830, London 2002
Ruddock, T, *Arch bridges and their builders* 1735–1835, Cambridge 1979
Scott, J. R., *The Galway Hookers: Sailing work boats of Galway Bay*, Limerick 1983
Semple, Maurice, *Reflections on Lough Corrib*, Galway 1974
Semple, Maurice, *Some Galway Memories*, Galway 1973
Smiles, Samuel, *Life of Telford*, London 1867
Symonds, Thomas E., *Observations on the Fisheries of the west coast of Ireland*, London 1855
Thackeray, W. M., *An Irish Sketch-book, 1842*, London 1875
Villiers-Tuthill, Kathleen, *History of Clifden 1810–1860*, Galway 1981
Villiers-Tuthill, Kathleen, *Beyond The Twelve Bens: A History of Clifden and District 1860-1923*, Galway 1986
Villiers-Tuthill, Kathleen, *Patient Endurance: The Great Famine in Connemara*, Dublin 1997
Villiers-Tuthill, Kathleen, *History of Kylemore Castle & Abbey*, Galway 2002
Villiers-Tuthill, Kathleen, Rags to Riches in Clifden 'Society', in *Connemara, Journal of the Clifden and Connemara Heritage Group* Vol. 1, No. 1, Clifden 1993
Wakefield, Edward, *An Account of Ireland Statistical and Political*, Vol. 1, London 1812
White, James, & Kevin Bright (eds.), *Treasures of The Royal Dublin Society: A Summary Catalogue of the works of art in the collection of The Royal Dublin Society*, Dublin 1998
Wilde, Sir W. *Lough Corrib, Its Shores and Islands*, Galway 2002
Young, Arthur, *A Tour in Ireland 1776–1779* Vol. 1 & 2, ed. A. W. Hutton. Shannon 1970

Newspapers

Connaught Journal, 1830 - 1832
Galway Advertiser, 1831 - 1832
The Times, 26 January 1832

Index